Strategic Planning and Control in Complex Project Management

Strategic Planning and Control in Complex Project Management

Guest Editors

Wenxin Shen
Wenzhe Tang
Jin Xue

MDPI

Basel • Beijing • Wuhan • Barcelona • Belgrade • Novi Sad • Cluj • Manchester

Guest Editors

Wenxin Shen
Department of Construction Management
Beijing Jiaotong University
Beijing
China

Wenzhe Tang
State Key Laboratory of Hydroscience and Engineering
Institute of Project Management and Construction Technology
Tsinghua University
Beijing
China

Jin Xue
School of Project Management
The University of Sydney
Sydney, NSW
Australia

Editorial Office
MDPI AG
Grosspeteranlage 5
4052 Basel, Switzerland

This is a reprint of the Special Issue, published open access by the journal *Buildings* (ISSN 2075-5309), freely accessible at: https://www.mdpi.com/journal/buildings/special_issues/54BNE4QK5K.

For citation purposes, cite each article independently as indicated on the article page online and as indicated below:

Lastname, A.A.; Lastname, B.B. Article Title. *Journal Name* **Year**, *Volume Number*, Page Range.

ISBN 978-3-7258-3865-3 (Hbk)
ISBN 978-3-7258-3866-0 (PDF)
https://doi.org/10.3390/books978-3-7258-3866-0

Contents

About the Editors

Wenxin Shen

Dr. Shen is an associate professor at the School of Economics and Management, Beijing Jiaotong University, Beijing, China. Her research areas include program management, interface management, and inter-organizational relationships in construction management. She is the winner of the 2025 IPMA Young Researcher Award. Her works have been published in top journals such as the *International Journal of Project Management*, the *Journal of Management in Engineering*, the *Journal of Construction Engineering and Management*, the *Journal of Civil Engineering and Management*, and *Renewable Energy*. She has been actively involved in various research projects funded by the National Natural Science Foundation of China (NSFC), the China Postdoctoral Science Foundation, and other prestigious institutions. As a Principal Investigator, she has led projects on multi-project coordination networks in megaprojects and has overseen the role of information technology in managing complex interfaces in megaprojects. She has received academic awards such as the Merit Paper Award (2020, 2021) of the 26th International Symposium on Advancement of Construction Management and Real Estate.

Wenzhe Tang

Prof. Tang is a professor at the Department of Hydraulic Engineering, Tsinghua University, where he also serves as the Deputy Director of the Department of Construction Management. His research focuses on hydropower engineering management, energy management, digital construction management, EPC project management, and risk management. He is particularly interested in partnership theory, performance incentives, and international engineering management. Currently, he leads several key projects, including National Natural Science Foundation of China (NSFC) projects. He also collaborates with industry partners like the China Three Gorges Corporation and Huaneng Lancang River Hydropower on clean energy management and technological innovation. Prof. Tang has received numerous awards, such as the Power Technology Innovation Award (2022), China Electric Power Science and Technology Award (2020), and the Outstanding Education Contribution Award from the China Project Management Research Committee (2019). His scholarly contributions include over 150 publications in top-tier journals like *Nature*, *Energy*, *Renewable Energy*, the *International Journal of Project Management*, the *Journal of Management in Engineering*, the *Journal of Construction Engineering and Management*, and the *Journal of Civil Engineering and Management*. His work bridges theoretical research and practical applications, advancing sustainable infrastructure development globally.

Jin Xue

Dr. Xue is a lecturer working in the prestigious John Grill Institute for Project Leadership (JGIPL), School of Project Management, University of Sydney. He is an academic lead of the theme "Project Models and Digital Transformation" at the JGIPL. He is a promising researcher specializing in pioneering research on stakeholder management in the dynamic and complex environment of data-driven megaprojects. His current research interest is focused on AI-driven project management studies. He has undertaken a series of AI and big data-driven project management research projects funded by the Association for Project Management, Sydney Environment Institute, Commonwealth Scientific and Industrial Research Organisation, and the Research Grants Council of Hong Kong. He has strong multi-disciplinary research capabilities proven by the data-driven publications in the leading ABDC (Australian Business Deans Council) management journals. His pioneering data-driven research has also been recognized by world-leading project management associations (Royal Institution of Chartered Surveyors and American Society of Civil Engineers).

buildings

MDPI

Article

The Interactive Effects of Communication Network Structure and Organizational Size on Task Performance in Project-Based Organizations: The Mediating Role of Bootleg Innovation Behavior

Xue Ding [1], Long Feng [2,*], Yao Huang [3] and Wei Li [4]

1 School of Financial Technology, Shanghai Lixin University of Accounting and Finance, Shanghai 201209, China; dingxue@lixin.edu.cn
2 Shanghai New Jing An (Group) Co., Ltd., Shanghai 200040, China
3 Department of Engineering Management, Changsha University of Science & Technology, Changsha 410114, China; yaohuang@csust.edu.cn
4 School of Information Science and Technology, Sanda University, Shanghai 201209, China; liwei@sandau.edu.cn
* Correspondence: fl@jatz-sh.com; Tel.: +86-135-8560-8156

Abstract: A PBO is a temporary organization formed by assembling members with diverse experiences and backgrounds, aimed at achieving specific innovation goals. Constructing a reasonable communication network structure and enhancing organizational synergy are effective ways to promote the sustainable development of the system. This study, based on the Input–Process–Output (IPO) model and social network analysis, utilized a group collaboration platform to conduct a three-stage communication experiment on 685 construction project managers. Under two organizational sizes, the internal mechanism of how communication networks with two levels of centralization influence task performance were tested. The results indicate that in the case of a smaller organizational size, PBOs using a decentralized communication network tend to achieve higher task performance. However, as the organizational size expands, PBOs employing a centralized communication network may surpass in task performance. Additionally, we found that with the expansion of organizational size, bootleg innovation behaviors of organizational members are continually stimulated, further enhancing collective task performance. This study, based on the evolution of communication network parameters, explores the structural characteristics of organizational communication networks and the mechanisms underlying the emergence of bootleg innovation behaviors. It delineates the key pathways for improving collective task performance. The findings can provide a scientific reference for the organizational evolution and development of engineering project management.

Keywords: project-based organizations (PBOs); communication network structure; organizational size; organizational evolution; bootleg innovation behavior; task performance

Citation: Ding, X.; Feng, L.; Huang, Y.; Li, W. The Interactive Effects of Communication Network Structure and Organizational Size on Task Performance in Project-Based Organizations: The Mediating Role of Bootleg Innovation Behavior. *Buildings* **2024**, *14*, 98. https://doi.org/10.3390/buildings14010098

Academic Editors: Wenxin Shen, Wenzhe Tang and Jin Xue

Received: 15 November 2023
Revised: 27 December 2023
Accepted: 28 December 2023
Published: 29 December 2023

1. Introduction

In recent years, project-based organizations (PBOs) have flourished in the construction industry. Their activities typically revolve around projects that encompass infrastructure such as airports, bridges, venues, and exhibitions [1–3]. The development of these constructions has a significant impact on the sustainable development of a city's economy, society, and environment. Research on the topic has branched out into different definitions of PBOs. For instance, some studies consider a PBO as a permanent structure encompassing multiple projects. In contrast, others view a PBO as a temporary legal project-based enterprise or firm created around a specific project outcome [4]. To avoid confusion, this article refers to the latter definition. In such a context, one PBO usually contains diverse participants with different knowledge and professional backgrounds, as well as interactive connections [5,6]. From the standpoint of social network analysis, the members of a PBO and their interrelationships serve as the foundation of its communication network [7]. In fact, the establishment of a PBO was aimed at achieving satisfactory project or task performances [8]. Recent research consistently highlights the significant impact of the communication network structure of a PBO on enhancing construction project performance. For instance, some studies, from the perspective of singular or interactive social network indicators, have explored the influence of the communication network structure on project performance [9,10]. Conversely, other research has analyzed project performance variances within established network structures from the viewpoint of information communication and knowledge-sharing coordination [11,12]. These investigations, through the lens of social network analysis, delve into the effects of organizational communication structures on task performance or project success, providing valuable insights.

The communication networks for PBOs can be categorized by their structural feather—centralization, which is defined as the degree to which communication flows disproportionately through organization members rather than being more equally distributed [13]. Recently, substantial research indicates that decentralized structures, in which organizational members are connected to most or all other members, adapt faster to unique information flows and have increased knowledge-based project performance [14,15]. In contrast, centralized structures, which contain a core leader or a leadership team with groups of peripheries, could improve collective learning and problem-solving ability by providing access to all critical information in a shifting task environment [13]. When further exploring the relationship between these communication networks and task performances, is there significant difference in project performance for PBOs which have adopted different network structures? What type of communication network structure would improve task performance? Further, are there conditions for the desired outcomes? Unfortunately, the answers to these questions remain inconsistent.

Furthermore, the size of an organization, typically measured by the number of employees or total assets, is also considered an important factor affecting task performance [16]. Taking the construction industry as an example, large PBOs often face problems of cost overruns and project delays [17], and there may be a negative correlation between organizational size and task completion [18]. Further, existing research suggests that organizational size and communication network structure are key elements influencing organizational activities [19], which have been presented to be critical factors for task completion outcomes. However, how the interaction mechanism between the communication network structure and organizational size impacts task performance for PBOs remains understudied. Additionally, the relationship between organizational size and technological innovation has always been a matter of debate. The earliest studies believed that the larger the organization size, the stronger its R&D capability and risk resistance, which promotes technological innovation [20]. However, some studies argue that as the size increases, organizational inertia tends to rise, reducing the willingness for technological innovation [21]. Recent studies suggest that the larger the organizational size, the more redundant resources it possesses, which might trigger bootleg innovative behavior among its members. This refers to members of the organization innovating outside of formal channels. In recent years, such bootleg innovative behavior has been considered a motivating factor for improving task performance [22–24].

Thus, effective communication channels facilitate knowledge sharing and knowledge innovation [25,26]. Innovative behavior has been regarded as a key factor in enhancing the vitality of a PBO and in promoting project success [27,28]. However, for construction PBOs with temporary characteristics and engaged in creative production, will the interaction between communication networks of different centralization and different organizational sizes stimulate members' deviant innovative behaviors and further enhance task performance? These discussions have left a series of open questions for traditional research and theoretical perspectives.

Based on the relational contract theory in construction engineering, we identified potential one-to-one correspondences between representative communication network structures and project contract delivery modes. Depending on the position and connections of each node in the network, these three pairs of coupled relationships, from a complete clique structure/Integrated Project Delivery (IPD) mode, a locally clustered structure/Design–Build (DB) mode, to a core periphery structure/Design–Bid–Build (DBB) mode, show a gradual increase in network structure centralization [7,29,30]. At the same time, drawing on the Input–Process–Output (IPO) model [31,32], we examined the impact of communication networks with varying levels of centralization and organizational size on the task performance of PBOs. That is, our study aims to compare the performance differences brought by centralized and decentralized communication networks from the perspective of expanding organizational size, thus filling a research gap. This study also clarifies whether bootleg innovation behavior plays a mediating role in this process.

This study has three theoretical contributions. Firstly, building on recent studies that explore project performance improvement from the perspective of network structure metrics for PBOs, this study examines the interactive relationship between communication network structure and organizational size based on collective tasks, including individual task execution, internal information sharing, and collective decision making. This enriches the research results related to improving project performance. Secondly, this study explores the intrinsic mechanism of performance improvement in PBOs by bootleg innovation behavior, expanding the research of innovation management in the construction industry. Thirdly, the conclusion of this study explores the relationship between organizational size and bootleg innovation behavior from the perspective of organizational evolution in project organizations, expanding the scope of issues addressed in project management.

The next section of this paper presents a literature review and hypothesis formulation. Section 3 introduces the experimental design, procedure, and measurements. Section 4 presents data analysis and test results. Section 5 discusses the results, outlining theoretical contributions and managerial implications. Lastly, Section 6 is the conclusion, including discussions on limitations and suggestions for future research.

2. Literature Review and Hypotheses Development

2.1. Theoretical Foundation: Input–Process–Output Framework

McGrath proposed an IPO framework for analyzing collective behavior and performance [31]. The IPO model clarifies the impact of inputs on outcomes through the interaction process. Inputs are described as antecedents that enable and constrain organization interactions, which include individual-level factors (e.g., competencies and personalities), team-level factors (e.g., structure and size), and environmental-level factors (e.g., task features and environmental complexity). These antecedents from different levels combine to drive team interaction processes toward task accomplishment. Processes are used to describe the emergent states of how inputs are transformed into outcomes. There are multiple criteria for outcomes. They may include task performance (e.g., quality, speed to solution) and members' affective reactions (e.g., member satisfaction, cohesiveness, viability). The IPO model has served as a foundation for research on collective performance outcomes, and it has also been involved from perspective of composite performance measures [32].

2.2. Communication Network Structure and Centralization

The construction of engineering projects is fraught with uncertainties [33–35]. By establishing a PBO, members with diverse skills, experiences, and tools can be brought together [36–38]. Individuals offer support to others through innovative activities, ultimately achieving task objectives within specified time, cost, and quality constraints [28,39]. Throughout this process, organizational members need to make efforts to share private information with others, and the knowledge acquired and internalized by other members can be externalized as knowledge collectively held, shared, and agreed upon within the organization [40]. Processes of collective knowledge acquisition, sharing, integration, and updates will heavily rely on the communication structures. Therefore, how to measure a communication network structure has attracted more widespread attention. Recently, social network analysis has been described as a tool for analyzing the complex interrelations through network structures, whereas network attributes related to roles, interactions, linkages, and metrics are discussed [41].

Based on existing research, a key feature of a communication network structure is centralization [42,43]. For a PBO, as the inequality or disparity in the number of contacts a group member has with others increases, the degree of centralization also increases [44]. Therefore, from the perspective of centralization, communication network structures can be categorized into two types: centralized structures and decentralized structures. For instance, the core periphery structure is a commonly observed centralized structure in empirical studies. It consists of an internally fully connected core subnetwork and peripheral individuals connected to the core but not each other. Those situated within the core subnetwork wield substantial control over the network's overall functions. In contrast, the peripheral participants play a pivotal role in shaping project objectives by channeling essential resources to the core [30,45]. This type of communication structure is often associated with the DBB contract delivery model. Core enterprises (such as the owners) interface with peripheral enterprises (such as the designers, constructors, and supervisors). There are no contractual relationships between the peripheral enterprises, nor is there an official flow of information.

However, an increasing number of studies indicate that centralization may also have its flaws because a person or persons in the periphery cannot share ideas or information directly with each other, or the key information that must flow through cores may be a bottleneck for collective problem solving [46,47]. Moreover, the central nodes may lead the whole organization by their bias, should they adopt bad ideas or bad solutions [48]. In contrast, decentralized structures have largely attracted more favor in which members can directly communicate with each other rather than just with a core. the structure of a complete clique includes a group of members, all of whom are connected to each other, like a large workgroup or organization members using a single shared digital communication channel. For instance, a complete clique structure wherein individuals can freely interact with others. Organization members are able to learn each other's skills and expertise more fluently and coordinate the interdependent activities between members, where each member contributes to maximize the collective benefits [29,41]. This structure is often found in the PBOs of the IPD delivery model, where each member shares responsibility and risk. Maximization of project performance is achieved based on the insights and talents of each organizational member [49,50]. In addition, the locally clustered structure consists of interconnected subgroups. Each subgroup attends to a specific aspect of the project. Interactions, such as knowledge and resource transfer, are concentrated within the boundaries of subgroups and are relatively sparse between them [13,30]. For mega-projects, when the EPC general contractor cannot handle all the tasks, multiple DB subunits are established. Each DB is interconnected through schedule and cost interfaces. Contractual relationships also exist within the subgroups. Decentralized structures are presented to adapt to shifting environments with faster information flows and increased knowledge-based work [13].

To further improve performance from a communication network structure perspective, researchers move a step closer to resolving the questions above by integrating centralization with other structure metrics (e.g., network density) or external environmental factors (e.g., adaptability to shifting environment) which unite these disparate findings under an integrated theoretical roof. However, to the best of our knowledge, related research in the construction industry still needs to be expanded to explore the characteristics and trial conditions of varying communication networks from a more wide structural metrics perspective.

2.3. Organizational Size and Network Size

A PBO is constructed to create innovative products or services [5,51]. Organizational size for a PBO can be measured by various aspects, including number of employees, turnover, and total balance sheet [52]. It has been identified that organizational size will affect the management mechanisms adopted by the organization and the interests of members [53], the organizational financial capacity, and innovative performance [54]. It is worth noting that, on the one hand, a larger organizational size improves the ability of the organization to cope with complex and uncertain tasks [55]. On the other hand, a larger organizational size is accompanied by a more complex organizational structure [56], which, in turn, increases the difficulty of management and increases the uncertainty of task completion [57].

From a network structure perspective, we obtained similar findings. The network size has a significant impact on the effectiveness of network governance. A small size is beneficial in trust-building among participants because it is easier for a small group of members to interact and communicate directly than for a large number of participants. And a network with small size can help to allocate physical and human resources to projects in a timely and accurate manner, as well as increase responsiveness and allow for the discovery and utilization of dispersed knowledge [58]. Specifically for knowledge workers, a larger communication network is not better due to time and energy constraints [59]. However, if there are too many tasks, a large-scale network with abundant resources is indispensable [58]. Based on the discussions above, we suggest that the key impact of organizational size should be further explored to improve task performances of PBOs.

2.4. Communication Network Structure, Organizational Size, and Task Performance

In recent years, people have been discussing the differences in task performance brought about by adopting centralized communication networks versus decentralized communication networks. Many studies suggest that decentralized structures have more communication channels which facilitate adapting faster to information flow and innovative work based on increased knowledge [14,15]. However, some new research suggests that centralized structures have advantages in uncertain and complex external environments. For instance, when there is staff turnover, new members might find it challenging to quickly understand and participate in communication within a decentralized structure, which may harm the task performance. In contrast, within a centralized structure, new members can quickly discern the set communication channels and established coordination procedures, proactively offer fresh insights and data to central nodes, and aid in completing tasks [44]. Recent experiment-based research aligns with this finding, indicating that within a centralized structure, the core node exclusively upholds the somewhat autonomous peripheral nodes, without imposing uniformity demands on its members. Such a setup fosters the creation and linkage of varied approaches, facilitating the dissemination and adaptation of potent concepts in an ever-evolving context [13]. For PBOs in the construction industry, the limitation of the construction period has always been a key issue in project management, and the ability to handle tasks quickly is undoubtedly beneficial. On the other hand, from the perspective of organizational size, when the organization is smaller in size, decentralized structures achieve better task performance due to their numerous communication channels [7]. However, as the organization grows larger, centralized structures can enhance task performance by increasing the average communication frequency on available paths [44]. Different types of construction projects have varying scales, and they also face numerous uncertainties.

These streams of research for the IPO model and the previous discussions lead us to question: Does the interaction between the communication network structure and organizational size affect task performance in PBOs? Under which communication network structure, and at what level of organizational size, will a higher task performance be inspired? To answer these questions, we first need to clarify whether the interaction between the communication network structure and organizational size will have an impact on the project performance for PBOs. Following this, we present the differences in task performance between centralized and decentralized structures across two organizational sides. We posit that

H1. *Communication network structure and organizational size interact to affect task performance for PBOs in construction industry.*

H1a. *When the organizational size is smaller, communication networks with decentralized structures can lead to superior task performance.*

H1b. *When the organizational size is larger, communication networks with centralized structures can lead to superior task performance.*

2.5. The Mediator of Bootleg Innovation Behavior

In an organizational environment characterized by diversity, boundarylessness, and openness, innovation is incrementally considered to be a critical way to improve performance for most types of organizations [60]. The activities, structures, and rules to achieve innovation will affect an organization's innovation speed and outcome [61]. To break deliberate strategies which may hinder organization innovation [62,63], bootleg innovation is presented as a portfolio approach, improving organizations' innovation tendency and ability [22]. Bootleg innovation is defined as bottom-up or unplanned ideation activities through which innovation ideas are initiated and elaborated without authorization but to the benefit of the organization [64]. The previous research mostly focused on bootleg innovation behavior and innovation performance from an individual perspective. For instance, the relationship between innovation climate, costs, and benefits and the willingness and behavior of organization members [24,65]. It is noteworthy that the success of bootleg innovation lies in the organization situation, not the bootlegging behavior itself [66]. Hence, a few studies have tried to explore the role of bootleg innovation from a collective perspective. For instance, the bootleg innovation tendency at an organizational level can be intentionally regulated and implemented, contributing to organizational renewal and innovation [67]. However, if an organization has constructed an appropriate innovation management structure, excessive bootlegging may dismantle its efficiency and effectiveness [68].

As an organization that aims to innovate products or services, an effective organizational structure can reduce management complexity by facilitating communication and innovation [69]. PBOs in the construction industry are usually operated within an established network structure and communication channels and facilitate the innovation process through knowledge sharing. On the one hand, organization innovation will improve task performance as well as project success [70]; on the other hand, the innovation process will take time, which is in contradiction with the time limitations for task completion of a construction project [71]. Could bootleg innovation be a strategy to resolve this paradox? Based on our literature review, research on bootleg innovation in PBOs still requires further exploration. Furthermore, as the structure of the communication network and organizational size are key parameters reflecting organizational characteristics, they will also influence the collective inclination towards innovation [72]. Therefore, examining PBOs that adopt various communication network structures and identifying at which organizational sizes they are most likely to initiate bootleg innovation, and subsequently assisting organizations to break through challenges in enhancing project performance, is of paramount significance. These arguments lead us to expect that

H2. *Bootleg innovation behavior mediates the impact of the communication network structure (centralized vs. decentralized) on task performance for PBOs in the construction industry when the organizational size is larger.*

Therefore, based on these three hypotheses, this study proposes the theoretical model shown in Figure 1.

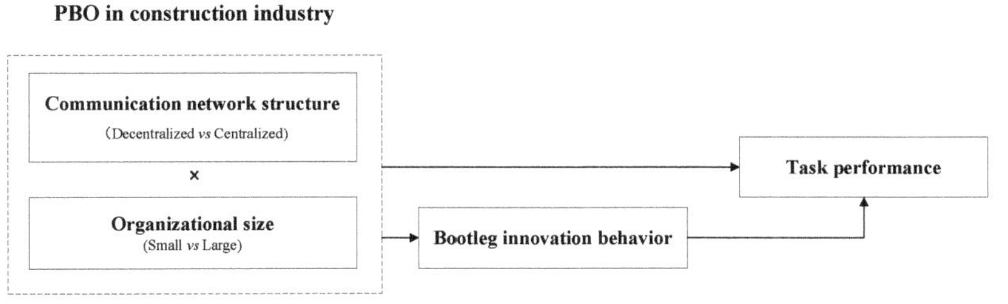

Figure 1. Theoretical model.

3. Methods

3.1. Research Overview

The aim of this study is to explore the impact of communication network structures and organizational size on task performance for PBOs in construction industry (H1). Furthermore, by studying the behavior of organizational members, we seek to determine which organizational sizes best support performance improvements within communication networks of varying centralizations. In addition, we attempt to demonstrate the conditions for the mediating effect in bootleg innovation behavior (H2).

As each PBO is unique, such as in their organizational size, mission characteristics, and environmental factors [73,74], it is difficult to accurately extract the impact of specific communication network features on project completion by directly observing and collecting data via scenario surveys. Current research has utilized gaming experiments to collect data, simulating the actual challenges faced in the work environment [44,75]. Participants are able to immerse themselves in these particular situations and give authentic feedback to accomplish the given tasks, allowing researchers to examine their actions and outcomes [76,77].

Based on the Bavelas–Leavitt–Guetzkow series of experiments [78–80] and the latest communication network experimental framework [13,44], this study designed an online communication and collaboration experiment platform to simulate the activities of PBOs in the construction industry. By observing and recording the individual work and collaborative performance of participants under specific communication conditions and task requirements, we aim to validate our hypotheses. The main features of this experimental scenario are as follows: First, the realism in the replication of the scenario, where the communication network structure is derived from existing case studies. Second, the accessibility of task execution, where the collective task execution process is similar to real-life scenarios, and task requirements are easily understood and executed by participants with construction project experience. Third, the feasibility of variable measurements, where participant performance, task completion status, and other variable measurement data can be monitored and recorded through the system platform.

3.2. Experiment Treatment

The basic setup of the experiment originates from the fundamental characteristics of PBOs, which are that various professionals achieve task objectives through communication and collaboration. Before the main experiment, we conducted a pilot study to verify the experimental design and ensure appropriate measurement of the relevant variables. In the experiment, we manipulated two variables: communication network structure and organizational size. First, the communication network structure is described through network centralization. We set up three communication structures: (1) complete clique, where all members can communicate with each other; (2) locally clustered, where subcliques can communicate internally with certain members connecting externally; (3) core periphery, where core members are interlinked and peripheral members connect with only one core member. We built the communication network using communication clients and recommended communication partners to participants based on network features. Second, based on the basic setup in a PBO where all members are part of the communication network, the organizational size is reflected by the number of members in the communication network. Based on existing research [13], this experiment establishes communication networks of two organizational sizes, with node counts of 9 and 12, respectively. Each node corresponds to a member, who play the roles of owner, designer, and contractor. They possess information specific to their professional fields. Yellow node members have more decision-making authority than blue node members. Communication network structures are formed through pairwise communication and collaboration among members. Therefore, communicating and collaborating with non-recommended members is considered as bootleg innovation behavior. Essentially, this study employed a 3 (communication network structure: complete clique, locally clustered, core periphery) × 2 (organizational size: small, large) between-subject design (Figure 2).

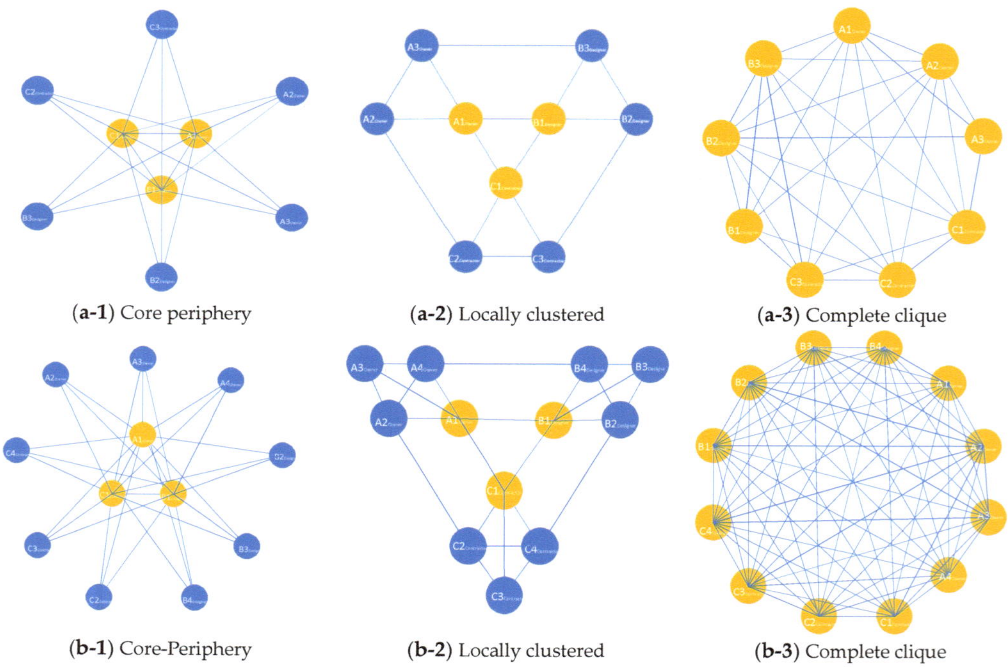

(**a-1**) Core periphery (**a-2**) Locally clustered (**a-3**) Complete clique

(**b-1**) Core-Periphery (**b-2**) Locally clustered (**b-3**) Complete clique

Figure 2. Network visualizations. (**a-1–a-3**) Network within smaller organizational size (9 points); (**b-1–b-3**) Network within larger organizational size (12 points).

Participants, after being randomly assigned roles, will obtain an exclusive information library containing ten pieces of private information. This means they have different communication and information access rights. Only roles represented by the orange dots have the authority to integrate information. Connection relationships indicate that they are recommended communication collaboration partners. Each step in the process includes specific task requirements and key hints. Under the guidance of these hints, participants search for crucial information in their respective information libraries and share it with recommended or non-recommended partners. Members with integration capabilities combine three different pieces of information to complete each process step. Notably, multiple process steps can be executed simultaneously. Within 30 min, the groups formed by the participants need to analyze tasks, extract vital data, pair up to share information, and combine the three key pieces of information to ultimately complete a program. These participants need to complete a total of 40 programs in a specific order. An example of the experimental interface can be seen in Figure 3.

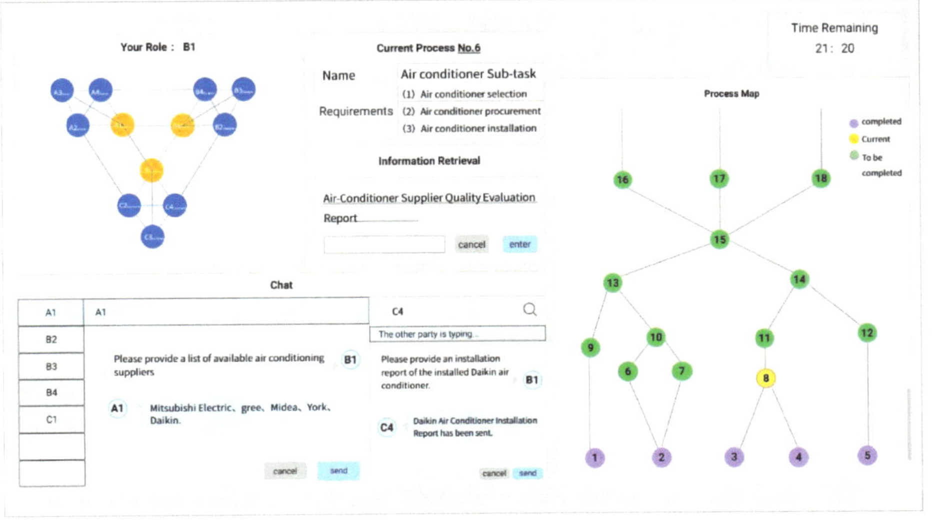

Figure 3. Screenshot of the experimental interface.

3.3. Measures

Both the mediating variables and the outcome variables in this study were measured through behavioral measures. Following a suggestion from [81], we took the number of operations completed by the group within 30 min as the task performance. Referring to the manipulative materials on pro-social rule violations [82] and based on the definition of bootleg innovation, this article takes the communication collaboration behavior between non-recommended communication partners as a measure of bootleg innovation behavior. In this context, when one member sends a message to another member, it is counted as one instance of bootleg innovation behavior.

3.4. Sample and Procedure

Utilizing expert interviews and targeted recommendations, a total of 685 participants were enlisted from Shanghai and Changsha, China. These individuals are either currently engaged in or have previous experience in engineering project management. Participants who complete the formal experiment will receive a ¥50 shopping card. Additionally, the best-performing group in each scenario will receive an extra ¥100 shopping card per person. Following the recommendations of Meade and Craig [83], as well as Shore, Bernstein, and Jang [13], a three-phase participant pipeline is established, as shown in Figure 4.

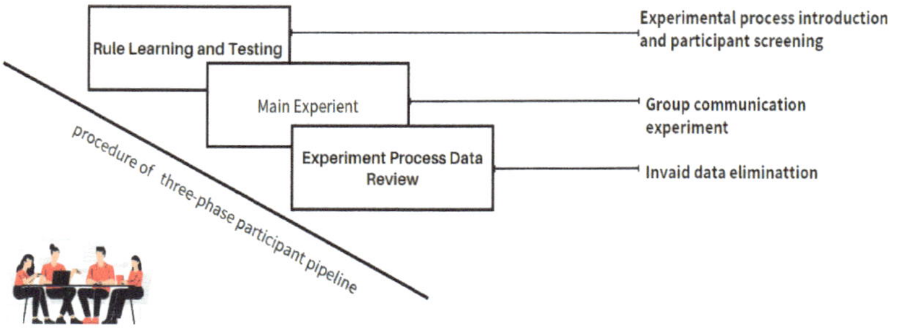

Figure 4. Experimental procedure.

First Phase: Rule Learning and Testing. Upon entering the lab, each participant sat in front of a computer in a private booth, watching a video explaining communication rules and the operating interface. After the video, participants answered seven questions about experimental rules based on a given scenario (See Appendix A). Participants who failed the test could attempt up to 3 more times.

Second Phase: Main Experiment. Participants who passed the first phase of testing entered the main experiment. They were randomly assigned to groups of 9 or 12 people and given specific roles. Under the recommended communication partners and specified role permissions, they collaborated to complete a task within 30 min through information extraction, information sharing, collective decision-making. Finally, participants filled out a survey about their basic personal information, such as gender, age, and work experience, which may influence their cognition and behavior in communication and collaboration, to serve as control variables [76,84].

Third Phase: Experiment Process Data Review. We checked each participant's message sending frequency to determine which participants sent messages significantly fewer times than the team average or sent repeated or invalid messages more than three times. Those with insufficient participation were excluded from their respective group's data. The implementation of the first phases and third phase aimed to identify and eliminate participants who did not follow instructions, lacked concentration, or were unable to complete tasks correctly. To minimize potential confounding factors, like uneven practice or fatigue effects, each participant could only join the main experiment once [85].

As this experiment was conducted in groups, we had to tolerate situations where the experiment could not commence due to an insufficient number of participants in each group. We provided opportunities for those participants who were not grouped the first time but were willing to continue to wait and join the next experimental group. To achieve a balanced sample, we repeated the procedure of the aforementioned three stages until we could no longer recruit more participants to maintain group balance, at which point data collection was halted. Ultimately, seven individuals could not partake in the main experiment due to either not passing the first-stage test or failing to be teamed up. Additionally, we excluded invalid data from three groups filtered out in the third stage and data from two groups that exceeded the sample balance. In the end, we recorded data from 10 groups in each scenario, totaling 60 groups of data. Based on the valid data from 630 participants, the average age was 32, with 57.3% male and 42.7% female participants. Furthermore, 36.2% had one to three years experience; 40.6% had three to five years experience; and 23.2% had more than five years experience.

4. Results

4.1. Manipulation Check

Referring to the research of Argote, Aven, and Kush [44], through the analysis of experimental process data, we conducted a manipulation check on the communication network structure. By examining the communication targets of group members, we observed the following: First, the communication frequency with recommended communication partners was significantly higher than with non-recommended ones ($p < 0.1$). Second, under the core periphery structure, the usage rate of available communication paths by the core nodes was slightly higher than that of other members ($p < 0.1$). In contrast, under the locally clustered structure and complete clique structure, there was no significant difference in the usage rates among members ($p > 0.1$). This suggests that the manipulation over the communication network structure from the centralization perspective is effective.

4.2. Testing of the Main Effect

First, a two-way ANOVA was carried out using the communication network structure and organizational size, taking task performance as predictors. The results indicated a significant interaction between the communication network and organizational size (F (1.54) = 25.498, $p = 0.000$, $\eta^2{}_p = 0.486$). Following this, a one-way ANOVA was used for simple effect analysis. As shown in Figure 5, when the organization size was smaller, the task performance of the locally clustered structure (M = 34.10, SD = 2.644) was significantly higher than the core periphery structure (M = 25.60, SD = 2.633; F (1.18) = 20.909, $p = 0.000$, $\eta^2{}_p = 0.742$). Additionally, the task performance of the complete clique structure (M = 31.40, SD = 3.026) was significantly higher than the core periphery structure (M = 25.60, SD = 2.633; F (1.18) =20.909, $p = 0.000$, $\eta^2{}_p = 0.573$). Thus, H1a was supported. On the other hand, when the organization size was larger, the task performance of the core periphery structure (M = 24.70, SD = 3.889) was significantly higher than the locally clustered structure (M = 20.70, SD = 1.947; F (1.18) = 8.461, $p = 0.000$, $\eta^2{}_p = 0.320$) and the complete clique structure (M = 21.40, SD = 2.675; F (1.18) = 4.888, $p = 0.004$, $\eta^2{}_p = 0.214$). Thus, H1b was supported. Therefore, H1 was validated.

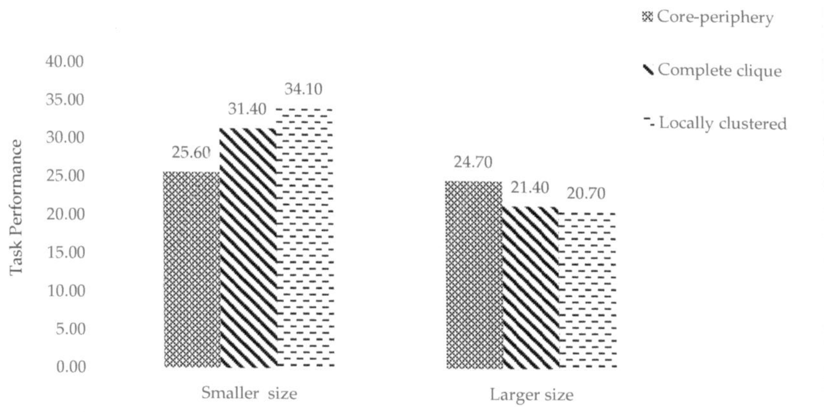

Figure 5. Task performance of interaction between communication networks and organizational size.

4.3. Testing of the Mediation Effect

The PROCESS was used to test the mediation effect. Model 4 was chosen with a sample size of 5000 and a 95% confidence interval, according to Hayes [86]. Due to the characteristic of the complete clique structure with fully interconnected members, bootleg innovation behavior does not exist and is not included as a research object. The results showed that, under a smaller organizational size, Effect (core periphery vs. locally clustered) = −0.416, 95%CI = [−0.3076, 0.0924]. The interval includes 0, indicating that the mediation effect of bootleg innovation behavior is not significant at this time. However, under a larger organizational size, Effect (core periphery vs. locally clustered) = −1.4170, 95%CI

= [−2.9063, −0.3162]. The interval does not include 0, indicating a significant mediation effect of bootleg innovation behavior at this time. Hypothesis H2 is supported.

5. Discussion

5.1. Main Findings

Although there has been an increasing number of studies using social network analysis metrics to explore the relationship between organizational collaboration and project performance for PBOs in the construction industry, research examining the role of communication network characteristics on member's bootleg innovative behavior and their impact on task performance is still scarce. Our study, which involved communication experiments based on individual tasks, information sharing, and collective decision making, with 685 construction project management professionals, yielded the following findings:

For a PBO, the impact of the communication network structure on task performance is moderated by organizational size. Specifically, when the organization is smaller in size, a decentralized structure can achieve better task performance than a centralized one. Conversely, in larger organizations, the task performance of centralized structures surpasses that of decentralized structures. This means that the influence of organizational size leads to opposite outcomes in the effects of communication network structures. Multiple factors might explain this phenomenon. In a decentralized structure, there are more central node members spread out, resulting in dispersed power. When the organization is smaller, it benefits members with various experiences and capabilities to fully utilize their potential. While completing individual innovative tasks, they use numerous communication channels to share information externally, prompting a vast amount of effective collaboration within the PBO, swiftly enhancing task performance. For instance, the decentralized structure represented by the IPD contract model promotes strategic alliances among all members throughout a project's life cycle. This can achieve efficient and rapid collaborative work, especially for highly innovative tasks. This strategy meets all organizational member interests and requirements, aiding in achieving strategic objectives [87,88]. The strength of a decentralized structure is that it encourages more members to share knowledge and contribute insights to collective decision making. However, as the organizational size grows, quickly concentrating manpower and material resources often becomes challenging, exposing the issue of incomplete information or resources under a single dispersed channel [89,90]. At this point, the central node in the centralized structure can leverage its positional advantage, helping to form Simmelian ties of extremely high strength and stickiness [91]. That is, coordinating production resources, mobilizing appropriate organizational member activities, focusing on key issues, and improving the production efficiency for a PBO. A case from the 2010 Shanghai World Expo shows that, under urgent tasks and difficult coordination, establishing a project command composed of experienced senior executives from relevant government departments can integrate various resources. With efficient communication and coordination, the project progress can be accelerated, improving completion quality [92]. Research on the mega-project of the 2012 London Olympics also supports this conclusion, suggesting that project managers play a crucial role in strengthening organizational coordination and integrating management teams [93,94].

Furthermore, this study explored the evolutionary mediating effect of bootleg innovative behavior between communication network structure and task performance across different organizational sizes. Specifically, in smaller PBOs, the mediating effect of bootleg innovative behavior between communication network structure and task performance is not significant. However, as the organization grows, bootleg innovative behavior has been validated as the intrinsic mechanism through which communication network structure affects task performance. In smaller organizations, members have relatively smooth channels to collaborate with others, establishing collective decisions based on knowledge sharing and rapidly enhancing task performance [84,95]. As organizational size increases, communication between members and core members becomes increasingly challenging. Especially within limited task timeframes, when innovation plans based on formal communication channels encounter obstacles, members might opt for bootleg innovative actions to generate more profits for the organization and demonstrate their value. The goal is to ensure the organization achieves its innovation targets, subsequently benefiting the project [23,24]. This situation mirrors the triggering scenarios of bootleg innovative behavior in traditional organizations, emphasizing members' autonomy, concealment, and informality in innovative activities [96].

5.2. Theoretical Contributions

This article describes the impact of the interplay between communication network structure and organizational size on task performance for PBOs in the construction industry. This allows us to contribute to the current literature on project management, mainly reflected in the following aspects.

Firstly, based on recent studies that explore the collaborative relationships of project organization members from a social network analysis perspective [41,97], this research advances our understanding of the relationships in organizational structures for PBOs by revealing the fundamental role of communication network structures and how communication networks with different centralizations affect task performance moderated by organizational size. Especially for large-scale PBOs, which are easily affected by unstable, complex, uncertain, and ambiguous environments over the lengthy project construction cycle, a centralized structure can play a significant organizational synergy role from the individual to the collective level, thus improving project performance [2,98].

Secondly, this study expands the literature on the application of innovative behavior in construction project management. The experimental results of this study confirm the driving force of bootleg innovation behavior on task performance under certain communication network structures. The occurrence of this behavior is closely related to the duration requirements of construction projects, reflecting the behavior of organizational members creating more profits for the organization and the desire to realize their self-worth [99,100].

Thirdly, since organizations are embedded in social systems, their strategic actions are largely influenced by the social environment. Due to the contextual and strategic nature of construction projects, their size has always been a focal point [23,101]. This study expands the element of organizational size from the conventional perspective of a control variable to an independent variable, exploring the evolutionary development relationship between organizational size and innovation behavior, thereby enriching the research boundaries of management in PBOs.

5.3. Managerial and Practical Implications

This study offers significant insights for decision makers and leaders in construction project management. Within PBOs in the construction industry, there is a need for an adaptive communication organizational structure based on the behavior and interaction coordination of organization members, which is crucial for organizational performance management.

It is worth noting that communication and coordination between organizational members play a vital role in enhancing project outcomes. When executing projects, decision makers should gather workers with diverse knowledge and backgrounds based on the project type and environmental factors and establish an appropriately centralized communication structure. On the one hand, using the 2020 Dubai Expo as an example, appointing a central coordinator as the hub is essential for large-scale PBOs. It is recommended to implement coordination among sub-projects by building management teams and coordinating resources. This approach aims to mitigate potential negative impacts brought about by daily changes and disturbances, thus improving collective outcomes [94]. The solution is approached from a central collaborative perspective. On the other hand, in the context of medium and small organizations, members with different professional expertise in a decentralized communication structure have clearer rights and responsibilities. This promotes knowledge exchange and sharing, forming efficient task decisions based on lean construction concepts. This type of organizational structure also provides adaptive collaboration for the work coordination problems that often arise during project progress [102,103].

However, leaders should also recognize that, under certain relational structures, even if formal task interdependence channels are formed between members, interactions among them may not effectively occur due to various challenges. Therefore, managers should pay attention to employees' deviant innovation behaviors and focus on individual employee autonomy in innovation and creative performances. Our research suggests that, especially in larger-size PBOs with tight project timelines, deviant innovative behaviors have a positive effect on enhancing project performance. Leaders should foster innovation among employees by strengthening their psychological capital and work enthusiasm, thereby enhancing their innovative capabilities [104–106].

6. Conclusions

This study, grounded in the IPO model and combined with relational contract management practices, explores the relationship between the communication network structure, organizational size, and task performance of PBOs in the construction industry. We did this by focusing on key indicators of communication networks—centralization and size—using a communication experimental platform. We summarized organizational structure features across different task scenarios, identifying key factors to elucidate internal organizational relational mechanisms. The number of people in the organization, the roles they play, structural hierarchies, experimental scenarios, as well as methods to measure task and goal variables, are representative both in theory and practice. Our experimental findings revealed that smaller-sized organizations achieved improved task performance when PBOs utilized a decentralized communication network. As the organizational size expanded, centralized communication-network PBOs demonstrated superior task outcomes. Furthermore, we observed that when the organizational size increases, it could stimulate deviant innovation behaviors among employees, subsequently aiding in enhancing task performance.

However, due to the diversity of types of PBOs in the construction industry, this study cannot precisely present and measure all the structures in reality. For instance, some mega-projects are more suited to a combination of multiple organizational structures. Additionally, due to the limitations of experimental conditions, the design of organizational size is referenced from classical experiments and limited to two levels. In the future, trend characteristics can be obtained by increasing measurement levels. The measurement of bootleg innovation behavior can also be expanded in multiple dimensions, including the structure to which it belongs and the manner of behavior. In the future, field experiments can be combined with laboratory experiments. Based on the findings of this study, case studies corresponding to specific scenarios can be added to validate or expand the conclusions of this research. Furthermore, our study, conducted exclusively in China, did not account for the influence of cultural variations on organizational member behavior. Future research could benefit from incorporating diverse backgrounds and participant viewpoints to broaden the scope of understanding.

Author Contributions: Conceptualization, Methodology, Investigation and Data Curation, Writing—Original Draft, X.D.; Writing—Review and Editing, L.F., Y.H., and W.L.; Funding Acquisition, X.D. and Y.H. All authors have read and agreed to the published version of the manuscript.

Funding: This research was funded by Shanghai University Teacher Training Program and the Teacher Professional Development Funding, grand number(ZZ202220047) and Science Foundation of the Hunan Provincial Department of Education, grant number (21B0295).

Institutional Review Board Statement: Not applicable.

Informed Consent Statement: Not applicable.

Data Availability Statement: The original data are held by all four authors. If there are relevant research needs, the data can be obtained by sending an email to the corresponding author. The data are not publicly available due to privacy restrictions.

Acknowledgments: The authors would like to express their sincere gratitude to the Editor and the anonymous reviewers for their insightful and constructive comments.

Conflicts of Interest: Author Long Feng was employed by Shanghai New Jing An (Group) Co., Ltd. The remaining authors declare that the research was conducted in the absence of any commercial or financial relationships that could be construed as a potential conflict of interest.

Appendix A

Table A1. Questionnaire in first phase of participant filter.

Item	Measure
1	How many people are in your team?
2	What is the total duration of your tasks?
3	Who are the recommended communication targets for you?
4	What is the ultimate goal of your team?
5	What are the requirements for completing each process?
6	Do you have the authority to assemble information to complete the process?
7	How many tasks did you finally complete?

References

1. Miterev, M.; Turner, J.R.; Mancini, M. The organization design perspective on the project-based organization: A structured review. *Int. J. Manag. Proj. Bus.* **2017**, *10*, 527–549. [CrossRef]
2. Eliwa, H.K.; Jelodar, M.B.; Poshdar, M. Information and communication technology (ICT) utilization and infrastructure alignment in construction organizations. *Buildings* **2022**, *12*, 281. [CrossRef]
3. Zhang, X.; Denicol, J.; Chan, P.W.; Le, Y. Designing the transition to operations in large inter-organizational projects: Strategy, structure, process, and people. *J. Oper. Manag.* **2023**, *in press*. [CrossRef]
4. Miterev, M.; Mancini, M.; Turner, R. Towards a design for the project-based organization. *Int. J. Proj. Manag.* **2016**, *35*, 479–491. [CrossRef]
5. Oliveira, N.; Lumineau, F. How coordination trajectories influence the performance of interorganizational project networks. *Organ. Sci.* **2017**, *28*, 1029–1060. [CrossRef]
6. Mishra, A.; Sinha, K.K.; Thirumalai, S.; Van de Ven, A. Sourcing structures and the execution efficiency of information technology projects: A comparative evaluation using stochastic frontier analysis. *J. Oper. Manag.* **2020**, *66*, 281–309. [CrossRef]
7. Du, J.; Zhao, D.; Zhang, O. Impacts of human communication network topology on group optimism bias in Capital Project Planning: A human-subject experiment. *Constr. Manag. Econ.* **2019**, *37*, 44–60. [CrossRef]
8. Bechky, B.A. Gaffers, gofers, and grips: Role-based coordination in temporary organizations. *Organ. Sci.* **2006**, *17*, 3–21. [CrossRef]
9. Wang, H.; Lu, W.; Soderlund, J.; Chen, K. The interplay between formal and informal institutions in projects: A social network analysis. *Proj. Manag. J.* **2018**, *49*, 20–35. [CrossRef]
10. Li, X.; Li, H.; Cao, D.; Tang, Y.; Luo, X.; Wang, G. Modeling dynamics of project-based collaborative networks for BIM implementation in the construction industry: Empirical study in Hong Kong. *J. Constr. Eng. Manag.* **2019**, *145*, 05019013. [CrossRef]
11. Ren, X.; Deng, X.; Liang, L. Knowledge transfer between projects within project-based organizations: The project nature perspective. *J. Knowl. Manag.* **2018**, *22*, 1082–1103. [CrossRef]
12. Zhao, D.; Duva, M.; Mollaoglu, S.; Frank, K.A.; Garcia, A.J.; Tait, J.R. Integrative collaboration in fragmented project organizations: Network perspective. *J. Constr. Eng. Manag.* **2021**, *147*, 04021115. [CrossRef]
13. Shore, J.; Bernstein, E.; Jang, A.J. Network centralization and collective adaptability to a shifting environment. *Organ. Sci.* **2022**, *34*, 2064–2096. [CrossRef]
14. Bernstein, E.; Bunch, J.; Canner, N.; Lee, M. Beyond the holacracy hype. *Harv. Bus. Rev.* **2016**, *94*, 38–49.
15. Lee, M.Y.; Edmondson, A. Self-managing organizations: Exploring the limits of less-hierarchical organizing. *Res. Organ. Behav.* **2017**, *37*, 35–58. [CrossRef]
16. Petruzzelli, A.M. The impact of technological relatedness, prior ties, and geographical distance on university–industry collaborations: A joint-patent analysis. *Technovation* **2011**, *31*, 309–319. [CrossRef]
17. Alkaissy, M.; Arashpour, M.; Zeynalian, M.; Li, H. Worksite accident impacts on construction and infrastructure: Nondeterministic analysis of subsectors and organization sizes. *J. Constr. Eng. Manag.* **2022**, *148*, 04022023. [CrossRef]
18. Xu, J.; Lu, W.; Papadonikolaki, E. Human-Organization-Technology Fit Model for BIM Adoption in Construction Project Organizations: Impact Factor Analysis Using SNA and Comparative Case Study. *J. Manag. Eng.* **2022**, *38*, 04022004. [CrossRef]

19. Reagans, R.E.; McEvily, B. Network structure and knowledge transfer: The effects of cohesion and range. *Adm. Sci. Q.* **2003**, *48*, 240–267. [CrossRef]
20. Villard, H.H. Competition, oligopoly, and research. *J. Polit. Econ.* **1958**, *66*, 483–497. [CrossRef]
21. Rothaermel, F.T.; Deeds, D.L. Exploration and exploitation alliances in biotechnology: A system of new product development. *Strateg. Manag. J.* **2004**, *25*, 201–221. [CrossRef]
22. Augsdorfer, P. Bootlegging and path dependency. *Res. Policy* **2005**, *34*, 1–11. [CrossRef]
23. Mainemelis, C. Stealing fire: Creative deviance in the evolution of new ideas. *Acad. Manag. Rev.* **2010**, *35*, 558–578.
24. Huang, D.; Zhu, T.; Wu, Y.; Sun, T. A Study on Paradoxical Leadership and Multiple Path Mechanisms of Employees' Bootleg Innovation. *Psychol. Res. Behav. Manag.* **2022**, *15*, 3391–3407. [CrossRef]
25. Liu, H.; Yu, Y.; Sun, Y.; Yan, X. A system dynamic approach for simulation of a knowledge transfer model of heterogeneous senders in mega project innovation. *Eng. Constr. Archit. Manag.* **2020**, *28*, 681–705. [CrossRef]
26. Zhou, Q.; Deng, X.; Hwang, B.-G.; Ji, W. Integrated framework of horizontal and vertical cross-project knowledge transfer mechanism within project-based organizations. *J. Manag. Eng.* **2020**, *36*, 04020062. [CrossRef]
27. Bhatti, S.H.; Zakariya, R.; Vrontis, D.; Santoro, G.; Christofi, M. High-performance work systems, innovation and knowledge sharing: An empirical analysis in the context of project-based organizations. *Empl. Relat. Int. J.* **2021**, *43*, 438–458. [CrossRef]
28. Korhonen, T.; Jääskeläinen, A.; Laine, T.H.; Saukkonen, N. How performance measurement can support achieving success in project-based operations. *Int. J. Proj. Manag.* **2023**, *41*, 102429. [CrossRef]
29. Zhang, L.; He, J.; Zhou, S. Sharing tacit knowledge for integrated project team flexibility: Case study of integrated project delivery. *J. Constr. Eng. Manag.* **2013**, *139*, 795–804. [CrossRef]
30. Garcia, A.J.; Mollaoglu, S.; Frank, K.A.; Duva, M.; Zhao, D. Emergence and evolution of network structures in complex interorganizational project teams. *J. Manag. Eng.* **2021**, *37*, 04021056. [CrossRef]
31. Hackman, J.R.; Morris, C.G. Group tasks, group interaction process, and group performance effectiveness: A review and proposed integration. In *Advances in Experimental Social Psychology*; Elsevier: Amsterdam, The Netherlands, 1975; Volume 8, pp. 45–99.
32. Mathieu, J.; Maynard, M.T.; Rapp, T.; Gilson, L. Team effectiveness 1997–2007: A review of recent advancements and a glimpse into the future. *J. Manag.* **2008**, *34*, 410–476. [CrossRef]
33. McGrath, J.E. *Social Psychology: A Brief Introduction*; Holt, Rinehart and Winston: New York, NY, USA, 1964.
34. Arashpour, M.; Abbasi, B.; Arashpour, M.; Hosseini, M.R.; Yang, R. Integrated management of on-site, coordination and off-site uncertainty: Theorizing risk analysis within a hybrid project setting. *Int. J. Proj. Manag.* **2017**, *35*, 647–655. [CrossRef]
35. Walker, D.H.; Davis, P.R.; Stevenson, A. Coping with uncertainty and ambiguity through team collaboration in infrastructure projects. *Int. J. Proj. Manag.* **2017**, *35*, 180–190. [CrossRef]
36. Bartol, K.M.; Srivastava, A. Encouraging knowledge sharing: The role of organizational reward systems. *J. Leadersh. Organ. Stud.* **2002**, *9*, 64–76. [CrossRef]
37. Bresnen, M.; Goussevskaia, A.; Swan, J. Embedding new management knowledge in project-based organizations. *Organ. Stud.* **2004**, *25*, 1535–1555. [CrossRef]
38. Thiry, M.; Deguire, M. Recent developments in project-based organisations. *Int. J. Proj. Manag.* **2007**, *25*, 649–658. [CrossRef]
39. Zhang, X.; Le, Y.; Liu, Y.; Liu, M. Fostering ambidextrous innovation in infrastructure projects: Differentiation and integration tactics of cross-functional teams. *J. Constr. Eng. Manag.* **2021**, *147*, 04021046. [CrossRef]
40. Grand, J.A.; Braun, M.T.; Kuljanin, G.; Kozlowski, S.W.J.; Chao, G.T. The dynamics of team cognition: A process-oriented theory of knowledge emergence in teams. *J. Appl. Psychol.* **2016**, *101*, 1353. [CrossRef]
41. James, O.K.; Christofer, M.H. Social networks and construction teams: Literature review. *J. Constr. Eng. Manag.* **2019**, *145*, 03119001.
42. Rulke, D.L.; Galaskiewicz, J. Distribution of knowledge, group network structure, and group performance. *Manag. Sci.* **2000**, *46*, 612–625. [CrossRef]
43. Katz, N.; Lazer, D.; Arrow, H.; Contractor, N. Network theory and small groups. *Small Group Res.* **2004**, *35*, 307–332. [CrossRef]
44. Argote, L.; Aven, B.L.; Kush, J. The effects of communication networks and turnover on transactive memory and group performance. *Organ. Sci.* **2018**, *29*, 191–206. [CrossRef]
45. Borgatti, S.P.; Everett, M.G. Models of core/periphery structures. *Soc. Netw.* **2000**, *21*, 375–395. [CrossRef]
46. Huang, S.; Cummings, J. When critical knowledge is most critical: Centralization in knowledge-intensive teams. *Small Group Res.* **2011**, *42*, 669–699. [CrossRef]
47. Becker, J.; Brackbill, D.; Centola, D. Network dynamics of social influence in the wisdom of crowds. *Proc. Natl. Acad. Sci. USA* **2017**, *114*, 5070–5076. [CrossRef]
48. Flyvbjerg, B. Delusions of success: Comment on Dan Lovallo and Daniel Kahneman. *Harvard Bus. Rev.* **2003**, *81*, 121–122.
49. Kent, D.C.; Becerikgerber, B. Understanding Construction Industry Experience and Attitudes toward Integrated Project Delivery. *J. Constr. Eng. Manag.* **2010**, *136*, 815–825. [CrossRef]
50. Bygballe, L.E.; Dewulf, G.; Levitt, R.E. The interplay between formal and informal contracting in integrated project delivery. *Eng. Proj. Organ. J.* **2015**, *5*, 22–35. [CrossRef]
51. DeFillippi, R.J. Organizational models for collaboration in the new economy. *Hum. Resour. Plan.* **2002**, *25*, 7.
52. Núñez-Pomar, J.; Prado-Gascó, V.J.; Sanz, V.A.; Hervás, J.C.; Moreno, F.C. Does size matter? Entrepreneurial orientation and performance in Spanish sports firms. *J. Bus. Res.* **2016**, *69*, 5336–5341. [CrossRef]

53. De Rooij, M.M.G.; Janowicz-Panjaitan, M.; Mannak, R.S. A configurational explanation for performance management systems' design in project-based organizations. *Int. J. Proj. Manag.* **2019**, *37*, 616–630. [CrossRef]
54. Sung, S.Y.; Choi, J.N. Multiple dimensions of human resource development and organizational performance. *J. Organ. Behav.* **2014**, *35*, 851–870. [CrossRef]
55. Ben Abdallah, S.; El-Boukri, S.; Floricel, S.; Hudon, P.A.; Brunet, M.; Petit, M.-C.; Aubry, M. A process–oriented framework to measure development performance and success of megaprojects. *Int. J. Proj. Manag.* **2022**, *40*, 685–702. [CrossRef]
56. Haveman, H.A. Organizational size and change: Diversification in the savings and loan industry after deregulation. *Adm. Sci. Q.* **1993**, *38*, 20–50. [CrossRef]
57. Shenoy, D.; Mahanty, B. Measuring the readiness of a megaproject. *Int. J. Manag. Proj. Bus.* **2021**, *14*, 999–1022. [CrossRef]
58. Yi, H.; Liu, W.; Li, F. Network structure and low-carbon governance performance: A qualitative comparative analysis. *Int. Public Manag. J.* **2020**, *24*, 23–46. [CrossRef]
59. Chen, L.; Gable, G.G. Larger or Broader: Performance Implications of Size and Diversity of the Knowledge Worker's Egocentric Network. *Manag. Organ. Rev.* **2015**, *9*, 139–165. [CrossRef]
60. Bartel, C.A.; Garud, R. The Role of Narratives in Sustaining Organizational Innovation. *Organ. Sci.* **2009**, *20*, 107–117. [CrossRef]
61. Kavadias, S.; Hutchison-Krupat, J. A framework for managing innovation. In *Informs Tutorials in Operations Research*; INFORMS: Catonsville, MD, USA, 2020; pp. 202–228.
62. Kopmann, J.; Kock, A.; Killen, C.P.; Gemünden, H.G. The role of project portfolio management in fostering both deliberate and emergent strategy. *Int. J. Proj. Manag.* **2017**, *35*, 557–570. [CrossRef]
63. Maniak, R.; Midler, C. Multiproject lineage management: Bridging project management and design-based innovation strategy. *Int. J. Proj. Manag.* **2014**, *32*, 1146–1156. [CrossRef]
64. Augsdorfer, P. Forbidden Fruit: An Analysis of Bootlegging, Uncertainty, and Learning in Corporate R&D. Ph.D. Thesis, University of Sussex, Sussex, UK, 1994.
65. Criscuolo, P.; Salter, A.J.; Wal, A.L.J.T. Going Underground: Bootlegging and Individual Innovative Performance. *Organ. Sci.* **2014**, *25*, 1287–1305. [CrossRef]
66. Lyu, L.; Zhang, H.; Gao, K. Why does distributed leadership foster or hamper bootlegging behavior of employees: The role of exploratory-exploitative learning tension and paradox mindset. *Math. Probl. Eng.* **2022**, *2022*, 3093641. [CrossRef]
67. Globocnik, D.; Peña Häufler, B.; Salomo, S. Organizational antecedents to bootlegging and consequences for the newness of the innovation portfolio. *J. Prod. Innov. Manag.* **2022**, *39*, 717–745. [CrossRef]
68. Burgelman, R.A.; Grove, A.S. Let chaos reign, then rein in chaos—Repeatedly: Managing strategic dynamics for corporate longevity. *Strateg. Manag. J.* **2007**, *28*, 965–979. [CrossRef]
69. Zhang, W.; Jiang, Y.; Zhang, W. Antecedents of online knowledge seeking of employees in technical R&D team: An empirical study in China. *IEEE Trans. Eng. Manag.* **2021**, *70*, 523–532.
70. Nawaz, A.; Tian, R. The impact of authentic leadership on project success: The mediating effect of organizational learning and innovation. *Int. J. Manag. Proj. Bus.* **2022**, *15*, 960–982. [CrossRef]
71. Iacono, M.P.; Esposito, V.; Berni, A. Temporary project network and innovation: Research on the Italian regional wine industry. *Int. J. Manag. Proj. Bus.* **2013**, *6*, 274–292. [CrossRef]
72. Lijauco, F.; Gajendran, T.; Brewer, G.J.; Rasoolimanesh, S.M. Impacts of Culture on Innovation Propensity in Small to Medium Enterprises in Construction. *J. Constr. Eng. Manag.* **2020**, *146*, 04019116. [CrossRef]
73. Zhou, Q.; Deng, X.; Hwang, B.G.; Yu, M. System dynamics approach of knowledge transfer from projects to the project-based organization. *Int. J. Manag. Proj. Bus.* **2022**, *15*, 324–349. [CrossRef]
74. Zhang, D.; Guo, P.; Zhao, J. The motives system for developing project-based inter-organizational cooperation. *Int. J. Proj. Manag.* **2021**, *40*, 167–180. [CrossRef]
75. Liu, J.; Cui, Z.; Yang, X.; Skitmore, M. Experimental investigation of the impact of risk preference on construction bid markups. *J. Manag. Eng.* **2018**, *34*, 04018003. [CrossRef]
76. Wang, Z.; Yang, X.; Chan, A.P.C.; Liu, J. Are project-based organizations willing to learn compliance lessons from sanctioned organizations close to them? The moderating effect of knowledge base compatibility and strength of the event. *Int. J. Proj. Manag.* **2021**, *39*, 672–682. [CrossRef]
77. Zhang, M.; Hou, C.; Zhang, M.; Niu, J.; Lai, Y.; Fu, H. Leveraging User Comments for the Construction of Recycled Water Infrastructure—Evidence from an Eye-Tracking Experiment. *Behav. Sci.* **2022**, *13*, 29. [CrossRef] [PubMed]
78. Guetzkow, H.S.; Simon, H.A. The impact of certain communication nets Upon organization and performance in task-oriented groups. *Manag. Sci.* **1955**, *1*, 233–250. [CrossRef]
79. Bavelas, A. Communication patterns in task-oriented groups. *J. Acoust. Soc. Am.* **1950**, *22*, 725–730. [CrossRef]
80. Leavitt, H.J. Some effects of certain communication patterns on group performance. *J. Abnorm. Soc. Psychol.* **1951**, *46*, 38. [CrossRef] [PubMed]
81. Lu, S.; Wang, L.; Ni, D.; Shapiro, D.L.; Zheng, X. Mitigating the harms of abusive supervision on employee thriving: The buffering effects of employees' social-network centrality. *Hum. Relat.* **2022**, *76*, 1441–1473. [CrossRef]
82. Morrison, E.W. Doing the job well: An investigation of pro-social rule breaking. *J. Manag. Acc. Res.* **2006**, *32*, 5–28. [CrossRef]
83. Meade, A.W.; Craig, S.B. Identifying careless responses in survey data. *Psychol. Methods* **2012**, *17*, 437. [CrossRef]

84. Xiao, Y.; Zhang, H.; Basadur, T.M. Does information sharing always improve team decision making? An examination of the hidden profile condition in new product development. *J. Bus. Res.* **2016**, *69*, 587–595. [CrossRef]
85. Charness, G.; Gneezy, U.; Kuhn, M.A. Experimental methods: Between-subject and within-subject design. *J. Econ. Behav. Organ.* **2012**, *81*, 1–8. [CrossRef]
86. Hayes, A.F. *Introduction to Mediation, Moderation, and Conditional Process Analysis: A Regression-Based Approach*; Guilford Publications: New York, NY, USA, 2017.
87. Matthews, O.; Howell, G.A. Integrated project delivery an example of relational contracting. *Lean Constr. J.* **2005**, *2*, 46–61.
88. Kaufmann, C.; Kock, A. Does project management matter? The relationship between project management effort, complexity, and profitability. *Int. J. Proj. Manag.* **2022**, *40*, 624–633. [CrossRef]
89. Cattani, G.; Ferriani, S. A core/periphery perspective on individual creative performance: Social networks and cinematic achievements in the Hollywood film industry. *Organ. Sci.* **2008**, *6*, 824–844. [CrossRef]
90. Englmaier, F.; Grimm, S.; Schindler, D.; Schudy, S. *The Effect of Incentives in Non-Routine Analytical Team Tasks-Evidence from a Field Experiment*; Working Paper No. 6903; CESifo Gmbh: Munich, Germany, 2018.
91. Jayaraj, S.; Doerfel, M.L.; Williams, T. Clique to win: Impact of cliques, competition, and resources on team performance. *J. Constr. Eng. Manag. Decis.* **2022**, *148*, 04022047. [CrossRef]
92. Zhai, Z.; Ahola, T.; Le, Y.; Xie, J. Governmental governance of megaprojects: The case of EXPO 2010 Shanghai. *Proj. Manag. J.* **2017**, *48*, 37–50. [CrossRef]
93. Davies, A.; Mackenzie, I. Project Complexity and Systems Integration: Constructing the London 2012 Olympics and Paralympics Games. *Int. J. Proj. Manag.* **2014**, *32*, 773–790. [CrossRef]
94. Shen, W.; Ying, W. Large-scale construction programme resilience against creeping disruptions: Towards inter-project coordination. *Int. J. Proj. Manag.* **2022**, *40*, 671–684. [CrossRef]
95. Lightle, J.P.; Kagel, J.H.; Arkes, H.R. Information exchange in group decision making: The hidden profile problem reconsidered. *Manag. Sci.* **2009**, *55*, 568–581. [CrossRef]
96. Lin, B.; Mainemelis, C.; Kark, R. Leaders' responses to creative deviance: Differential effects on subsequent creative deviance and creative performance. *Leadersh. Q.* **2016**, *27*, 537–556. [CrossRef]
97. Shi, Q.; Chen, X.; Xiao, C.; Han, Y. Network Perspective in Megaproject Management: A Systematic Review. *J. Constr. Eng. Manag.* **2022**, *148*, 03122002. [CrossRef]
98. Naderpajouh, N.; Matinheikki, J.; Keeys, L.A.; Aldrich, D.P.; Linkov, I. Resilience and projects: An interdisciplinary crossroad. *Proj. Leadersh. Soc.* **2020**, *1*, 100001. [CrossRef]
99. Lo, J.T.; Kam, C. Innovation of organizations in the construction industry: Progress and performance attributes. *J. Manag. Eng.* **2022**, *38*, 04022064. [CrossRef]
100. Li, H.; Zhang, Y. How Does Secure-Base Leadership Affect Employees' Taking-Charge Behavior: The Role of Psychological Availability and Independent Self-Construal. *Behav. Sci.* **2023**, *13*, 853. [CrossRef] [PubMed]
101. Fergusson, H.; Langford, D. Strategies for managing environmental issues in construction organizations. *Eng. Constr. Archit. Manag.* **2006**, *13*, 171–185. [CrossRef]
102. Wen, Q.; Qiang, M. Coordination and knowledge sharing in construction project-based organization: A longitudinal structural equation model analysis. *Autom. Constr.* **2016**, *72*, 309–320. [CrossRef]
103. Steen, J.; De Fillippi, R.; Sydow, J.; Pryke, S.; Michelfelder, I. Projects and networks: Understanding resource flows and governance of temporary organizations with quantitative and qualitative research methods. *Proj. Manag. J.* **2018**, *49*, 3–17. [CrossRef]
104. Rego, A.; Sousa, F.; Marques, C.; e Cunha, M.P. Authentic leadership promoting employees' psychological capital and creativity. *J. Bus. Res.* **2012**, *65*, 429–437. [CrossRef]
105. Paterson, T.A.; Luthans, F.; Jeung, W. Thriving at work: Impact of psychological capital and supervisor support. *J. Organ. Behav.* **2014**, *35*, 434–446. [CrossRef]
106. Osman, A.M.; Liu, Y.; Wang, Z. Influence of Organizational Culture on Construction Firms' Performance: The Mediating Roles of Innovation and Marketing Capabilities. *Buildings* **2023**, *13*, 308. [CrossRef]

buildings MDPI

Review

Extended Reality (XR) Training in the Construction Industry: A Content Review

Shixian Li [1], Qian-Cheng Wang [2,*], Hsi-Hsien Wei [1] and Jieh-Haur Chen [3]

[1] Department of Building and Real Estate, Hong Kong Polytechnic University, Hung Hom 999077, Hong Kong; hhwei@polyu.edu.hk (H.-H.W.)
[2] Department of Land Economy, University of Cambridge, Cambridge CB3 9EP, UK
[3] Department of Civil Engineering, Research Center of Smart Construction, National Central University, Jhongli, Taoyuan 320317, Taiwan
* Correspondence: qw250@cam.ac.uk

Abstract: As modern information technology advances and equipment devices update, extended reality (XR) technologies, including virtual reality (VR), augmented reality (AR), and augmented virtuality (AV) have witnessed an increasing use and application in construction training. This review aims to comprehensively examine the evolution of XR training in the construction domain. To achieve this, a systematic literature review of 74 journal papers from the Scopus database was conducted. This paper outlines the progression of XR training from 2009 to 2023, detailing related technologies like development platforms, display devices, and input devices. The literature review reveals that XR application in construction training spans five main areas: (1) safety management, (2) skill/knowledge acquisition, (3) equipment operation, (4) human–computer collaboration, and (5) ergonomics/postural training. Additionally, this review explores the impact of trainee roles on XR training outcomes and identifies the challenges faced by XR technology in construction training applications. The findings of this literature review are hoped to assist researchers and construction engineering trainers in understanding the latest advancements and challenges in XR, thereby providing valuable insights for future research.

Keywords: extended reality (XR); virtual reality (VR); augmented reality (VR); augmented virtual reality (AV); construction training

Citation: Li, S.; Wang, Q.-C.; Wei, H.-H.; Chen, J.-H. Extended Reality (XR) Training in the Construction Industry: A Content Review. *Buildings* **2024**, *14*, 414. https://doi.org/10.3390/buildings14020414

Academic Editors: Jurgita Antucheviciene and Osama Abudayyeh

Received: 30 November 2023
Revised: 16 January 2024
Accepted: 31 January 2024
Published: 3 February 2024

1. Introduction

The construction industry, often referred to as the backbone of urban development [1], necessitates a workforce with skills and adaptability to effectively navigate the intricate challenges inherent in powering social advancement [2] and achieving sustainable development goals [3,4]. Training plays a crucial role in the construction industry, serving as a vital link between theoretical knowledge and practical application. Notably, the advancements in modern information technology have facilitated the rise and widespread adoption of virtual reality (VR), augmented reality (AR), and related virtual technology in construction training [5]. Some studies also indicate that these technologies, compared to traditional training methods, enhance the effectiveness of training [6–8].

The terminology associated with virtual technologies lacks clarity in both academic and industrial contexts, encompassing terms such as virtual reality (VR), augmented reality (AR), mixed reality (MR), and extended reality (XR) [5]. For example, Wedel et al. [9] stated that MR combines VR and AR, while Zhao et al.'s literature review [10] positions MR alongside VR and AR, collectively denoted as XR. To enhance clarity and distinguish these perplexing concepts for the readers, this study incorporates the concept of a "virtuality continuum" proposed by Milgram and Kishino in 1994 [11]. As shown in Figure 1, this conceptual framework effectively distinguishes between AR, augmented virtuality (AV), and VR. Specifically, the concept of AR can be interpreted as overlaying computer-generated digital information onto the real-world environment, creating a seamless integration [12,13]. VR refers to a technology that creates computer-generated simulated environments that users can explore and interact with [13]. In addition, AV blending real-world elements into VR employs real objects as input devices for interaction within the virtual environment during construction training [14]. In this review, XR is utilized as an umbrella term encompassing virtual technologies, including VR, AR, and AV [11].

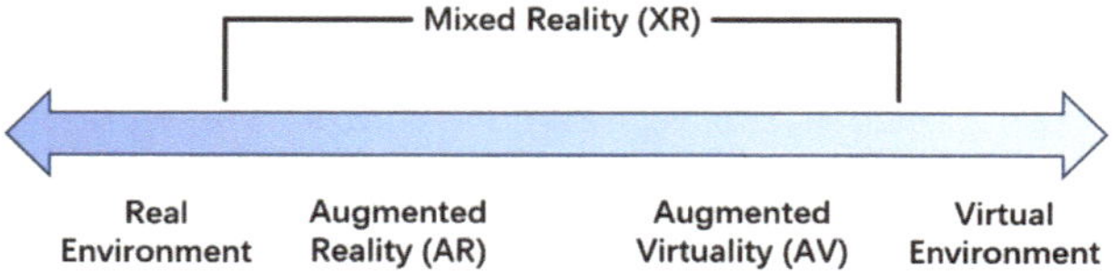

Figure 1. Reality-virtuality continuum.

The benefits of employing extended reality (XR) training in the construction industry are diverse. For example, VR simulations provide a risk-free setting for trainees to immerse themselves in realistic construction scenarios, refining their skills without compromising safety [15–17]. These technologies also facilitate collaborative training experiences, allowing professionals to participate in shared virtual environments and overcome geographical distances and disciplinary barriers [18,19]. Furthermore, VR can increase a trainees' motivation and decrease the cost compared to traditional training methods [19–21]. Additionally, AR can aid construction workers in task completion, contributing to reduced task duration and minimizing the likelihood of errors [22].

While XR training offers some advantages compared to traditional methods, it is undeniable that these technologies are still in their early developmental stages and have certain limitations. The creation of training scenarios and the achievement of realism encounters challenges. For example, current devices limit the trainee's visual perspective [23] with sensory feedback primarily focused on visual and auditory aspects, lacking tactile, olfactory, and other stimuli [24]. Additionally, developing high-quality training content demands a significant investment of both resources and time. Most VR training relies on head-mounted displays (HMD), allowing only one person to undergo training at a time, creating difficulties in scaling up simultaneous training for multiple participants on a larger scale. It is important to note that VR, AR, and AV technologies are different methods during training, although they are often grouped together. Therefore, a comprehensive literature review of XR technology in the construction field is essential. This will provide scholars and practitioners in related domains with background knowledge on XR training, enhancing their understanding of the current development status and the challenges faced in its implementation.

Furthermore, literature reviews pertaining to similar topics are also examined and evaluated in this study. It was revealed that there was a lack of research on the application of XR in the field of construction training. In terms of the review scope, certain reviews take a broader scope, offering a comprehensive examination of research within the entire architecture, engineering, and construction (AEC) field, maintaining a general focus but lacking a specific review dedicated to the construction sector [25–28]. For instance, Zhang et al. [26] utilized a mixed quantitative–qualitative review method, analyzing 206 journal articles to explore research trends and opportunities for VR application in the AEC industry. Other reviews delve into the intersection of education and construction training [29,30]. While education and construction training share a close relationship, they differ in their objectives, with education emphasizing knowledge transfer to students and focusing more on theoretical foundations, while construction training targets workers tasked with complex and hazardous construction site activities, emphasizing practical orientation. Furthermore, some reviews concentrate on construction safety training [31–34], but construction training extends beyond safety training, encompassing areas such as construction equipment operation [35]. A research gap emerges from these literature reviews, specifically the absence of a critical review focusing on XR technology in the realm of construction training.

To address the research gaps mentioned above, this review aims to provide a comprehensive overview of extended reality (XR) research pertaining to construction training, encompassing virtual reality (VR), augmented reality (AR), and augmented virtuality (AV). The specific objectives include as follows: (1) delineating the current state of development in XR training and associated technology applications; (2) examining the implementation areas of these studies within construction training; (3) investigating the influence of XR training participants on training outcomes; (4) and exploring challenges encountered in the domain of VR, AR, and AV technologies in construction training. The subsequent sections of this paper are structured as follows: Section 2 outlines the review protocol and methodology, including research questions, a systematic literature search strategy, and the data analysis of selected publications. In Section 3, the current research status of XR in construction training is provided, covering an analysis of the identified publications and specific techniques. Section 4 categorizes the application of XR in construction training. Section 5 focuses on the participants in the XR training experiments. Additionally, Section 6 delineates the challenges inherent in XR training. Finally, Section 7 concludes this literature review.

2. Review Protocol and Methodology

The literature review protocol adheres to the guidelines outlined by [36], comprising four essential phases for evaluating a quality literature review: (1) design, (2) conduct, (3) data abstraction and analysis, (4) and structuring and writing the review. As stated by Snyder [36], the design phase necessitates the formulation of a clear and well-motivated research question, followed by the selection of an appropriate review methodology tailored to the research query. It is imperative to establish a transparent search strategy, incorporating relevant search terms, explicit inclusion, and exclusion criteria. This section provides research questions to be addressed and the search strategy to be utilized, in detail.

2.1. Research Questions

As mentioned before, this paper aims to comprehensively understand the research of XR training in the construction domain. The specific research questions that this review paper intends to address are the following:

- Research Question 1 (Q1): What is the status of XR training in the construction industry?
- Research Question 2 (Q2): What are the applications of XR training in the construction industry?
- Research Question 3 (Q3): How do the participants of the XR training experiment have an impact on training results?
- Research Question 4 (Q4): What are the challenges of XR training faced in the construction industry?

2.2. Search Strategy

In this review, Scopus was utilized for literature searches, which is managed by Elsevier Publishing company and contains the metadata for over 82 million documents and more than 1.7 billion references [37]. A preliminary research search was conducted on selected sources using relevant search terms. Since this paper seeks to study the application of extended reality technology in construction training, the search terms were divided into three parts: research technology, target research areas, and target research purpose. The specific search string used to query Scopus was as follows:

TITLE-ABS ("VR" OR "virtual reality" OR "AR" OR "augmented reality" OR "XR" OR "extended reality" OR "MR") AND TITLE-ABS (construction OR "construction site" OR "construction project" OR "construction management" OR "construction engineering" OR "construction industry") AND TITLE-ABS (training OR "construction training" OR operation).

The initial search identified 2390 studies, which were then filtered based on publication date, with only studies published between January 2009 and December 2023 being considered. The document type was restricted to journal articles, the publication stage was final, and the language was limited to the English language. To further refine the selected articles, the study also narrowed the scope of subjects to engineering, computer science, social sciences, energy, environmental science, psychology, multidisciplinary, and decision sciences.

Following the initial screening, 506 articles were obtained. The researchers then conducted a cursory full-text reading of the 506 articles to exclude duplicates, articles outside the field of construction industry, articles irrelevant to the training, and articles where the full text was unavailable. Ultimately, a total of 74 articles were selected for an in-depth literature analysis. The database literature screening process is illustrated in Figure 2.

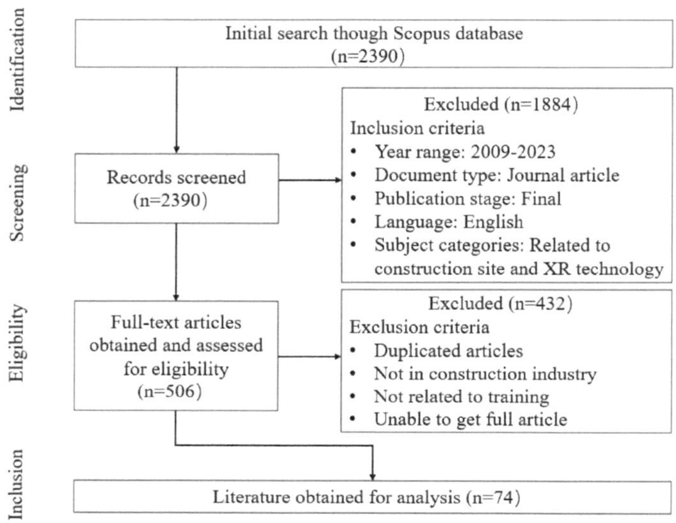

Figure 2. The literature screening process.

2.3. Data Analysis

The ultimate inclusion of 74 publications undergoes a comprehensive content review. To address the aforementioned research questions, specific data are selected from the chosen articles during the examination of their contents. The data utilized in this study are shown in Table 1.

Table 1. Data adopted for content analysis.

Data	Descriptions
Publication year	The year of publications, from 2009 to 2023
Publication source	The Journals that feature the chosen publications
Country	The first author's country where the selected publication originated
The adoption of XR technology	The XR technology employed in the selected publications, including VR, AR, AV
Keyword	The author keywords of the selected publications
XR technology	The XR technology in the selected publications, including development platforms, display devices, and input devices
XR application	Categories of XR application in construction training in the selected publications
Participants	The roles of participants in XR training experiments in the selected publications
Challenges	The challenges stated in XR training in the selected publications

3. Current Development of XR Training in Construction Industry

3.1. Overview of Identified Publications

Figure 3 illustrates the number of published articles about XR in construction training from 2009 to 2023. The trend in article publications was basically stable from 2009 to 2019, consistently featuring fewer than five articles annually. This pattern may be attributed to the construction industry's slower adoption of XR technology, trailing behind fields like manufacturing and automated driving. Commencing in 2020, a notable surge in published studies is observed, showing incremental growth throughout the years 2020 to 2023. This substantial increase is postulated to be linked to the emergence of the COVID-19 pandemic at the close of 2019. The global outbreak impacted the construction industry [38,39], disrupting normal operations and contributing to the surge in remote and online technologies, which thrived amid restrictions on large-scale offline gatherings. The escalating number of publications post-2020 underscores the sustained academic interest in investigating XR training within the construction domain.

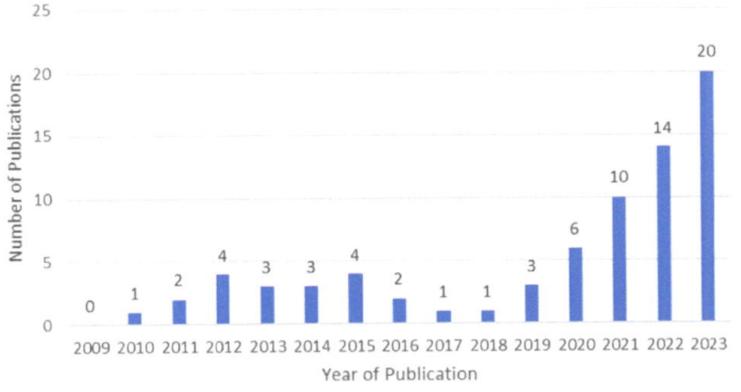

Figure 3. Number of publications from 2009 to 2023.

Table 2 presents the distribution of the 74 selected journal articles by publication source. Over 33 journals featured articles on XR technology in construction training. The top five journals, ranked by the number of selected articles, were Automation in Construction (14.9%), Journal of Computing in Civil Engineering (12.2%), Advanced Engineering Informatics (8.1%), Journal of Construction Engineering and Management (6.8%), and Safety Science (5.4%). These five journals collectively accounted for approximately half of the total selected articles. The distribution shows that most XR training articles were published in technology-related journals.

Table 2. Distribution of the selected journal articles by publication source.

Journal Title	Number of Selected Articles
Automation in Construction	11
Journal of Computing in Civil Engineering	9
Advanced Engineering Informatics	6
Journal of Construction Engineering and Management	5
Safety Science	4
Engineering, Construction, and Architectural Management	3
Construction Innovation	3
Buildings	3
Applied Sciences	2

Table 2. *Cont.*

Journal Title	Number of Selected Articles
Computer Applications in Engineering Education	2
International Journal of Computers Communications and Control	2
Sustainability	2
Accident Analysis and Prevention	2
Applied Ergonomics	1
CivilEng	1
Construction Management and Economics	1
Developments in the Built Environment	1
Education Sciences	1
Electronic Journal of Information Technology in Construction	1
i-com	1
IEEE Transactions on Learning Technologies	1
IEEE Transactions on Visualization and Computer Graphics	1
International Journal of Computational Methods and Experimental Measurements	1
International Journal of Injury Control and Safety Promotion	1
Journal of Architectural Engineering	1
Journal of Civil Engineering and Management	1
Journal of Intelligent and Robotic Systems: Theory and Applications	1
Journal of Robotics and Mechatronics	1
Journal of Safety Research	1
Journal of Surveying Engineering	1
Scientific World Journal	1
Virtual Reality	1
Visual Computer	1
Total	74

As shown in Figure 4, regarding the geographical affiliation of the first author, the United States leads with 38% of published papers. Subsequent to this, South Korea emerges as the second-highest contributor at 11%, followed by China (9%) and Australia (8%). Other countries have made comparatively modest contributions, with none exceeding three publications over the past 15 years.

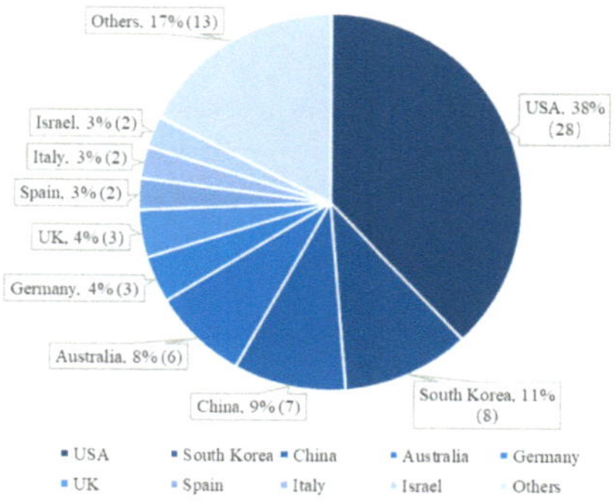

Figure 4. Distribution of publications by country.

The percentage of the research on XR technology in construction training is elucidated in Figure 5. VR technology takes the lead, constituting 89% of the applications, followed by AR at 8%, and AV at the lowest, with a mere 3%. This distribution indicates a predominant reliance on VR technology within XR applications for construction training, underscoring its substantial growth over the past 15 years.

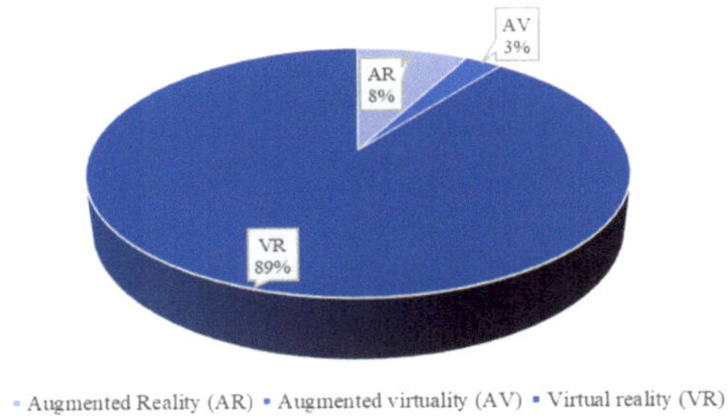

Figure 5. The percentage of VR/AR/AV applications in construction training.

Figure 6 depicts the co-occurrence network of author keywords, generated through VOSviewer 1.6.20, providing insights into the thematic content within this research domain. To improve visualization, the minimum occurrence threshold for author keywords was set to two. It is crucial to note that synonymous keywords were consolidated into representative terms. For instance, "VR" encompasses "virtual reality" and "virtual reality (VR)", and "hazard recognition" is used for both "hazard identification" and "hazard perception". After this consolidation process, 32 keywords were identified. In Figure 6, the size of the nodes corresponds to the frequency of keyword occurrences, and the thickness of the link between nodes indicates the strength of the relationship between node terms, with thicker lines representing stronger associations. Notably, XR technology in construction training emerges with the highest frequency and total link strength, prominently featuring VR. This underscores VR's widespread adoption and relative maturity in the current construction training research landscape. Apart from "VR", prevalent themes include "safety training", "construction safety", and "hazard recognition", emphasizing the collective focus on enhancing construction safety and hazard awareness. Keywords at the periphery of the network graph, such as "human–robot collaboration", "construction worker", and "user experience", although represented by smaller nodes and weaker associations, suggest potential emerging research themes that may evolve in the future and warrant attention.

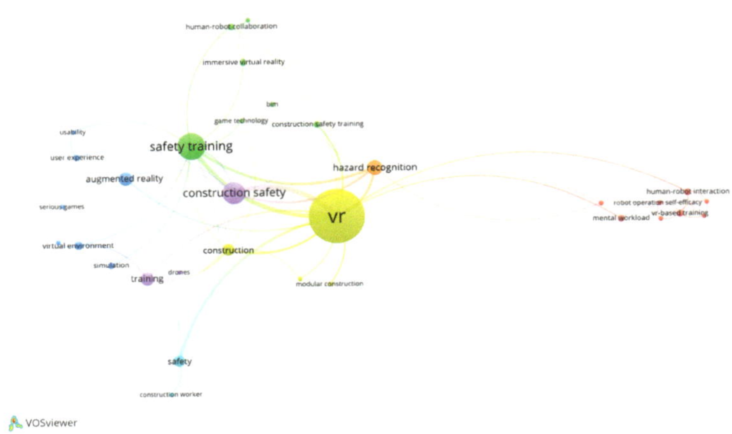

Figure 6. Author keyword co-occurrence network.

3.2. XR Training Technology

XR technology is evolving at an unprecedented rate, with tremendous benefits to the research and industry [40]. Access to new display and input devices for the consumer market, offering affordable pricing models, has accelerated the diffusion of XR technology [37]. This sub-section introduces the development platform, display device, and input device of XR technology in construction training.

3.2.1. Development Platform of XR

In the development of XR experience, a commonly employed approach involves the utilization of a game engine or a specialized development platform that has native support for XR. The game engine serves as the heart that encompasses nearly all of the functions and features required to develop a fully realized game [41,42]. It provides developers with a set of tools and functionality to assist and simplify the development, creation, and management of digital games, simulations, or other interactive applications. Modern game engines provide key features to create realistic and immersive virtual environment scenarios, including graphics rendering, physics simulation, scene graphs, audio systems, animation systems, scripting and programming interfaces, and networking functions [41,43]. For example, a graphics rendering system involves rendering objects, textures, lighting, and visual effects to enhance the fidelity of a virtual environment. In a study by Luo et al. [44], it was found that enhancing scenario fidelity significantly improves the sense of presence and usability in forklift safety training for workers. The audio system enables developers to seamlessly integrate and manage sound effects and other audio elements in construction training; Han et al. [45] introduced background sound to enhance immersion in scaffold-based safety training using VR.

Within the realm of XR training for the construction field, Unity and Unreal Engine stand out as the most widely utilized game engines. Unity3D, for example, features multi-platform system support, offering both free and commercial versions. Games developed using the Unity3D game engine can be exported as standalone applications for macOS and MS Windows, consoles such as Xbox and Wii, and smartphones running iOS, Android, Blackberry, and Windows [46]. Moreover, Unity's asset store comprises an extensive collection of images, 3D models, scripts, sound effects, and complete games, offering developers a time-saving resource [41]. Unreal Engine is renowned for its high-quality rendering effects, featuring a 64-bit color High Dynamic Range (HDR) rendering pipeline in Unreal Engine 4. This pipeline incorporates a diverse range of post-processing effects, including motion blur, depth of field, bloom, and screen space ambient occlusion [41]. In addition, both Unity and Unreal Engine have large communities; support is readily available from a diverse user base, comprising both professional and amateur developers, who actively contribute assistance in the community forums.

Other development platforms are less frequently used in the construction training field, such as WebXR, Oculus First Contact game, Second Life, and Torque 3D engine. WebXR, based on WebGL, enables the creation of XR experiences accessible through web browsers [47]. This approach allows trainees to access simulations with minimal hardware requirements. Dhalmahapatra et al. [48] employed the Oculus First Contact game to instruct users in comprehending the sequence of electric overhead crane operations and addressing potential hazards during work. Users can utilize the Oculus controllers to interact with the virtual agent within the environment. Second Life is not a game or game engine; it is a massive multiplayer 3D virtual world designed for social interactions, where users can meet, chat, play, explore, and build virtual spaces [41]. Hence, Second Life is regarded as a communication platform within the context of construction training, facilitating individuals from diverse geographical locations to engage in concurrent online learning [19]. As Cowan and Kapralos [41] pointed out, Torque 3D, available under the MIT open-source license, is free and supports Windows and major web browsers. It uses TorqueScript scripting language, featuring a syntax similar to C++, to accomplish the majority of the game engine's functionalities. Table 3 summarizes the advantages and disadvantages of the XR development platform.

Table 3. Advantages and disadvantages of the XR development platform.

Development Platform	Advantages	Disadvantages	Application Example
Unity Engine	• Large and user-friendly community support • Wide range of supported platforms • Offers pre-built assets and plugins by the Asset Store • Supports Cross-platform development	• Slightly lower graphical fidelity compared to Unreal Engine • Primarily uses C#, unfriendly to developers preferring other languages	[18,44,49]
Unreal Engine	• High-quality graphics and realistic rendering • Powerful visual scripting with Blueprints • Robust community and support • Includes built-in VR development tools	• Features a steeper learning curve compared to some other engines • Some tasks may require C++, posing challenges for non-programmers	[16,50]
WebXR	• Allows creating VR experiences accessible via web browsers • Compatible with VR headsets supporting WebXR	• Limited to certain browsers and devices	[47]
Oculus First Contact Game	• Designed to be an introductory experience, accessible for beginners in VR development. • Provides access to Oculus-specific features • Optimized performance for Oculus devices	• Exclusive to Oculus hardware	[48]
Second Life	• Excels in creating vast, user-generated virtual worlds • Designed to be user-friendly for content creation in its virtual environment	• Not as robust as professional game engines in complex virtual world	[19,51]
Torque 3D Engine	• Open-source (MIT), allowing developers to modify and customize the engine • Supports multiple platforms for greater flexibility	• Smaller community compared to Unity and Unreal, potentially resulting in fewer resources and support	[52,53]

3.2.2. Display Device

The effectiveness of XR technology in construction training is influenced by the immersive level of display devices. The immersive level determines the depth of engagement and realism experienced by trainees within the virtual construction environments. Accordingly, XR display devices can be classified into three categories based on their immersive levels: non-immersive display (e.g., 2D display), semi-immersive display (e.g., Cave Automatic Virtual Environment (CAVE), Microsoft HoloLens), and immersive display (e.g., Head-Mounted Display (HMD)).

Two-dimensional display devices, such as desktop monitors [54] and mobile screens [52], represent the entry level of XR immersion. While these devices lack the depth and immersive qualities of more advanced options, they remain valuable for certain construction training scenarios. For example, mobile screens, found on smartphones and tablets, offer a portable and accessible XR experience. Gomes Jr et al. [55] used a tablet to detect the equipment from various user perspectives within the industrial scenario. This approach facilitates the real-time incorporation of actual information about these pieces of equipment through AR annotations.

CAVE systems represent a semi-immersive solution for construction training, providing a more engaging experience than 2D displays. CAVEs consist of immersive projection environments where virtual content is displayed on multiple surfaces, creating a room-sized virtual experience [56]. This immersive projection enhances spatial awareness and allows trainees to interact with the virtual construction site on a larger scale [17]. The use of motion tracking and gesture-based input devices further enhance the realism of the training experience [57]. One notable advantage of the CAVE systems, is their ability to facilitate collaborative learning [58]. Multiple trainees can simultaneously participate in the same virtual construction scenario, fostering teamwork and communication. This collaborative aspect makes CAVEs well suited for group training sessions and team-building exercises within the construction industry. In addition, because of the fundamental characteristics of AR technology, which entails superimposing computer-generated content onto the user's real surroundings, all AR-related display devices cannot offer a completely immersive experience. Consider Microsoft HoloLens as an example. While wearing HoloLens, users can maintain a clear view of their surroundings while concurrently observing holographic images seamlessly integrated into the real-world environment [59].

VR HMDs offer a fully immersive experience by seamlessly integrating virtual content into the trainee's field of view [60]. These head-worn devices, resembling visors, completely envelop the user's eyes and ears, creating a deeply immersive environment that isolates users from external stimuli. Offering a first-person perspective, VR HMDs transport trainees to a virtual construction environment, fostering a sense of physical presence [61]. Moreover, VR HMDs enable realistic simulations by tracking the user's head movements and adjusting the virtual environment accordingly [40,62]. This responsiveness contributes to the authenticity of construction training, enabling trainees to inspect virtual structures, interact with virtual objects, and practice construction tasks with an elevated sense of presence. Therefore, numerous studies chose headset devices for construction training, such as Oculus Rift [6], HTC Vive [63], Samsung Odyssey [64], Oculus Quest [65], and HP Reverb [21] et al.

3.2.3. Input Device

Input devices are the bridge between the physical actions of trainees and their interactions within virtual construction environments, including keyboard and mouse, touch screen, controller/joystick, tracking devices, and specialized input devices. Traditional input devices like keyboards and mouses remain relevant in XR construction training, especially in desktop-based training cases [66]. Touchscreen interfaces extend the reach of XR training to devices such as tablets and interactive displays [35]. Trainees can directly manipulate virtual elements by tapping, swiping, and pinching touch-sensitive screens. Moreover, controllers and joysticks are handheld devices that often include buttons, triggers, and other input mechanisms to facilitate various interactions within the virtual environment [40]. In construction training applications, motion controllers simulate the handling of tools and equipment, allowing trainees to practice tasks ranging from bricklaying to operating heavy machinery [6,8,15]. This can be particularly useful for hands-free interactions and can enhance the overall user experience. Tracking devices, such as sensors and cameras, play a critical role in capturing the real-world movements of trainees. These devices enable accurate positional tracking, allowing XR systems to replicate the trainee's movements within the virtual environment [8,14,65,67]. In construction training, tracking devices contribute to a high level of realism by ensuring that virtual interactions closely mirror the trainee's physical actions. Some specialized input devices are also adopted to enhance the training experience, such as operation panels [23], steering wheels [44], and foot pedals [18].

4. XR Application in Construction Training

From the review, XR applications in construction training can be categorized into five groups, including (1) safety training; (2) skill/knowledge acquisition; (3) equipment operation; (4) human–robot collaboration; (5) and ergonomics/postural training. Table 4 represents the distribution of these publications.

Table 4. The distribution of publications characterized by XR application in construction training.

XR Training Application	Representative Studies	NO. of Studies
Safety management	[8,14,17,19,20,24,44–50,52–54,57,64–66,68–91]	44
Skill/knowledge acquisition	[21,22,51,55,92–98]	11
Equipment operation	[6,15,18,23,35,99–103]	10
Human-robot collaboration	[7,63,104–108]	7
Ergonomics/postural training	[16,67]	2
Total		74

4.1. Safety Management

In the realm of construction training, the predominant current research focuses on safety management. Over the past 15 years, 44 papers have been published, constituting 59.5% of the total selected papers. Construction sites pose various risks, such as fall hazards, electrical hazards, objects striking, and hazards related to collapse or caught-in/between situations [24,45]. Ensuring the safety of workers is paramount to the construction industry, and XR technology presents an innovative approach to training that surpasses the traditional methods. Researchers achieve this by devising construction tasks or scenarios that incorporate one or more hazards, allowing safety managers or workers to enhance hazard perception awareness, reduce accidents, or gain sufficient knowledge to respond effectively to hazardous situations [49,68].

XR technology facilitates the development of highly realistic safety simulations, enabling trainees to encounter hazardous scenarios within a controlled virtual environment. Moreover, Trainees can use XR devices to inspect the environment, identify hazards, and make informed decisions on how to address or avoid them. In a study conducted by Rey-Becerra et al. [20], VR was employed to replicate an overhead work scenario on a construction site. Participants, equipped with VR headsets (Pico Neo 3 Pro) and controllers, engaged in three virtual tasks: painting a façade using scaffolding, transporting two boxes on a platform, and installing a camera at the roof's corner. Throughout the simulation, participants were tasked with selecting the appropriate personal protective equipment and reporting unsafe situations and hazards using a virtual tablet. Another approach, proposed by Wolf et al. [14] involved an AV method to assess trainees' hazard awareness in a virtual work environment. AV technology allows for the tracking of the user's hand motion and details of all entities in the virtual world, enabling the user to interact with a virtual angle grinder (with real weight, shape, and function) using their hand. The virtual scenario tasked participants with operating the angle grinder to cut a pipe while ensuring compliance with the prerequisites for themselves and their virtual animated coworkers represented in the simulation.

XR technology also enables dynamic and lifelike emergency response exercises, giving trainees practical experience in responding to critical situations. Wang et al. [46] used BIM and Unity3D to develop a virtual fire training system aimed at enhancing safety evacuation awareness. In addition, following simulated scenarios, XR facilitates in-depth post-incident analysis. Trainees have the opportunity to review their actions, comprehend the repercussions of their decisions, and explore alternative courses of action [87].

4.2. Skill/Knowledge Acquisition

One purpose of construction training utilizing XR technology is to enhance the competencies and expertise of construction workers. In this review, 11 papers, accounting for 14.9% of the total, were dedicated to applications related to skill and knowledge acquisition—the second-largest focus area in construction training.

In construction training, VR can provide trainees with a three-dimensional virtual space, replicating diverse construction sites with different layouts, structures, and spatial challenges. Through VR devices, trainees can interact with and manipulate virtual objects, promoting a deeper understanding of spatial relationships. Conesa et al. [21] developed an immersive shared virtual scenario enabling multiple students to collaborate in building a model. This study's results demonstrated that VR has the potential to enhance the spatial skills of trainees.

The primary objective of AR training is to augment the skills and knowledge of construction professionals. Notably, in construction assembly tasks, AR proves valuable by offering workers graphical models enriched with contextualized information. This assistance enables them to carry out assembly work more quickly and accurately [109,110]. For instance, Gabajova et al. [96] introduced a virtual training task where trainees were tasked with assembling an industrial plug. Using AR technology through a tablet, the group could access each step of the assembly procedure, resulting in a reduction in the average assembly time compared to providing only a user manual. Similarly, Hou et al. [22] employed AR to embed digitized information into a real-world workspace displayed on a TV monitor. This approach furnished workers with the correct assembly procedure, leading to an enhanced accuracy in completing the pipe assembly by the trainees. AR proves to be a valuable tool in assisting and allowing trainees to practice and refine their assembly skills.

Other construction skills and knowledge can also be gained through XR technology. For example, Goulding et al. [97] employed VR interactive training, enabling trainees to experiment with offsite production work practices within a secure and controlled environment. This approach allows trainees to explore and grasp new methods, processes, and modes of thinking. Furthermore, Osti et al. [98] developed a virtual sector focused on timber-based construction, offering effective training for workers and enhancing the manual skills of young carpenters.

4.3. Equipment Operation

XR technology has also been implemented in construction equipment training, with a total of 10 studies (13.5%) reviewing its use. Primarily within VR environments, XR replicates the controls and functionalities of authentic construction machinery. This innovation allows trainees to engage in hands-on practice with equipment like excavators, forklifts, and cranes within a secure environment, eliminating the potential dangers associated with using actual machinery [18]. Virtual training not only provides a risk-free setting but also yields cost savings by mitigating expenses such as fuel consumption and equipment rental [30]. XR simulations enable trainees to acquaint themselves with equipment interfaces, hone precise maneuvers, and cultivate the skills essential for secure and efficient equipment operation. The training modules further allow for a progressive learning approach, enabling trainees to advance through various difficulty levels. This structured progression ensures that trainees master basic controls before tackling more intricate tasks, facilitating a systematic and effective skill development process.

Liu et al. [18] introduced a multi-user excavator teleoperating system that uses two joysticks and two pedals to simulate a real excavation experience. This system can facilitate collaborative work between an excavator operator and a signaler. Pooladvand et al. [100] created a crane simulation system within a virtual reality environment. This system automatically produced lifting objects and obstacles, incorporating comprehensive lift studies and a crane path planning system for real-time evaluation. It assessed the safety and feasibility of comprehensive lift planning in real time, providing crane operators and lift engineers with experience comparable to actual operation. Other small equipment like drones [101] and angle grinders [99] can also be trained using VR. These studies demonstrated the promising potential of XR for diverse equipment operating training.

4.4. Human–Robot Collaboration

As robotics become increasingly integrated into construction processes, XR facilitates training scenarios involving human–robot collaboration. Recent research has successfully merged VR with robotic systems, creating a seamless environment for data sharing and interaction [63]. In a study by Ye et al. [104], a comprehensive robot-assisted motor training system was introduced to enhance expert motor skills through a keyhole welding training task. The system recorded the movement process of expert motor skills using haptic feedback from the robotic system. The recorded data were then played back to the trainee through perceptual learning, enabling them to comprehend the movement pattern through proprioception. The system offered features such as repetition, pausing, and adjusting training speed, thereby aiding novices in mastering motor skills. Another instance of human–robot collaboration is highlighted in a study by Adami et al. [7]. The researchers investigated whether participation in virtual training with 25 trainees remote operating a demolition robot could enhance trust in the robot, self-efficacy, mental load, and situational awareness. The results demonstrated that VR training significantly improved cognitive factors compared to traditional face-to-face training. Consequently, the utilization of VR technology for training human–robot collaboration holds immense potential for the future.

4.5. Ergonomics/Postural Training

XR implementations in ergonomics/postural training are comparably limited, comprising only two journal articles. Nonetheless, in the construction industry, maintaining proper ergonomics and posture is critical to preventing musculoskeletal injuries, reducing the risk of injury, and improving overall health [111]. XR technology proves valuable in simulating the postures adopted by construction workers during their daily tasks. It not only evaluates these postures but also fosters proper ergonomic awareness and posture, thereby mitigating the risk of injuries associated with repetitive tasks or awkward postures. For example, Akanmu et al. [67] developed a cyber-physical postural training system that leverages VR technology to integrate virtual environments with physical architectural resources, such as wood. In this system, trainees utilized wearable sensors (IMUs) and a Vive tracker to ascertain their position and posture within the virtual environment. As trainees engaged in wood frame construction, the system captured data on body part rotations for analysis. Subsequently, it delivered instructional material aimed at promoting safe posture, thus enhancing the overall effectiveness of postural training. Although XR has minimal application in construction training, this technology provides a platform to create a safer and healthier working environment in the construction industry.

5. Participants of XR Training Experiments

XR training experiments encompass a diverse range of participants, primarily classified into two categories: students and construction professionals. Students typically come from disciplines associated with construction, such as construction engineering, civil engineering, mechanical engineering, and electrical engineering. On the other hand, construction professionals encompass various roles, including welders, concrete workers, rebar workers, carpenter workers, electric construction site workers, operators, safety managers, project managers, site managers, and safety inspectors. In Figure 7, among the 74 publications obtained for this review, 23 articles exclusively featured students, representing 31% of the total, while an equivalent number of articles concentrated on the involvement of construction professionals. Furthermore, 18 articles (24%) showcased a mixture pattern, with participants including both student and construction professional roles. Notably, the identity of the participants remained unspecified in 10 articles, constituting 14%.

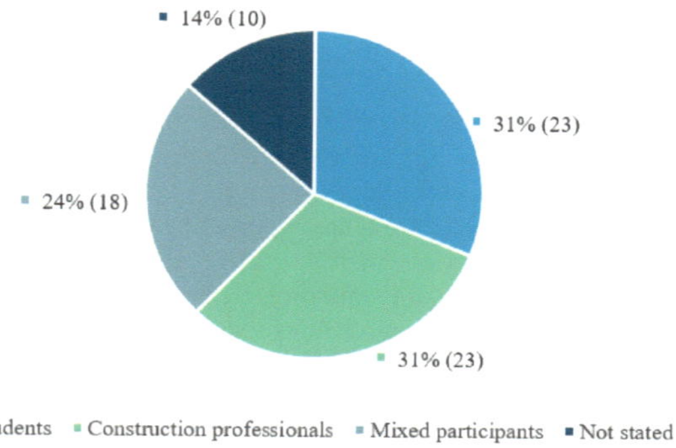

Figure 7. Percentage of participants involved in the experiments.

Numerous studies have observed variations in training outcomes for trainees who are either students or working professionals. For example, Eiris et al. [54] conducted an experiment involving 38 construction students and 38 construction professionals, comparing a 360 degree panorama with a virtual reality safety training platform featuring four hazards. The results revealed that students perceived the 360 degree panorama as more realistic than the virtual reality environment, whereas construction professionals did not discern a significant difference. In addition, Adami et al. [63] undertook a study with 25 construction workers and 25 graduated construction engineering students, focusing on demolition robot operation training. The findings indicated that students exhibited greater knowledge acquisition, while workers demonstrated higher levels of trust and self-efficacy in the robot. Consequently, the effectiveness of XR training appears to be role-specific within the construction domain, suggesting the need for customized training content and methodologies tailored to distinct roles.

The outcome of XR training is also influenced by the trainee's experience. For instance, in a safety training study by Yu et al. [74], 40 novice and 40 experienced workers underwent virtual construction site training involving 17 hazardous scenarios. The results indicated that novices exhibited significantly higher safety learning gains compared to their experienced counterparts. Such comparisons between novice and experienced workers offer valuable insights into the adaptability of XR training for individuals at various stages of their careers.

Some researchers have focused on the linguistic background of the trainees who participated in the XR training experiment. Afzal and Shafiq [50], for instance, employed BIM-based VR technology to conduct safety training focused on fall hazards in a multilingual setting in the United Arab Emirates. To mitigate the impact of language differences, participants in the training opted for a construction hierarchy, enabling them to discuss, address, and explain primary risks to on-site workers who did not share a common language. Understanding how participants communicate in virtual environments is crucial for designing XR training experiences that facilitate clear and effective interactions.

The diverse demographic composition of experiment participants engaged in XR training introduces a multifaceted dimension to the evaluation of its effectiveness. The characteristics of the study sample selected for XR training play a vital role in influencing the ultimate training effect and impacting the overall validity of the study's findings. To ensure the broader applicability of XR training outcomes, researchers should carefully consider participants' features throughout the evaluation process. In addition to the trainee's identity, work experience, and language background, other factors merit consideration, such as the gender and age of the trainees, their previous experience with XR training, etc. Additionally, it is important to consider the sample size, as a sample that is too small may fail to adequately represent a diverse group of individuals. Considering the characteristics of XR experimental participants can enhance the depth of analysis, providing a more nuanced understanding of factors influencing XR training.

6. Challenges of XR Training

This paper presents a comprehensive review of 74 articles on XR training in construction published between 2009 and 2023. Although the application of XR technology in the field of construction training is becoming more and more widespread and shows the advantages and great research potential of this technology, it is undeniable that it is still in the early stage of XR technology development. Therefore, analyzing the main obstacles and challenges that exist in current XR training technology can help researchers understand the issues that they may face when conducting research on XR technology in the construction training domain.

As shown in Table 5, this paper describes the challenges faced by VR, AR, and AV technologies in the application of construction engineering training. VR consists of nine main aspects:

- Visual interface design: The current equipment constrains the trainee's visual perspective, and overcoming this limitation to attain a more extensive and panoramic view poses a challenge.
- Motion capture technique: Current motion capture technologies come with calibration challenges, and the precision of captured positions is not sufficient.
- Content development: Creating high quality and relevant VR training content specific to the construction industry can be time consuming and resource intensive. Developing realistic simulations, 3D models, and interactive scenarios that accurately represent construction processes and hazards requires expertise and collaboration between subject matter experts, instructional designers, and VR developers.
- Multi-sensory: The current sensory feedback primarily involves visual and auditory stimuli; incorporating additional sensory feedback can enhance the alignment of the virtual world with the real world.
- Health and safety: Ensuring the safety of trainees during VR simulations is crucial. Designing VR training experiences that accurately replicate real-world construction hazards and safety protocols while maintaining a safe training environment can be challenging. It is essential to strike a balance between realistic training scenarios and minimizing the risk of physical or psychological harm to trainees.
- Assessment method: Most VR training assessments rely on questionnaires to gauge effectiveness, yet this evaluation method depends on the subjective sentiments of the trainees.
- Scalability and accessibility: Ensuring accessibility for workers with diverse backgrounds, abilities, and language proficiency can be a challenge that requires careful consideration during VR training development.
- Long-term effect: Most VR training evaluates short-term effectiveness, whereas the ultimate goal of training is for trainees to acquire a skill or knowledge over the long term.
- Skill requirement: Mastering VR technology demands highly specialized skills and can prove challenging for many individuals.

Challenges of AR technology application in the field of construction include two aspects: (1) Health and safety: AR training should prioritize safety by accurately representing construction hazards and safety protocols. Prolonged use of head-mounted displays can cause visual discomfort for trainees. (2) Tracking technique: It is challenging to track motion in unprepared environments. Regarding challenges in AV technology, creating a high-fidelity environment with multiple participants and well-designed training assistant content proves to be difficult.

Table 5. Challenges of XR training application in the construction industry.

XR Technology	XR Training Application Challenges	Description	References
VR	Visual interface design	The design of the visual interface may prioritize situational comprehensiveness over realism in recent studies	[18]
		Trainees' view is limited, and they cannot accurately understand their situation.	[23]
	Motion capture technique	Motion capture technique requires cumbersome calibration before each new task trial.	[18]
		Moton tracking is not accurate enough.	[83]
	Content development	Training contents inconsistent with real-time life cause confusion among participants.	[99]
		Factors like wind, weather, temperature, lighting, and visibility, are not adopted.	[100,101]
		The problem remains of how to measure the task complexity and its impact.	[15]
	Multi-sensory	Improve the training experience by adding haptic, locomotion, and auditory feedback, etc.	[24,106]
	Health and safety	VR-based training may have physical side effects	[8,24,48,107]
		Tasks requiring dexterity, involving high force, sudden changes in force, large accelerations, or rapid movements carry a risk of injury.	[104]
	Assessment method	The measurement of the questionnaire is subjective.	[20,44,80]
		Negative emotions are not measured during training.	[45]
	Scalability and accessibility	Ensuring training accessibility for workers with diverse backgrounds, abilities, and language proficiency can be a challenge	[20,72]
	Long-term effect	Lack of research on the long-term effect of training results	[45,53,70,74,81,87,106]
	Skill requirement	The development process requires special skills and extra effort.	[19,99]
AR	Health and safety	Prolonged use of head-mounted displays can cause visual discomfort for trainees	[55]
	Tracking technique	Markerless tracking and 3D feature-based tracking are not yet fully developed for application, particularly on mobile devices.	[22]
AV	Content development	Lack of multi-user virtual environment with varying degrees of assistance in achieving training objectives	[14]

7. Conclusions

The purpose of this literature review was to provide a comprehensive overview of research on extended reality (XR) technology training in the construction industry. This study was conducted by systematically searching the relevant literature over a period of 15 years, from 2009 to 2023, resulting in 74 journal articles. This literature review first describes the selected publications as a whole and then shows the technology of XR including its development platform, display device, and input device. Regarding the application of XR training in the construction industry, it can be categorized into the following five categories: (1) safety management (2) skill/knowledge acquisition (3) equipment operation (4) human–robot collaboration (5) ergonomics/postural training. In addition, this study also investigated the challenges faced in the implementation of XR in construction training applications. The specific challenges encountered in VR, AR, and AV training applications were presented separately.

This study contributes to three main areas. Firstly, it provides an overview of the current state of XR technology (VR/AR/AV) and its applications in construction training, offering insights into the latest developments. Secondly, this review also delves into the identity characteristics of XR trainees, marking the first such analysis in the literature and exploring their potential impact on XR training experimental results. Lastly, through a comprehensive analysis of the existing literature, this study identifies challenges encountered in XR training within the construction industry. This information provides audiences with insights into the current obstacles in XR training development and hopes to offer guidance for future advancements in this field.

It is essential to acknowledge its limitations. First, the focus on XR technologies exclusively within the realm of construction training may inadvertently neglect broader applications within the construction industry. Additionally, the sourcing of 74 journal articles solely from the Scopus database may have resulted in the omission of pertinent literature from other databases, potentially limiting the comprehensiveness of the findings. Moreover, while the three facts of XR technology are introduced, it is imperative to recognize that this may not capture the entirety of the expansive XR landscape. While recognizing certain limitations, it is expected that the findings presented in this review serve as a valuable reference for fellow researchers in their respective fields.

Author Contributions: Conceptualization, S.L., H.-H.W. and Q.-C.W.; methodology, S.L.; software, S.L.; validation, S.L., Q.-C.W., H.-H.W. and J.-H.C.; resources, S.L. and Q.-C.W.; data curation, S.L.; writing—original draft preparation, S.L.; writing—review and editing, S.L., Q.-C.W., H.-H.W. and J.-H.C.; visualization, S.L.; supervision, H.-H.W. and J.-H.C. All authors have read and agreed to the published version of the manuscript.

Funding: The authors acknowledge the support for this research by the Department of Building and Real Estate of The Hong Kong Polytechnic University under Project: P0044823 (1-WZ2E). Any opinions, findings, conclusions, and recommendations expressed in this paper are those of the authors and do not necessarily reflect the views of The Hong Kong Polytechnic University.

Conflicts of Interest: The authors declare no conflicts of interest.

References

1. Niazi, G.A.; Painting, N. Significant factors causing cost overruns in the construction industry in Afghanistan. *Procedia Eng.* **2017**, *182*, 510–517. [CrossRef]
2. Moshood, T.D.; Adeleke, A.Q.; Nawanir, G.; Ajibike, W.A.; Shittu, R.A. Emerging challenges and sustainability of industry 4.0 era in the Malaysian construction industry. *Int. J. Recent Technol. Eng.* **2020**, *4*, 1627–1634. [CrossRef]
3. Wang, Q.C.; Yu, S.N.; Chen, Z.X.; Weng, Y.W.; Xue, J.; Liu, X. Promoting additive construction in fast-developing areas: An analysis of policies and stakeholder perspectives. *Dev. Built Environ.* **2023**, *16*, 100271. [CrossRef]
4. Liu, X.; Yang, D.; Arentze, T.; Wielders, T. The willingness of social housing tenants to participate in natural gas-free heating systems project: Insights from a stated choice experiment in the Netherlands. *Appl. Energy* **2023**, *350*, 121706. [CrossRef]
5. Rauschnabel, P.A.; Felix, R.; Hinsch, C.; Shahab, H.; Alt, F. What is XR? Towards a framework for augmented and virtual reality. *Comput. Hum. Behav.* **2022**, *133*, 107289. [CrossRef]
6. Albeaino, G.; Eiris, R.; Gheisari, M.; Issa, R.R. DroneSim: A VR-based flight training simulator for drone-mediated building inspections. *Constr. Innov.* **2022**, *22*, 831–848. [CrossRef]
7. Adami, P.; Rodrigues, P.B.; Woods, P.J.; Becerik-Gerber, B.; Soibelman, L.; Copur-Gencturk, Y.; Lucas, G. Impact of VR-based training on human–robot interaction for remote operating construction robots. *J. Comput. Civ. Eng.* **2022**, *36*, 04022006. [CrossRef]
8. Jacobsen, E.L.; Solberg, A.; Golovina, O.; Teizer, J. Active personalized construction safety training using run-time data collection in physical and virtual reality work environments. *Constr. Innov.* **2022**, *22*, 531–553. [CrossRef]
9. Wedel, M.; Bigné, E.; Zhang, J. Virtual and augmented reality: Advancing research in consumer marketing. *Int. J. Res. Mark.* **2020**, *37*, 443–465. [CrossRef]
10. Zhao, X.; Zhang, M.; Fan, X.; Sun, Z.; Li, M.; Li, W.; Huang, L. Extended Reality for Safe and Effective Construction Management: State-of-the-Art, Challenges, and Future Directions. *Buildings* **2023**, *13*, 155. [CrossRef]
11. Milgram, P.; Kishino, F. A taxonomy of mixed reality visual displays. *IEICE Trans. Inf. Syst.* **1994**, *77*, 1321–1329.
12. Dunleavy, M.; Dede, C. Augmented reality teaching and learning. In *Handbook of Research on Educational Communications and Technology*; Springer: New York, NY, USA, 2014; pp. 735–745.
13. Ahmed, S. A review on using opportunities of augmented reality and virtual reality in construction project management. *Organ. Technol. Manag. Constr. Int. J.* **2018**, *10*, 1839–1852. [CrossRef]

14. Wolf, M.; Teizer, J.; Wolf, B.; Bükrü, S.; Solberg, A. Investigating hazard recognition in augmented virtuality for personalized feedback in construction safety education and training. *Adv. Eng. Inform.* **2022**, *51*, 101469. [CrossRef]
15. Su, X.; Dunston, P.S.; Proctor, R.W.; Wang, X. Influence of training schedule on development of perceptual–motor control skills for construction equipment operators in a virtual training system. *Autom. Constr.* **2013**, *35*, 439–447. [CrossRef]
16. Dias Barkokebas, R.; Li, X. VR-RET: A Virtual Reality–Based Approach for Real-Time Ergonomics Training on Industrialized Construction Tasks. *J. Constr. Eng. Manag.* **2023**, *149*, 04023098. [CrossRef]
17. Sacks, R.; Perlman, A.; Barak, R. Construction safety training using immersive virtual reality. *Constr. Manag. Econ.* **2013**, *31*, 1005–1017. [CrossRef]
18. Liu, D.; Kim, J.; Ham, Y. Multi-user immersive environment for excavator teleoperation in construction. *Autom. Constr.* **2023**, *156*, 105143. [CrossRef]
19. Le, Q.T.; Pedro, A.; Park, C.S. A social virtual reality based construction safety education system for experiential learning. *J. Intell. Robot. Syst.* **2015**, *79*, 487–506. [CrossRef]
20. Rey-Becerra, E.; Barrero, L.H.; Ellegast, R.; Kluge, A. Improvement of short-term outcomes with VR-based safety training for work at heights. *Appl. Ergon.* **2023**, *112*, 104077. [CrossRef]
21. Conesa, J.; Mula, F.J.; Bartlett, K.A.; Naya, F.; Contero, M. The Influence of Immersive and Collaborative Virtual Environments in Improving Spatial Skills. *Appl. Sci.* **2023**, *13*, 8426. [CrossRef]
22. Hou, L.; Wang, X.; Truijens, M. Using augmented reality to facilitate piping assembly: An experiment-based evaluation. *J. Comput. Civ. Eng.* **2015**, *29*, 05014007. [CrossRef]
23. Ki, J.S. Interactive training simulator for aerial working platform in a virtual environment. *Comput. Appl. Eng. Educ.* **2011**, *19*, 733–738. [CrossRef]
24. Li, W.; Huang, H.; Solomon, T.; Esmaeili, B.; Yu, L.-F. Synthesizing personalized construction safety training scenarios for VR training. *IEEE Trans. Vis. Comput. Graph.* **2022**, *28*, 1993–2002. [CrossRef]
25. Tastan, H.; Tong, T. Immersive virtual reality in AECO/FM to enhance education: Systematic literature review and future directions. *Comput. Appl. Eng. Educ.* **2023**, *31*, 1465–1491. [CrossRef]
26. Zhang, Y.; Liu, H.; Kang, S.-C.; Al-Hussein, M. Virtual reality applications for the built environment: Research trends and opportunities. *Autom. Constr.* **2020**, *118*, 103311. [CrossRef]
27. Wen, J.; Gheisari, M. Using virtual reality to facilitate communication in the AEC domain: A systematic review. *Constr. Innov.* **2020**, *20*, 509–542. [CrossRef]
28. Spitzer, B.O.; Ma, J.H.; Erdogmus, E.; Kreimer, B.; Ryherd, E.; Diefes-Dux, H. Framework for the use of extended reality modalities in AEC Education. *Buildings* **2022**, *12*, 2169. [CrossRef]
29. Tan, Y.; Xu, W.; Li, S.; Chen, K. Augmented and Virtual Reality (AR/VR) for Education and Training in the AEC Industry: A Systematic Review of Research and Applications. *Buildings* **2022**, *12*, 1529. [CrossRef]
30. Wang, P.; Wu, P.; Wang, J.; Chi, H.-L.; Wang, X. A critical review of the use of virtual reality in construction engineering education and training. *Int. J. Environ. Res. Public Health* **2018**, *15*, 1204. [CrossRef] [PubMed]
31. Zoleykani, M.J.; Abbasianjahromi, H.; Banihashemi, S.; Tabadkani, S.A.; Hajirasouli, A. Extended reality (XR) technologies in the construction safety: Systematic review and analysis. *Constr. Innov.* **2023**. [CrossRef]
32. Salinas, D.; Muñoz-La Rivera, F.; Mora-Serrano, J. Critical analysis of the evaluation methods of extended reality (XR) experiences for construction safety. *Int. J. Environ. Res. Public Health* **2022**, *19*, 15272. [CrossRef]
33. Moore, H.F.; Gheisari, M. A review of virtual and mixed reality applications in construction safety literature. *Safety* **2019**, *5*, 51. [CrossRef]
34. Kassem, M.; Benomran, L.; Teizer, J. Virtual environments for safety learning in construction and engineering: Seeking evidence and identifying gaps for future research. *Vis. Eng.* **2017**, *5*, 16. [CrossRef]
35. Zhang, Z.; Wong, M.O.; Pan, W. Virtual reality enhanced multi-role collaboration in crane-lift training for modular construction. *Autom. Constr.* **2023**, *150*, 104848. [CrossRef]
36. Snyder, H. Literature review as a research methodology: An overview and guidelines. *J. Bus. Res.* **2019**, *104*, 333–339. [CrossRef]
37. Trindade, N.V.; Ferreira, A.; Pereira, J.M.; Oliveira, S. Extended reality in AEC. *Autom. Constr.* **2023**, *154*, 105018. [CrossRef]
38. Muñoz-La Rivera, F.; Mora-Serrano, J.; Oñate, E. Virtual reality stories for construction training scenarios: The case of social distancing at the construction site. *WIT Trans. Built Environ.* **2021**, *205*, 37–47.
39. Ogunnusi, M.; Hamma-Adama, M.; Salman, H.; Kouider, T. COVID-19 pandemic: The effects and prospects in the construction industry. *Int. J. Real Estate Stud.* **2020**, *14*, 120–128.
40. Anthes, C.; García-Hernández, R.J.; Wiedemann, M.; Kranzlmüller, D. State of the art of virtual reality technology. In Proceedings of the 2016 IEEE Aerospace Conference, Big Sky, MT, USA, 5–12 March 2016; pp. 1–19.
41. Cowan, B.; Kapralos, B. An overview of serious game engines and frameworks. In *Recent Advances in Technologies for Inclusive Well-Being: From Worn to Off-Body Sensing, Virtual Worlds, and Games for Serious Applications*; Springer: Cham, Switzerland, 2017; pp. 15–38.
42. Thorn, A. *Game Engine Design and Implementation*; Jones & Bartlett Publishers: Sudbury, MA, USA, 2011.
43. Gregory, J. *Game Engine Architecture*; CRC Press: Boca Raton, FL, USA, 2018.

44. Luo, Y.; Ahn, S.; Abbas, A.; Seo, J.; Cha, S.H.; Kim, J.I. Investigating the impact of scenario and interaction fidelity on training experience when designing immersive virtual reality-based construction safety training. *Dev. Built Environ.* **2023**, *16*, 100223. [CrossRef]
45. Han, Y.; Yang, J.; Diao, Y.; Jin, R.; Guo, B.; Adamu, Z. Process and outcome-based evaluation between virtual reality-driven and traditional construction safety training. *Adv. Eng. Inform.* **2022**, *52*, 101634. [CrossRef]
46. Wang, B.; Li, H.; Rezgui, Y.; Bradley, A.; Ong, H.N. BIM based virtual environment for fire emergency evacuation. *Sci. World J.* **2014**, *2014*, 589016. [CrossRef] [PubMed]
47. Bao, L.; Tran, S.V.-T.; Nguyen, T.L.; Pham, H.C.; Lee, D.; Park, C. Cross-platform virtual reality for real-time construction safety training using immersive web and industry foundation classes. *Autom. Constr.* **2022**, *143*, 104565. [CrossRef]
48. Dhalmahapatra, K.; Maiti, J.; Krishna, O. Assessment of virtual reality based safety training simulator for electric overhead crane operations. *Saf. Sci.* **2021**, *139*, 105241. [CrossRef]
49. Rokooei, S.; Shojaei, A.; Alvanchi, A.; Azad, R.; Didehvar, N. Virtual reality application for construction safety training. *Saf. Sci.* **2023**, *157*, 105925. [CrossRef]
50. Afzal, M.; Shafiq, M.T. Evaluating 4d-bim and vr for effective safety communication and training: A case study of multilingual construction job-site crew. *Buildings* **2021**, *11*, 319. [CrossRef]
51. Arain, F.M.; Burkle, M. Learning construction project management in the virtual world: Leveraging on second life. *J. Inf. Technol. Constr. (ITcon)* **2011**, *16*, 243–258.
52. Zhao, D.; McCoy, A.; Kleiner, B.; Feng, Y. Integrating safety culture into OSH risk mitigation: A pilot study on the electrical safety. *J. Civ. Eng. Manag.* **2016**, *22*, 800–807. [CrossRef]
53. Zhao, D.; Lucas, J. Virtual reality simulation for construction safety promotion. *Int. J. Inj. Control Saf. Promot.* **2015**, *22*, 57–67. [CrossRef]
54. Eiris, R.; Gheisari, M.; Esmaeili, B. Desktop-based safety training using 360-degree panorama and static virtual reality techniques: A comparative experimental study. *Autom. Constr.* **2020**, *109*, 102969. [CrossRef]
55. Gomes, D.L., Jr.; dos Reis, P.R.J.; de Paiva, A.C.; Silva, A.C.A.; Braz, G., Jr.; Gattass, M. An approach for construction of augmented reality systems using natural markers and mobile sensors in industrial fields. *Int. J. Comput. Commun. Control* **2017**, *12*, 507–518. [CrossRef]
56. Manjrekar, S.; Sandilya, S.; Bhosale, D.; Kanchi, S.; Pitkar, A.; Gondhalekar, M. CAVE: An emerging immersive technology—A review. In Proceedings of the 2014 Uksim-Amss 16th International Conference on Computer Modelling and Simulation, Cambridge, UK, 26–28 March 2014; pp. 131–136.
57. Perlman, A.; Sacks, R.; Barak, R. Hazard recognition and risk perception in construction. *Saf. Sci.* **2014**, *64*, 22–31. [CrossRef]
58. de Back, T.T.; Tinga, A.M.; Nguyen, P.; Louwerse, M.M. Benefits of immersive collaborative learning in CAVE-based virtual reality. *Int. J. Educ. Technol. High. Educ.* **2020**, *17*, 51. [CrossRef]
59. Coppens, A. Merging real and virtual worlds: An analysis of the state of the art and practical evaluation of Microsoft Hololens. *arXiv* **2017**, arXiv:1706.08096.
60. Concannon, B.J.; Esmail, S.; Roduta Roberts, M. Head-mounted display virtual reality in post-secondary education and skill training. *Front. Educ.* **2019**, *4*, 80. [CrossRef]
61. John, N.W.; Pop, S.R.; Day, T.W.; Ritsos, P.D.; Headleand, C.J. The implementation and validation of a virtual environment for training powered wheelchair manoeuvres. *IEEE Trans. Vis. Comput. Graph.* **2017**, *24*, 1867–1878. [CrossRef] [PubMed]
62. Norouzi, N.; Bölling, L.; Bruder, G.; Welch, G. Augmented rotations in virtual reality for users with a reduced range of head movement. *J. Rehabil. Assist. Technol. Eng.* **2019**, *6*, 2055668319841309. [CrossRef] [PubMed]
63. Adami, P.; Singh, R.; Rodrigues, P.B.; Becerik-Gerber, B.; Soibelman, L.; Copur-Gencturk, Y.; Lucas, G. Participants matter: Effectiveness of VR-based training on the knowledge, trust in the robot, and self-efficacy of construction workers and university students. *Adv. Eng. Inform.* **2023**, *55*, 101837. [CrossRef]
64. Shringi, A.; Arashpour, M.; Golafshani, E.M.; Dwyer, T.; Kalutara, P. Enhancing Safety Training Performance Using Extended Reality: A Hybrid Delphi–AHP Multi-Attribute Analysis in a Type-2 Fuzzy Environment. *Buildings* **2023**, *13*, 625. [CrossRef]
65. Abotaleb, I.; Hosny, O.; Nassar, K.; Bader, S.; Elrifaee, M.; Ibrahim, S.; El Hakim, Y.; Sherif, M. An interactive virtual reality model for enhancing safety training in construction education. *Comput. Appl. Eng. Educ.* **2023**, *31*, 324–345. [CrossRef]
66. Fernández, A.; Muñoz La Rivera, F.; Mora Serrano, J. Virtual reality training for occupational risk prevention: Application case in geotechnical drilling works. *Int. J. Comput. Methods Exp. Meas.* **2023**, *11*, 55–63.
67. Akanmu, A.A.; Olayiwola, J.; Ogunseiju, O.; McFeeters, D. Cyber-physical postural training system for construction workers. *Autom. Constr.* **2020**, *117*, 103272. [CrossRef]
68. Choi, D.; Seo, S.; Park, H.; Hong, T.; Koo, C. Forecasting personal learning performance in virtual reality-based construction safety training using biometric responses. *Autom. Constr.* **2023**, *156*, 105115. [CrossRef]
69. Patil, K.R.; Bhandari, S.; Agrawal, A.; Ayer, S.K.; Perry, L.A.; Hallowell, M.R. Analysis of YouTube Comments to Inform the Design of Virtual Reality Training Simulations to Target Emotional Arousal. *J. Constr. Eng. Manag.* **2023**, *149*, 04023077. [CrossRef]
70. Cheng, J.-Y.; Gheisari, M.; Jeelani, I. Using 360-Degree Virtual Reality Technology for Training Construction Workers about Safety Challenges of Drones. *J. Comput. Civ. Eng.* **2023**, *37*, 04023018. [CrossRef]
71. Shringi, A.; Arashpour, M.; Dwyer, T.; Prouzeau, A.; Li, H. Safety in Off-Site Construction: Simulation of Crane-Lifting Operations Using VR and BIM. *J. Archit. Eng.* **2023**, *29*, 04022035. [CrossRef]

72. Yoo, J.W.; Park, J.S.; Park, H.J. Understanding VR-based construction safety training effectiveness: The role of telepresence, risk perception, and training satisfaction. *Appl. Sci.* **2023**, *13*, 1135. [CrossRef]
73. Shringi, A.; Arashpour, M.; Golafshani, E.M.; Rajabifard, A.; Dwyer, T.; Li, H. Efficiency of VR-Based Safety Training for Construction Equipment: Hazard Recognition in Heavy Machinery Operations. *Buildings* **2022**, *12*, 2084. [CrossRef]
74. Yu, W.-D.; Wang, K.-C.; Wu, H.-T. Empirical comparison of learning effectiveness of immersive virtual reality–based safety training for novice and experienced construction workers. *J. Constr. Eng. Manag.* **2022**, *148*, 04022078. [CrossRef]
75. Getuli, V.; Capone, P.; Bruttini, A.; Sorbi, T. A smart objects library for BIM-based construction site and emergency management to support mobile VR safety training experiences. *Constr. Innov.* **2022**, *22*, 504–530. [CrossRef]
76. Huang, D.; Wang, X.; Liu, J.; Li, J.; Tang, W. Virtual reality safety training using deep EEG-net and physiology data. *Vis. Comput.* **2022**, *38*, 1195–1207. [CrossRef]
77. Seo, H.J.; Park, G.M.; Son, M.; Hong, A.-J. Establishment of virtual-reality-based safety education and training system for safety engagement. *Educ. Sci.* **2021**, *11*, 786. [CrossRef]
78. Comu, S.; Kazar, G.; Marwa, Z. Evaluating the attitudes of different trainee groups towards eye tracking enhanced safety training methods. *Adv. Eng. Inform.* **2021**, *49*, 101353. [CrossRef]
79. Xu, Z.; Zheng, N. Incorporating virtual reality technology in safety training solution for construction site of urban cities. *Sustainability* **2020**, *13*, 243. [CrossRef]
80. Nykänen, M.; Puro, V.; Tiikkaja, M.; Kannisto, H.; Lantto, E.; Simpura, F.; Uusitalo, J.; Lukander, K.; Räsänen, T.; Heikkilä, T. Implementing and evaluating novel safety training methods for construction sector workers: Results of a randomized controlled trial. *J. Saf. Res.* **2020**, *75*, 205–221. [CrossRef] [PubMed]
81. Jeelani, I.; Han, K.; Albert, A. Development of virtual reality and stereo-panoramic environments for construction safety training. *Eng. Constr. Archit. Manag.* **2020**, *27*, 1853–1876. [CrossRef]
82. Choi, M.; Ahn, S.; Seo, J. VR-Based investigation of forklift operator situation awareness for preventing collision accidents. *Accid. Anal. Prev.* **2020**, *136*, 105404. [CrossRef] [PubMed]
83. Shi, Y.; Du, J.; Ahn, C.R.; Ragan, E. Impact assessment of reinforced learning methods on construction workers' fall risk behavior using virtual reality. *Autom. Constr.* **2019**, *104*, 197–214. [CrossRef]
84. Lu, X.; Davis, S. Priming effects on safety decisions in a virtual construction simulator. *Eng. Constr. Archit. Manag.* **2018**, *25*, 273–294. [CrossRef]
85. Bosché, F.; Abdel-Wahab, M.; Carozza, L. Towards a mixed reality system for construction trade training. *J. Comput. Civ. Eng.* **2016**, *30*, 04015016. [CrossRef]
86. Albert, A.; Hallowell, M.R.; Kleiner, B.; Chen, A.; Golparvar-Fard, M. Enhancing construction hazard recognition with high-fidelity augmented virtuality. *J. Constr. Eng. Manag.* **2014**, *140*, 04014024. [CrossRef]
87. Teizer, J.; Cheng, T.; Fang, Y. Location tracking and data visualization technology to advance construction ironworkers' education and training in safety and productivity. *Autom. Constr.* **2013**, *35*, 53–68. [CrossRef]
88. Li, H.; Chan, G.; Skitmore, M. Multiuser virtual safety training system for tower crane dismantlement. *J. Comput. Civ. Eng.* **2012**, *26*, 638–647. [CrossRef]
89. Guo, H.; Li, H.; Chan, G.; Skitmore, M. Using game technologies to improve the safety of construction plant operations. *Accid. Anal. Prev.* **2012**, *48*, 204–213. [CrossRef] [PubMed]
90. Shelton, B.E.; Scoresby, J.; Stowell, T.; Capell, M.R.; Alvarez, M.A.; Coats, K.C.C. A Frankenstein approach to open source: The construction of a 3D game engine as meaningful educational process. *IEEE Trans. Learn. Technol.* **2010**, *3*, 85–90. [CrossRef]
91. Alzarrad, A.; Miller, M.; Chowdhury, S.; McIntosh, J.; Perry, T.; Shen, R. Harnessing Virtual Reality to Mitigate Heat-Related Injuries in Construction Projects. *CivilEng* **2023**, *4*, 1157–1168. [CrossRef]
92. Qin, Y.; Bulbul, T. Electroencephalogram-based mental workload prediction for using Augmented Reality head mounted display in construction assembly: A deep learning approach. *Autom. Constr.* **2023**, *152*, 104892. [CrossRef]
93. Beh, H.J.; Rashidi, A.; Talei, A.; Lee, Y.S. Developing engineering students' capabilities through game-based virtual reality technology for building utility inspection. *Eng. Constr. Archit. Manag.* **2022**, *29*, 2854–2877. [CrossRef]
94. Bolkas, D.; Chiampi, J.D.; Fioti, J.; Gaffney, D. First assessment results of surveying engineering labs in immersive and interactive virtual reality. *J. Surv. Eng.* **2022**, *148*, 04021028. [CrossRef]
95. Habibnezhad, M.; Shayesteh, S.; Jebelli, H.; Puckett, J.; Stentz, T. Comparison of ironworker's fall risk assessment systems using an immersive biofeedback simulator. *Autom. Constr.* **2021**, *122*, 103471. [CrossRef]
96. Gabajová, G.; Furmannová, B.; Medvecká, I.; Grznár, P.; Krajčovič, M.; Furmann, R. Virtual training application by use of augmented and virtual reality under university technology enhanced learning in Slovakia. *Sustainability* **2019**, *11*, 6677. [CrossRef]
97. Goulding, J.; Nadim, W.; Petridis, P.; Alshawi, M. Construction industry offsite production: A virtual reality interactive training environment prototype. *Adv. Eng. Inform.* **2012**, *26*, 103–116. [CrossRef]
98. Osti, F.; de Amicis, R.; Sanchez, C.A.; Tilt, A.B.; Prather, E.; Liverani, A. A VR training system for learning and skills development for construction workers. *Virtual Real.* **2021**, *25*, 523–538. [CrossRef]
99. Jelonek, M.; Fiala, E.; Herrmann, T.; Teizer, J.; Embers, S.; König, M.; Mathis, A. Evaluating Virtual Reality Simulations for Construction Safety Training: A User Study Exploring Learning Effects, Usability and User Experience. *i-com* **2022**, *21*, 269–281. [CrossRef]

100. Pooladvand, S.; Taghaddos, H.; Eslami, A.; Nekouvaght Tak, A.; Hermann, U. Evaluating mobile crane lift operations using an interactive virtual reality system. *J. Constr. Eng. Manag.* **2021**, *147*, 04021154. [CrossRef]
101. Sakib, M.N.; Chaspari, T.; Behzadan, A.H. Physiological data models to understand the effectiveness of drone operation training in immersive virtual reality. *J. Comput. Civ. Eng.* **2021**, *35*, 04020053. [CrossRef]
102. Ito, M.; Funahara, Y.; Saiki, S.; Yamazaki, Y.; Kurita, Y. Development of a cross-platform cockpit for simulated and tele-operated excavators. *J. Robot. Mechatron.* **2019**, *31*, 231–239. [CrossRef]
103. Kim, B.; Kim, C.; Kim, H. Interactive modeler for construction equipment operation using augmented reality. *J. Comput. Civ. Eng.* **2012**, *26*, 331–341. [CrossRef]
104. Ye, Y.; Zhou, T.; Du, J. Robot-assisted immersive kinematic experience transfer for welding training. *J. Comput. Civ. Eng.* **2023**, *37*, 04023002. [CrossRef]
105. Shayesteh, S.; Ojha, A.; Liu, Y.; Jebelli, H. Human-robot teaming in construction: Evaluative safety training through the integration of immersive technologies and wearable physiological sensing. *Saf. Sci.* **2023**, *159*, 106019. [CrossRef]
106. Wang, X.; Liang, C.-J.; Menassa, C.C.; Kamat, V.R. Interactive and immersive process-level digital twin for collaborative human–robot construction work. *J. Comput. Civ. Eng.* **2021**, *35*, 04021023. [CrossRef]
107. Adami, P.; Rodrigues, P.B.; Woods, P.J.; Becerik-Gerber, B.; Soibelman, L.; Copur-Gencturk, Y.; Lucas, G. Effectiveness of VR-based training on improving construction workers' knowledge, skills, and safety behavior in robotic teleoperation. *Adv. Eng. Inform.* **2021**, *50*, 101431. [CrossRef]
108. Gogouvitis, X.V.; Vosniakos, G.-C. Construction of a virtual reality environment for robotic manufacturing cells. *Int. J. Comput. Appl. Technol.* **2015**, *51*, 173–184. [CrossRef]
109. Kwiatek, C.; Sharif, M.; Li, S.; Haas, C.; Walbridge, S. Impact of augmented reality and spatial cognition on assembly in construction. *Autom. Constr.* **2019**, *108*, 102935. [CrossRef]
110. Yang, Z.; Shi, J.; Jiang, W.; Sui, Y.; Wu, Y.; Ma, S.; Kang, C.; Li, H. Influences of augmented reality assistance on performance and cognitive loads in different stages of assembly task. *Front. Psychol.* **2019**, *10*, 1703. [CrossRef]
111. Ahankoob, A.; Charehzehi, A. Mitigating ergonomic injuries in construction industry. *IOSR J. Mech. Civ. Eng.* **2013**, *6*, 36–42. [CrossRef]

buildings

MDPI

Article

A Conceptual Framework for Planning Road Digital Twins

Munkhbaatar Buuveibaatar [1,2], Ioannis Brilakis [2], Matt Peck [3], George Economides [4] and Wonhee Lee [5,*]

[1] Research Institute of Artificial Intelligent Diagnosis Technology for Multi-Scale Organic and Inorganic Structure, Kyungpook National University, Sangju 37224, Republic of Korea; munkh@knu.ac.kr
[2] Department of Engineering, University of Cambridge, 7a JJ Thomson Avenue, Cambridge CB3 0FA, UK; ib340@cam.ac.uk
[3] AtkinsRealis, 2 Chamberlain Square, Birmingham B3 3AX, UK; matt.peck@atkinsrealis.com
[4] Department for Transport, Great Minster House, 33 Horseferry Road, London SW1P 4DR, UK; george.economides@dft.gov.uk
[5] Department of Location-Based Information System, Kyungpook National University, Sangju 37224, Republic of Korea
* Correspondence: wlee33@knu.ac.kr; Tel.: +82-054-530-1492

Abstract: Digital twin (DT) is an emerging technology gaining traction across various industries. However, its development and application in the architecture, engineering, and construction (AEC) industry remain in their early stage, lagging considerably behind other sectors. This is primarily attributed to the challenges facing the AEC industry, including digital transformation and the lack of formal standards for DT implementation. This study aims to contribute to the conceptualization of DT planning—the early stage of the DT lifecycle—focusing on the road transportation sector, particularly road physical twin planning within the AEC industry. To achieve this, we reviewed the relevant literature defining DT planning. We also examined stakeholders' relevant guidelines and documents from national bodies that roadmap the road DT planning process to understand the scope and identify knowledge gaps. Based on the findings, mapping the existing road planning process to the constituents of road DT planning was performed for the applicable planning steps. Finally, we proposed a five-layered road DT planning framework that will roadmap future implementations comprising data acquisition, data processing, data modeling and algorithms, data analysis and control, and a service layer plus users. In addition, a case study is incorporated to validate the feasibility of the framework toward applying it further in practice.

Keywords: digital twin; road digital twin; digital twin planning; road digital twin planning framework; digital roads; future roads; digital assets; digital transformation

Academic Editors: Wenxin Shen, Wenzhe Tang and Jin Xue

Received: 16 December 2024
Revised: 14 January 2025
Accepted: 17 January 2025
Published: 21 January 2025

Citation: Buuveibaatar, M.; Brilakis, I.; Peck, M.; Economides, G.; Lee, W. A Conceptual Framework for Planning Road Digital Twins. *Buildings* **2025**, *15*, 316. https://doi.org/10.3390/buildings15030316

1. Introduction

Digital twin (DT) is an emerging technology gaining traction across multiple industries. DT is a virtual representation of a physical asset or object, which enables a cyber-physical connection that facilitates real-time monitoring of assets, various simulations and analyses, and decision-making processes [1]. The most prominent DT applications include asset twinning, component twinning, system twinning, and process and workflow twinning [2].

The global DT market was estimated at USD 16.75 billion in 2023, with an expected compound annual growth rate (CAGR) of 35.7% from 2024 to 2030 [3]. Another report projects the global DT market to grow from USD 10.1 billion in 2023 to USD 110.1 billion by 2028, at a CAGR of 61.3% [4].

As estimated by Gartner, over 40% of large companies worldwide will be using DT alongside other computing technologies (e.g., metaverse) to increase revenue by 2027 [5].

Additionally, a 2024 report indicates that approximately 47% of executives across a wide range of industry verticals understand the benefits of using DTs, with 63% of these executives planning to integrate them into their operations by 2029 [2].

Despite these good figures, the development and application of DT in the architecture, engineering, and construction (AEC) industry remain in their early stage, lagging considerably behind other sectors [6,7]. Although the application of DTs has increased in certain areas of the AEC industry, their wider adoption is restricted by challenges, such as digital transformation, digitalization, and the lack of formal standards for DT implementation [1,6]. Regarding these challenges, the question of when and how to start applying DTs in the industry, particularly at which lifecycle stage of the physical twin (PT) (assets) it should be used first, needs to be addressed. This naturally leads researchers and practitioners to become well-acquainted with the DT lifecycle stages, as in the PT lifecycle.

In this study, we approach the DT lifecycle (concept), particularly emphasizing the beginning of the lifecycle—DT planning. However, some underlying concerns exist regarding the concept and application of DT planning. First, the definition of DT planning is unclear, making its practical implementation challenging. Second, there is a lack of guidelines or roadmaps necessary for creating DT planning. This study aims to contribute to the conceptualization of DT planning, which is the early stage of the DT lifecycle, focusing on the road transportation sector, particularly the planning of road PT within the AEC industry.

To achieve this objective, we reviewed the relevant literature on defining DT planning to contribute to the conceptualization of road DT planning. Additionally, we examined stakeholders' relevant guidelines and documents from national bodies that roadmap the road DT planning process to understand the scope and identify knowledge gaps. Finally, we derived/proposed a road DT planning process/scope based on the findings on planning PTs. Figure 1 shows the study flow from the literature review to the proposed framework development.

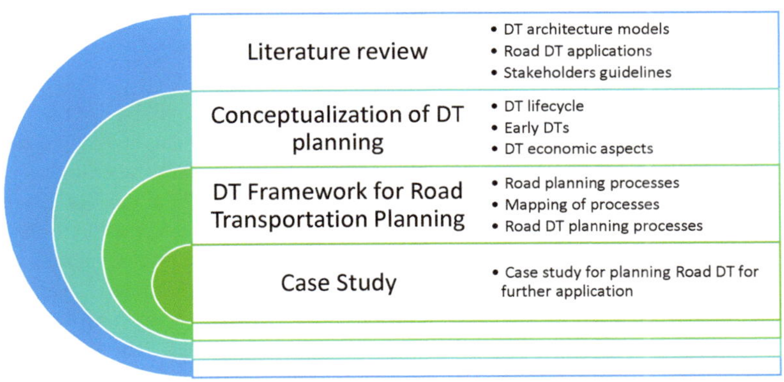

Figure 1. Research methodology used in the study.

The rest of this paper is organized as follows: Section 2 provides background on the architecture model of DT and describes its applications in the road sector and the identified research gaps. In Section 3, a conceptualization of DT planning is elaborated from the aspects of the DT lifecycle and early DTs with their economic consideration. Section 4 presents the proposed conceptual framework for road DT planning. This section consists of several subsections, including the requirements capturing and derivation of road DT planning from the road transportation planning processes. Also, a framework for road DT planning is proposed with the relevant case study to foster its implementation in practice. Finally, Section 5 summarizes the significance of the research findings and highlights the benefits of the proposed framework for further development. This section also discusses the limitations, challenges, and directions for future research.

2. Background

2.1. Digital Twin Architecture Model

Fundamentally, DT consists of three major components: a physical element (asset or object) that currently exists or will exist in the real world, referred to as the PT; a digital representation of that element that exists in the digital/cyber world, referred to as the DT; and the connections that enable the exchange of data and information between the two worlds [8].

These major components are further broken down and represented as layers within DT architecture models. DT system architecture typically consists of physical space (instrumental physical assets), digital space (data layer, data processing layer, models and algorithms layer, and analysis and control layer), user interface/applications, and users [9]. A more detailed architecture expands these layers to include physical space, cyber-physical data storage layer, primary processing layer, models and algorithms layer, analysis layer, and visualization and user interface layer [10].

A hierarchical DT architecture comprising five layers—a data acquisition layer, a transmission layer, a digital modeling layer, a data/model integration layer, and a service layer—has been introduced for building and city-level management [11]. A similar five-layer architecture has been proposed for building DT maturity measurements [1].

Based on the relevant literature, a five-layered DT architecture has been identified, as illustrated in Figure 2, which represents a common DT architecture model that will be applied in the subsequent development of road DT planning.

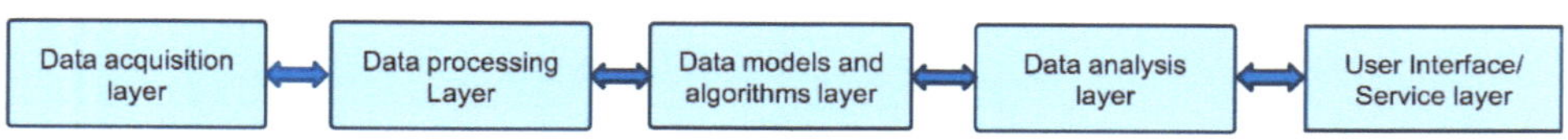

Figure 2. Five-layered digital twin model.

The data acquisition layer is the first layer of the five-dimensional DT architecture, serving as the foundation for all subsequent layers. The precise description of the physical object and its information is a fundamental and crucial step in creating a virtual model [10]. The data processing layer integrates all data resources according to the established data structure and converts data into usable information for further [10,11]. The third layer, the models and algorithms layer, focuses on developing the analysis models and algorithms for the components. The fourth layer, the data analysis layer, performs prediction, optimization, monitoring, and control. Furthermore, smart functions, such as AI and machine learning, support advanced decision-making [11]. The last layer, the application/service layer, presents the results to the users for decision-making and management purposes.

2.2. Road Digital Twin Applications

In recent years, DT-related research activities have gained momentum in various industries other than manufacturing. Among these sectors, the civil engineering field, particularly the road industry, is increasingly exploring the application of DT.

DT is applied throughout a physical asset's lifecycle, encompassing data acquisition, design, construction, operation and maintenance, and demolition, for multiple purposes, including asset monitoring and management, analysis and diagnosis, informed decision-making, and automatic control [12]. The operation and maintenance phase of the asset's lifecycle has the most applications of DTs [13], as performing maintenance tasks can significantly influence a company's business [14].

Consilvio et al. [15] introduced a DT-based decision support tool to aid road inspection and maintenance operations, thereby improving road (pavement) maintenance processes. This tool integrates different data sources and processes raw data into valuable information, which is then analyzed to derive optimal intervention plans.

Tanne et al. [16] developed a conceptual framework of the DT model to advance asset management and decision-making for city roads as a part of the Technology 4.0 implementation. This conceptual framework serves as a basis for implementing integrated road asset management and more efficient and improved community services.

Jiang et al. [17] proposed a DT-based sustainable urban road planning method for constructing new roads and widening existing roads. This framework enables the selection of road alignments based on geospatial map data while considering multiple factors to facilitate multi-criteria decision-making, including building displacement, land use, traffic congestion, driving route selections, air quality, and noise levels.

Jiang et al. [18] proposed a digital twin approach for checking the clearance of underpass roads in highway widening projects using open geospatial data. The underpass road digital twin with Building Information Modeling (BIM) was created to support redesigning and checking clearance of road infrastructure. The framework developed is effective and affordable for designing road widening projects by checking underpass clearance.

Furthermore, for existing assets, Nochta et al. [19] presented a city-scale DT as an emerging urban analytics tool to address the problems of traffic congestion, air quality, urban growth, and energy supply. This tool integrates existing data from various sectors to develop more comprehensive management plans.

For the current road networks, Machl et al. [20] introduced digital techniques to aid planners in designing agricultural core road networks. These networks consist of existing roads crucial for agricultural transportation and must meet both current and future agricultural needs. The design is based on the DT of the cultivated landscape.

On the other hand, regarding road transportation applications from the smart city planning or urban DT aspect, Malé and Lagier [21] reported an interactive planning platform as a decision support tool that enables the planning authority of Lyon to develop different future scenarios to innovate the traditional urban planning approach by engaging in linking different domains in urban planning in an integrated and inclusive way. The platform consists of four main functions including territory knowledge (e.g., spatial map), action (tool functions such as creating lanes), scenario (e.g., demography changes), and results (indicators calculation). One of those actions includes modifying transport infrastructure, such as transforming the highway into an urban roadway.

The Digital Twin of the City of Zurich for Urban Planning [22] used DT technology to support the decision-making of the City of Zurich. City themes, such as such buildings, bridges, vegetation, etc., are transformed into three-dimensional spatial data and models in a digital space, and they are updated when required to take advantage of digital space. Thus, digital prototypes support visualization and analysis, enabling the demonstration of interactions with the built environment, and digitally made scenarios can be used in decision-making bodies. Road as a street space is depicted in three-dimensional space to enable reducing the need for inspections and capturing data efficiently.

Dembski et al. [23] developed an urban digital twin prototype for a town in Germany covering aspects of a three-dimensional built environment model, road network model, urban mobility, and wind flow simulations, employing a participatory planning approach. The urban DT platform was visualized in a virtual reality platform for better understanding and inclusivity of public participation. This platform facilitates communications and decision support for relevant stakeholders such as urban planners, designers, and the public for the efficient creation of smart, sustainable cities and DTs. In addition, the prototype was able to test traffic-related interventions in terms of traffic impact and emissions. Particularly, the traffic planning scenarios allow for the possible reduction of congestion and pollution in the downtown.

Akroyd et al. [24] introduced a universal DT concept based on a dynamic knowledge graph that is implemented using the semantic web technology. The concept called the world avatar comprises concepts and instances that are linked together and agents that perform actions on both concepts and instances. From the urban planning aspect, the introduced platform would enable various functions and interoperability for constructing data, transportation flows, below-the-surface infrastructures, and climate data to improve the planning and design of cities by automating data processing and integrating notions from various planning aspects.

2.3. Gaps Identified in the Road Digital Twin Application

A literature review reveals that DT applications in the road sector have mainly focused on the maintenance and operation phase. In contrast, other lifecycle stages, such as planning and design, particularly road planning, are less considered. It may relate to the infrastructure's need for effective and efficient maintenance and operation support. Therefore, it can be understood that DT application in the road industry is still in its early stage, with a low level of maturity. Generally, the planning stage is often overlooked and not recognized as an independent lifecycle stage, as it is usually incorporated into the design stage.

In contrast, urban DTs or smart city applications do not significantly consider road planning aspects. Rather, they are focused on comprehensive aspects of cities and representing their facilities and city elements in multidimensional virtual space and underlining existing infrastructures to create future scenarios. Traffic simulation is one of the most related features in road planning in those smart urban planning applications. From this perspective, urban or city DTs are more closely linked to digital twin design, which is the next step after digital twin planning. This stage involves creating graphical representations of digital assets that correspond to planned physical assets, facilitating effective visualization and enabling various analytical processes. This can be accomplished using Geographic Information Systems (GIS), BIM, or by integrating both approaches.

While current methods for planning road alignments, whether for specific routes or for widening existing infrastructure, are becoming more advanced, there is still a notable absence of DT planning methods, especially for major infrastructure development projects such as national highways and trunk roads. These projects require significant funding and necessitate careful consideration of various factors before commencement.

However, the specific steps and processes involved in DT planning, particularly in the road infrastructure, are yet to be clarified, as this phase has received less attention than other lifecycle stages. Thus, it is necessary to be aware of and learn from the real road planning steps to derive its parallel to identify relevant requirements. It is also essential to reference existing DT architecture models to derive a DT architecture for planning road DTs, ensuring consistent implementation.

This study approaches from the road planning perspective, which serves as a use case for developing DT planning. Specifically, this study will take examples from road planning activities and processes mainly in the United Kingdom (UK).

3. Conceptualization of Digital Twin Planning

3.1. Digital Twin Lifecycle

DTs can be developed at any stage of a physical asset's lifecycle. However, they should be planned as early as possible, preferably during the pre-project or project stage, to maximize their benefits [25]. Ideally, DTs should begin at the inception of a new asset's lifecycle and continue throughout the entire lifecycle, even after the disposal phase when the PT no longer exists [8]. All DT developments and stages can occur even before the corresponding PT lifecycle begins, indicating that DT can occur even a few steps before any PT stage.

From DT planning to construction, there may be no existing physical assets. This implies that DTs can exist as digital assets before the physical assets are built. Digital assets are planned assets, such as roads, bridges, buildings, etc., in a digital format. Similar to a physical asset's lifecycle, the DT lifecycle comprises planning, design, construction, maintenance, operation, and disposal/recycling (Figure 3).

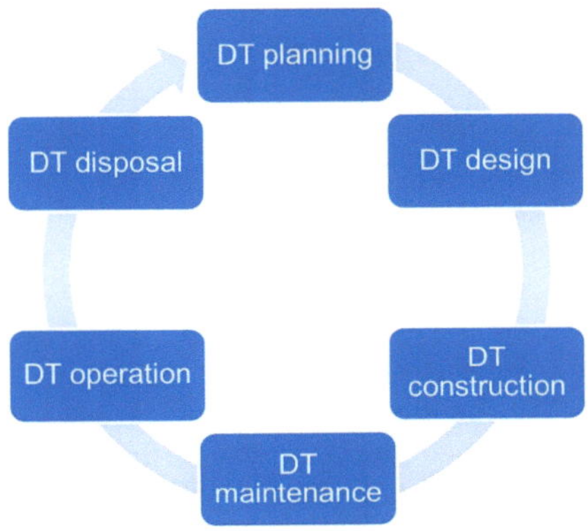

Figure 3. Digital twin lifecycle.

However, a common misconception is that DTs do not exist until a physical asset is constructed. In reality, DTs can exist even before their physical replicas are built. Therefore, a DT can be conceptualized before the availability of its physical asset. It is important to note that a DT's parallel can exist at various points in its lifecycle, and a DT can also continue to exist after the physical asset has been demolished [8].

Additionally, there is a fallacy that a DT created before a PT is simply a "digital model" that refers to a digital representation of planned physical assets, rather than a true DT. This distinction depends on the nature of its information exchange (e.g., manual or automatic) with the PT [26]. The key point is that if the digital model or DT created before PT is intended to be realized as a physical asset, it remains a DT. Otherwise, it is simply a digital model [8].

DT planning, which is the foremost stage, should start even before the planning of PT begins. From this aspect, DT planning can be considered the pre-planning stage of the PT. Therefore, DT planning is more about defining the requirements so that the DT design of the next step can be developed with a smooth handover. Defining DT requirements may refer to understanding the use purpose (functions and expected outcomes of DTs), data requirements (required data types), and constraints that restrict developing DTs (e.g., data accuracy and access) [27].

The next phase, DT Design, should be built on the DT planning, which delivers user requirements from a certain application domain. The main difference between DT planning and design is that while DT planning is to understand DT scope, feasibility, and costs by taking into consideration user and/or data requirements, DT design is rather to create DT asset classes, data structures, and cloud architectures based on the requirements captured. In addition, DT design can be visualized in multidimensional GIS or BIM environments, or possibly combinations of them for effective model development and interaction.

3.2. Early Digital Twins

The concept of DT planning is often overlooked in the existing literature. This study focuses on DT planning, particularly in the early stages of twinning. In this section, we will outline concepts related to DT planning and define its role and main capabilities.

Madni et al. [28] classified the levels of DT maturity/sophistication into four stages: Pre-DT, DT, Adaptive DT, and Intelligent DT. Pre-DT represents the earliest stage, where DT is created before the physical asset to aid decision-making during the conceptual and preliminary design. In this stage, a virtual prototype is typically used to validate some key decisions regarding the system and reduce specific technical risks early in the design process [28].

Haraguchi et al. [29] introduced a maturity model framework for city DTs consisting of eight stages, starting from the "preparation and planning stage", which is an initial stage involving planning and creating a framework for DTs. This stage, although not involving any data collection, focuses on raising awareness of DTs early in the process. The aim is to encourage political support for adopting DTs to enhance decision-making, ensuring alignment with the local authority's strategies related to the DT program [29].

As reported by [8], DT planning refers to the creation phase of a DT, i.e., a prototypical DT. All the products that will be built originate from this phase. Prototypical DT, in other words, DT Prototype (DTP), encompasses all the information needed to describe and construct a new physical asset.

Sacks et al. [30] classified the lifecycle of DTs and PTs through the concepts of fetal, child, and adult DT. Regarding DT planning, the fetal DT encompasses the conceptual design, including design, detailing, and planning activities, and ends with an as-planned process and as-designed product. The child DT begins during the fetal DT phase, at the start of prefabrication and construction, and ends with an as-built product and as-performed process. Based on these concepts, [31] compared the BIM fidelity of models, equating LOD 100—Conceptual Design—to the fetal DT and LOD 200—Preliminary Design—to the transition from fetal to child DT in the context of DT planning.

Jones et al. [32] proposed the concept of early DT, which encompasses engineering design tasks, such as task clarification, concept design, embodiment design, and detail design, to support effective design and information and solution evaluation. In the earlier stages, computational analyses are crucial for minimizing rework costs and ensuring sound decision-making, especially as these stages consume most of the project budget [32]. The concepts related to DT planning are outlined in Figure 4.

The planning phase is often overlooked, as it tends to be integrated into the design stage. However, some studies have reported specific tasks relevant to the DT planning stage. DTs are designed to support decision-making with well-thought-out and timely interventions and to deal with complicated and uncertain conditions.

Strategy and planning are critical components of early-stage twinning and represent significant use cases for DTs. In this sense, DTs can aid in formulating government policies and organizational strategies or planning the construction of new assets for future investments and transportation scenarios [33]. Moreover, DTs facilitate multi-criteria decision-making, serving as a collaborative decision-making tool for stakeholders to select the best options among various preferences and choices [33].

In the design phase, design data and digital assets, including a digital representation that facilitates simulation and further what-if analysis, may be required for new assets with the provision of DT during the project, design and construction, and operation for the entire lifecycle [25].

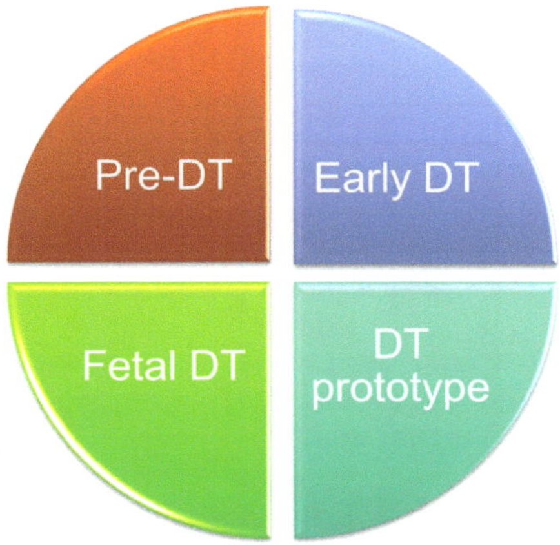

Figure 4. Digital twin planning-related concepts.

Singh et al. [34] identified various applications of DT in several industries, particularly regarding DT planning for smart city applications, resource planning and management, and decision-making process support, and for the construction section, improving decision-making for the feasibility of the project.

From the city DT aspect, a prominent research theme is planning and prediction, whose specific tasks for planning are policy evaluation, simulation, and "what-if scenarios" [35]. Based on the review, potential use cases and applications related to DT planning are illustrated in Figure 5.

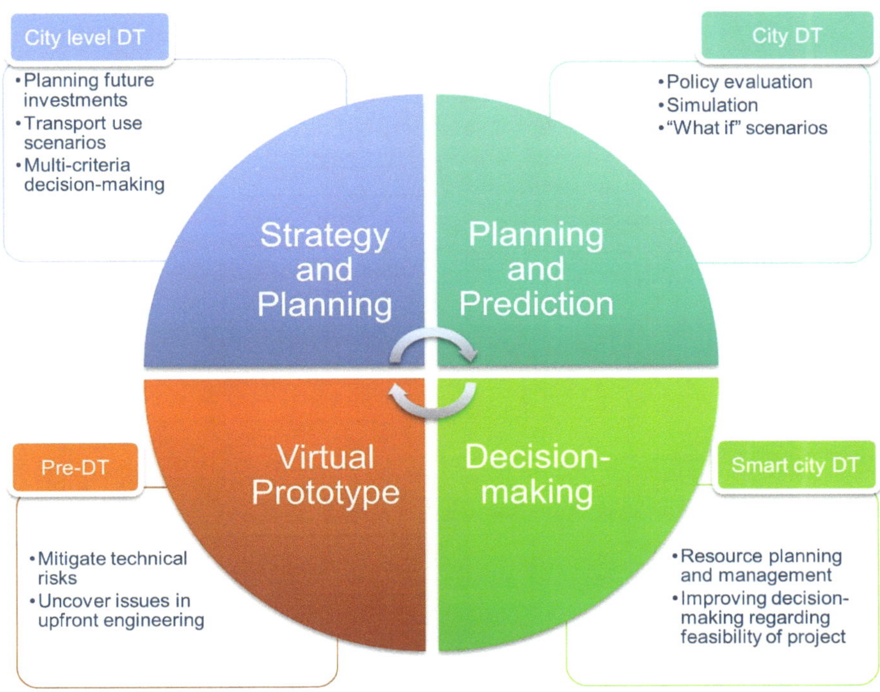

Figure 5. Digital twin planning-related potential use cases and applications.

3.3. Economic Aspect of (Early) Digital Twins

DTs were expensive and had restricted benefits when created in the past. However, it turned out to be a valuable thing owing to the increased availability of storage and computing costs as well as the use cases and potentials that enable DT implementation [36]. In addition, DTs may be cost-effective when applied to planned assets (e.g., new infrastructures) rather than existing assets during the lifecycle.

Although DTs require most of their digital resources at the time of creation, the entire cost of these prototypes—initial versions—reduces over time. Particularly, DTs can be valuable for cost and time in terms of their capability to provide a digital environment where relevant tests and experiments are performed without the actual input of physical materials, labor, and other resources [37].

There are not many identified KPIs for measuring DT performance except for works performed in the manufacturing domain. In this domain, KPIs such as performance (e.g., throughput, orders, lead time), quality (e.g., on-time delivery, order picking errors), or cost-related (e.g., materials, labor) ones are usually measured on cyber-physical systems or manufacturing, logistics equipment [38].

Similar to the categories in the manufacturing domain, road DT planning KPIs may consist of system performance metrics such as scalability (e.g., system performance vs. data volume), quality metrics such as the results or accuracy of prediction DT yields, and cost-related metrics such as software/platform cost, development cost, and labor cost, and all these features can be used to evaluate the feasibility of DTs for their further application.

Experts suggest that considering the return on investment, which is the ratio of profit over costs (investments), will be useful to decide whether to use DTs for businesses. However, the return on investment made during the implementation of DTs still remains to be quantified. Learning from the previously implemented DTs and use cases, as well as structuring benefit realization, will help to solve this problem [25].

When developing a strategy for structuring benefit realization, it is crucial to look into problems and issues that are faced in the enterprise, mid- and long-term performance, improved processes, and potential new revenue channels. The business values that DT offers can be categorized as follows [36]:

- Improved quality (e.g., overall quality);
- Improved services (e.g., customer experience);
- Operations cost (e.g., improved product design and engineering change);
- Record retention (e.g., better managing digital records);
- New product control (e.g., reduce overall time to market and cost);
- Revenue growth opportunities (e.g., product upgrade, improve efficiency).

Therefore, considering this cost-benefit aspect of DTs will be important when adopting DTs for businesses. This study provides one example of a DT use case—road transportation planning from the functional aspect so that the benefits of implementing it would be realized.

4. Digital Twin Framework for Road Transportation Planning

4.1. Capturing the Requirements for Road DT Planning from the Road Planning Process

4.1.1. Road Planning Processes

Road planning as a process is conducted before building every new road infrastructure. In practice, it usually requires considering multifactor in a complex project management environment with multi-stakeholders to make a feasible decision. The costs to build a new road are enormous, and it needs a systematic project management approach. The foremost stage is conventionally called a feasibility study, which is considered a (pre-)planning stage.

The process may differ from country to country and region to region; however, generally, these processes pertain to a similar flow from the beginning to the end of the study.

In the case of Korea [39], new and large-scale projects covering major infrastructures such as roads, railways, seaports, water-related, etc. are subject to preliminary feasibility study (PFS). The PFS here involves making a succinct evaluation in a short period to evaluate projects and make budgetary decisions. The result of this process is to yield a provisional evaluation and to conduct preceding research for a further detailed feasibility study.

An economic analysis, policy analysis, and technical feasibility analysis are the main steps of the PFC. The economic analysis is conducted to decide whether the next phase, a detailed feasibility study, is required. It covers the cost-benefit analysis (e.g., B/C ratio, NPV), transportation demand forecast (e.g., trip generation, distribution, split, and assignment), economic feasibility assessment (e.g., construction cost, operation and maintenance cost, user benefits), as well as fiscal feasibility assessments (e.g., profitability of projects). Whilst policy analysis involves examining the macro aspect of the project, i.e., evaluation from the national economic perspective, such as balanced regional development. Particularly, it excludes evaluation items that are not included in the economic analysis part and that are still important for assessing project feasibility, such as regional development concerns, regional economy development, consistency with relevant plans, and environmental impact assessments. Lastly, in the case of technical feasibility, the expert's consulting is performed, and a more detailed analysis is required in the detailed feasibility study stage.

In the case of the US [40,41], a feasibility study is conducted in the earlier stages of major project development. The feasibility study helps to decide whether to move forward to the next stages of a project, as it requires more detailed and advanced analysis, such as environmental analysis, public hearings, schematic design, and right-of-way surveys. Thus, feasibility studies are not intended to provide detailed design, environmental analysis, and cost calculations. It is rather to identify high-level engineering elements, the impact on the public, and the economic feasibility of new roads or improvement of the existing roads.

Generally, the feasibility study stages are concentrated on preliminary analysis, which helps determine whether to go forward, assessing environmental and engineering constraints, and identifying preliminary route options. After that, with input from the public opinion and additional technical analysis, route options that are further analyzed are refined and recommended. These route options are the starting point of future relevant project steps to be taken into consideration.

A checklist for conducting a feasibility study [42] suggests that various steps and relevant considerations for effect measurement should be performed not only in the planned event place but also in the hosting region. Particularly, these measures include travel forecast (e.g., modal split, traffic events), market analysis (e.g., travel time, distance analysis), parking demand analysis (e.g., on/offsite parking demand, parking occupancy), traffic demand analysis (e.g., directional distribution, event-generated traffic assignment), and roadway capacity analysis (e.g., screen line capacity and section/point capacity).

On the other hand, the UK has a very similar structure to the road transportation planning activities in other regions and countries and the feasibility study is performed in the foremost stage in terms of delivering a major road project. In this paper, we select the UK's road transportation planning practices as representative processes among other countries, and the next sections report more detailed steps and activities toward scoping the road DT planning.

4.1.2. Current Road Network and Digital Roads Initiative in the UK

The strategic road network (SRN) of England comprises 4500 miles of motorways and major A-roads. These roads connect the regions and nations of the UK, as well as major facilities, such as the Channel Tunnel, ports, and airports, enabling safe and free mobility of people and goods, keeping people connected, and supporting economic growth across the UK [43]. Consequently, the SRN is accountable for 90% of passenger travel and approximately 70% of freight transportation, serving as the backbone of the UK transportation system [44].

On the other hand, digital roads are revolutionizing England's important assets—its major roads—by catalyzing a technology-enabled network for better road infrastructure. These digital roads will exploit data, technology, and networks to improve the design, construction, operation, and maintenance of the SRN [45]. Additionally, they will support the long-term vision for the SRN, such as the "Strategic Vision 2050", which aims for an economy-supporting, greener, safer, more reliable and integrated, and smarter network [46].

Digital roads are structured around three core themes, one of which relates to the study, "Digital Design and Construction". This theme aims to facilitate digital design, including scheme designs and long-term planning through digital tools [45]. Currently, we are in the second Road Investment Strategy (RIS2) period (2020–2025), with RIS3 occurring from 2025 to 2030. Typically, these RISs provide plans, visions, etc., that lay the foundation for future road development. During these periods and beyond, the expectation is to develop detailed DT for all new schemes, leverage data and information for all business processes to make important business decisions, extend DT across the entire network, and combine these efforts into a national DT framework [45].

4.1.3. Road Planning Processes in the UK

The project control framework (PCF) is a methodology for managing and delivering major projects, such as road construction, through collaboration between the Department for Transport (DfT) and National Highways.

PCF consists of three major phases: options, development, and construction. The options phase, which includes the pre-project stage, explores the preferred road solution to the transportation problem. The pre-project stage involves a feasibility study focused on strategy development, shaping, and prioritization, which occurs before entering the PCF stages. After these steps, the development phase begins, aiming to design the preferred solution through the required statutory processes until the investment decision point for building the road solution. Finally, the construction phase focuses on building the road solution, handover for operation, and the project closure.

Each phase has implementation steps with clearly defined start and end points, along with key milestones (Figure 6). At the end of each stage, an approval procedure is required to start the next stage. Projects may drop out of the lifecycle at any point before the investment decision if they do not meet certain criteria, such as value for money and affordability of the road solution.

The PCF can be applied flexibly such that single projects can start from Phase 3—development—as these projects do not require land acquisition or an environmental assessment because they fall within the highway boundary. Examples of such projects include online widening and controlled or smart roadways, such as all-lane running [47].

This study aims to examine the road PT planning process to draw parallels from the relevant literature and stakeholder documents. Our focus is on identifying the road planning process referred to as the (pre-)planning stage. According to the project control framework from national bodies, the main activities in this pre-project phase include:

- Identifying and prioritizing potential transportation problems;
- Shaping, investigating, and assessing the effectiveness of transportation schemes and road network solutions;

- Initiating a major road project if it is considered the most feasible solution to the transportation problem.

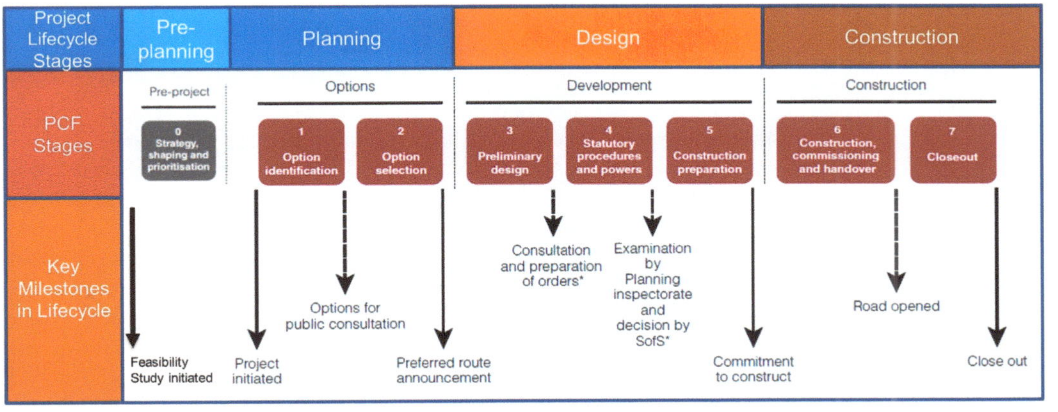

Figure 6. Project control framework stages with their key milestones (Adapted from [47]).

The options phase is divided into two distinct stages: option identification and option selection. The option identification stage deals with identifying options to be presented for (non-statutory) public consultation based on results from the previous stage/phase. Traffic modeling and economic assessment are conducted on options that could serve as potential road solutions to the transportation problem. Specifically, detailed traffic modeling is required to assess whether a scheme will be effective both now and in the future. The economic evaluation of transportation projects is crucial for helping decision-makers prioritize schemes and options to ensure the best use of public funds [47].

In the option selection stage, public consultations, including exhibitions, are held to gather and analyze public opinions and comments regarding the preferred options. Further refinements related to traffic modeling, economic assessment, and environmental impact are conducted. At the end of this stage, options to proceed further are determined, and a public announcement is made regarding the "preferred route" [47]. A summary of the PCF options phase (pre-project and planning steps) is provided in Figure 7.

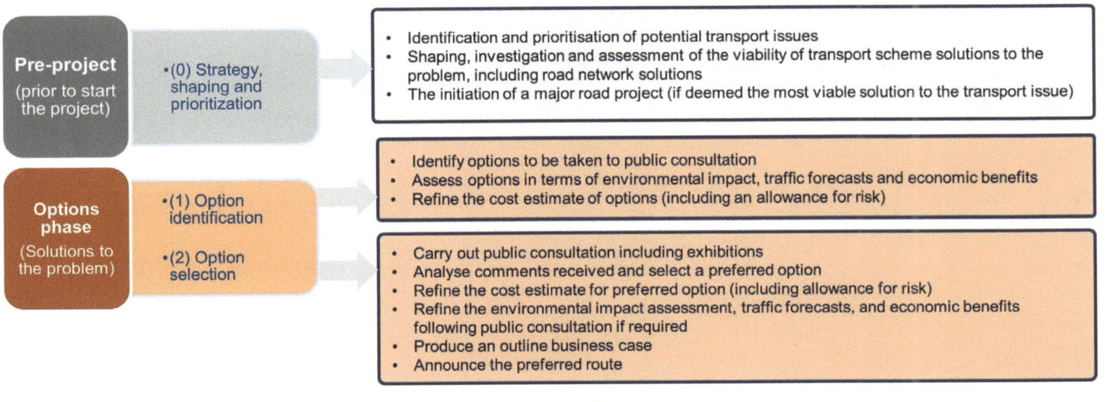

Figure 7. Road planning processes as part of the PCF.

As DT planning is the pre-planning step of PT planning, we consider the pre-project (feasibility study) stage, which occurs before the inception of PCF stages. This is also because the options phase (option identification and selection) focuses on further refining and elaborating the better-performing options identified in the pre-project stage. Additionally, this phase involves mandatory processes, such as public consultations on the refined options.

The primary purpose of road DT planning is to provide stakeholders with an initial idea about the project, offering insights without the need for immediate economic investment. Therefore, DT planning can serve as a reference model before starting any formal stage of the PCF. The existing frameworks and procedures mentioned in this paper, such as the PCF and Transport Analysis Guidance (WebTAG), should be followed.

The Transport Analysis Guidance, or Transport Appraisal Process, also called WebTAG, provided by the DfT [48], offers specific step-by-step guidance for implementing PCF stages. It consists of three stages, which are all relevant to a certain point/stage of PCF implementation: (1) option development, (2) further appraisal, and (3) implementation, monitoring, and evaluation. In this study, we examine the specific steps in planning road PT to draw parallels, as the PCF pre-project stage corresponds to a part of the option development stage.

The option development stage aims to identify intervention needs and create options to address the project's objectives. This stage is generally desk-based, using easily accessible data, backed by stakeholder involvement and public consultation. The resulting options are then assessed to obtain a small number of viable options for further detailed appraisal in Stage 2. The analysis aims to meticulously estimate the probable performance and impact of the interventions. WebTAG Stage 1 comprises nine steps (Figure 8), corresponding to the PCF 0—Pre-project stage. In practice, if any options are selected for further consideration after the pre-project or feasibility study, they proceed to the next stage (PCF 1—Options Identification), which begins once a project is initiated. Table 1 details these pre-planning steps (feasibility study), along with their main input and model information.

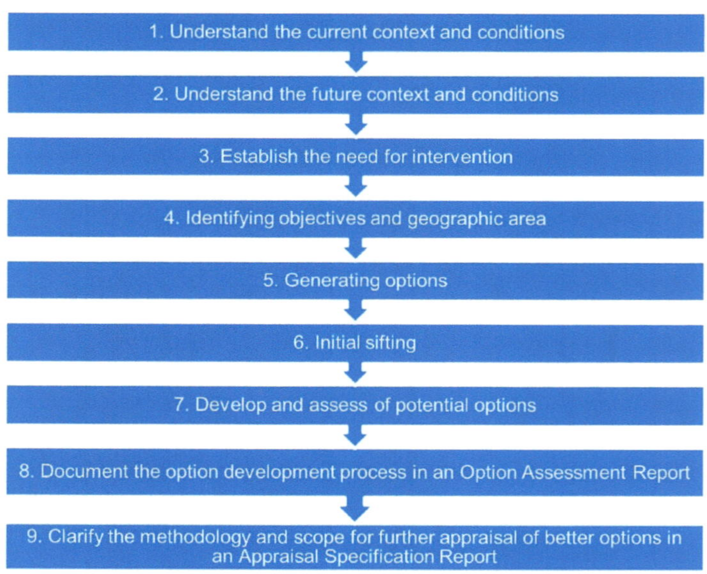

Figure 8. Feasibility study (pre-planning) steps of the road planning process.

Table 1. Road planning steps in detail.

Planning Steps	Detailed Steps	Main Considerations and Inputs (Models)	References
Step 1—Understanding the current situation	Current transportation and other policies Current travel demand and levels of service Current opportunities and constraints	Local, regional, and national policies of the target area and the transportation sector Travel demand, level of service, and capacity information Physical, legal, and institutional constraints and relevant opportunities	[48]
Step 2—Understanding the future situation	Future land uses and policies Future changes to the transportation system and demand	• Macro level: National planning policy framework, local and regional development plan • Micro level: Study-specific forecasts (transportation and land use) • TEMPRO database is used as a reference for forecasting population, household, and employment • Forecasts with and without scheme cases • Cost-benefit analysis • NTEM data are used and accessed by TEMPRO • Highway Assignment model • Public Transport Assignment model • Variable demand model	[48–52]
Step 3—Establishing the need for Intervention	Current transportation-related problems Future transportation-related problems Underlying causes	• Problem consultation information • Transportation system evaluation • Existing transportation data and tools • Other benchmarking performance • Future transportation problem without scheme case • Forecast future travel demand and level of service without the scheme Determination of the causes of the problems	[48,53]
Step 4a—Identifying Objectives	High-level or strategic outcomes	• Development of hierarchy of objectives: high-level or strategic outcomes, specific or intermediate objectives, and operational objectives • Setting targets for both qualitative and quantitative	[48]
Step 4b—Defining the Geographic Area of Impact to be addressed by the Intervention	The area of impact for generating options	The geographical extent of the travel market and the current and future issues	[48]
Step 5—Generating Options	Range of alternative measures or interventions aimed at addressing the identified objectives	• Opinions from local stakeholders, colleagues, consultants, authorities, and the public • Brainstorming sessions, workshops • Discussions with sponsoring organizations • Reconsider ideas from the previously discarded proposals • Benchmarking with appropriate comparator areas, not only in the UK • Thorough research, including professional journals, press, the internet, other projects, and other evidence databases	[48]

Table 1. *Cont.*

Planning Steps	Detailed Steps	Main Considerations and Inputs (Models)	References
Step 6—Initial Sifting	Discarding unpromising options	The Early Assessment and Sifting Tool (EAST) is used to assess options that have progressed or been discarded, based on evidence	[48,54]
Step 7—Development and Assessment of Potential Options	Identifying better-performing options	• Potential options are evaluated based on the "Transport Business Case" criteria using the Options Assessment Framework, which incorporates the best-practice Treasury five-case model (Strategic, Economic, Financial, Commercial, and Management cases) to identify better-performing options for further appraisal • The appraisal summary table is fulfilled as a result	[48]
Step 8—Produce an Options Assessment Report (OAR), or similar	Documenting the Stage 1 process of identifying the need for intervention and the process of option development and selection	• The OAR should align with Steps 1 to 8 of this framework • The framework should also include the viable options (e.g., a low-cost solution) to be considered for refined appraisal in Stage 2	[48]
Step 9—Clarify Modeling and Appraisal Methodology	Clarifying the methodology and scope of further appraisal	The methods should be recorded in an Appraisal Specification Report (ASR) or similar, including an Appraisal Specification Summary Table	[48]

Steps 1 to 4 constitute the basic stage, while Steps 5 through 8 focus on option generation and assessment/evaluation. Step 9, the last stage, is dedicated to determining a methodology essential for conducting subsequent stages/steps.

Step 1 involves understanding the current situation, including current transportation and other policies, current travel demand, levels of service, current opportunities, and constraints. This step's main inputs and models include policy-related document materials, currently available transportation data and information, and relevant constraints.

Step 2 focuses on assessing the current situation and understanding the future situation. This includes future land use and land policies, as well as future changes to the transportation system and transportation demand. It is essential to evaluate future land use and policies at the macro level, examine the interaction between transportation and planning decisions, and identify land-use planning decisions that may require study-specific forecasts. The Trip End Model Presentation Program (TEMPRO) database is used as a reference for forecasting population, household, and employment.

Future changes need to be calculated by considering two forecasts: one with the intervention (the "with-scheme" forecast) and another without the intervention (the "without-scheme" forecast). Typically, separate forecasts should be created for at least two years to account for changes in population and other transportation infrastructure over time. Additionally, cost-benefit analysis provides guidance on selecting the appropriate forecast years [55].

A transport model is a computerized tool that converts forecasting assumptions into a demand forecast (number of trips) and supply (level of service and cost of travel) on the transportation network. Figure 9 presents a highly generalized transport model component based on the standard transport model structure outlined in [55]. Consequently, the demand model relies on trip ends and travel costs. The trip end model is based on the National Trip End Model (NTEM) dataset, which reflects the DfT's standard assumptions about growth in demand, represented in units of trip ends, and the DfT's central assumption of travel demand growth between any two specified years. TEMPRO provides access to this dataset.

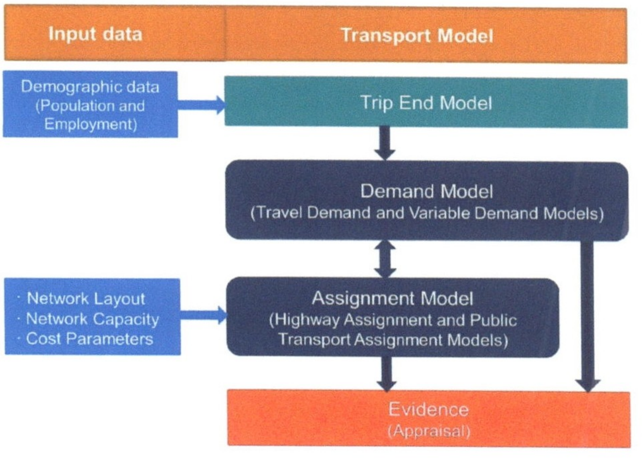

Figure 9. Highly generalized transport model structure.

The assignment model is used to develop a cost matrix estimated from the trip matrix (trip demand and supply) and a mathematical representation of the transport network model. Additionally, the assignment model allocates the trips by route and computes both the demand and cost for each link and junction within the network. Highway assignment models enable the analysis of social and environmental impacts through vehicle flows on links. In highway assignment, cost varies depending on congestion on the network, and the demand and cost for each route are interdependent. The public transport assignment model depicts how passengers select their routes through the transportation network. The network's complexity is determined by its physical appearance, the routes, and the number of service occurrences. The variable demand model is primarily used to model individual travel for highway and local transportation trips by car and bus schemes. The goal of variable demand modeling is to forecast and measure changes in transport conditions that influence demand.

Step 3 involves establishing the need for intervention for the current problem. This includes current and future transportation-related problems and their main causes. Key inputs and models for the current transportation-related issues are derived from public consultation information, transportation system evaluation, existing transportation data and tools, and performance benchmarking against other cases. Existing data may include the NTEM data (planning and demand data), highway traffic data, public transport data, road network data, and other data (such as census data, national travel survey data, and aviation data) [53]. Future transportation-related problems and their underlying causes are determined based on future travel demand forecasts without schemes. The causes of the issues must be examined before solutions are generated.

Step 4 consists of two main steps: (4a) identifying objectives regarding high-level or strategic outcomes, specific or intermediate objectives, and operational objectives; and (4b) defining a geographic area of impact to be improved by the intervention, particularly the area of impact for generating options. A set of intervention-focused objectives to solve the identified problems should be defined. These objectives must reflect the issues and context outlined in Steps 1 to 3, particularly the opportunities and constraints in Step 1 and the main causes depicted in Step 3. These objectives influence the direction of the appraisal methods detailed in Step 9 and their further refinement.

The area of impact affects the scope of options generated for the next step. Furthermore, determining the corresponding intervention's geographic boundary requires understanding the geographical extent of the travel market and key origins and destinations (Steps 1 and 2) and analyzing the geographical scope of current and future transportation problems and their underlying causes (Step 3).

Step 5 involves generating options by proposing various alternative measures or interventions to address the previously defined objectives. There are several requirements for generating options. In this step, all possible options should be generated, including all transportation modes, infrastructure, relevant regulations, pricing, and other factors affecting travel behavior. Specifically, for highway solutions, options should include varying sections or intersection standards and other alternatives to resolve the problems in the given area, such as providing public transportation, implementing demand management, and adopting traffic management methods. Additionally, options should include interventions that decrease or affect the need for travel and capital spending. Although measures included in options may not completely solve the problem, they can still help achieve the goals. At a later stage, some options may be considered either as a single or combined form that will enable possible interventions.

Ideas can be generated through various methods, including responses from local stakeholders, colleagues, consultants, nearby authorities, and the public. Additional methods include brainstorming sessions and workshops, discussions with sponsoring organizations, reconsidering ideas from previously discarded proposals, benchmarking with suitable comparable areas not only in the UK but also research-based methods, including professional journals/press, the internet, other projects, and other evidence databases.

Step 6 involves discarding options that do not meet certain intervention objectives and align with local, regional, and national programs and strategies, or government priorities. These options may not meet viability and acceptability criteria due to being impractical in specific economic, environmental, geographical, or social contexts, not technically good, not financially affordable, or unacceptable to stakeholders and the public. This step employs the Early Assessment and Sifting Tool (EAST), which facilitates the evaluation of options, determining which should be advanced or discarded (unpromising options) based on evidence. The EAST is an Excel-based decision support tool designed to rapidly summarize and present evidence on various transportation options. It can evaluate and compare several transportation options, packages, strategies, and plans across all modes and regions. This tool is versatile and suitable for early-stage development, but the reliability of its comparisons depends on the quality of the evidence used. The EAST is not designed to make recommendations or to be used for final funding decisions. Instead, it facilitates the consideration and advancement of options or elimination (unpromising options) based on evidence.

Step 7—In this step, the number of options is reduced to identify the best-performing ones. This involves developing potential options in detail and gathering robust evidence to compare their costs, benefits, and impacts. Potential options are assessed with the "Transport Business Case" criteria using the Options Assessment Framework, which incorporates the best-practice Treasury five-case business model (Strategic, Economic, Financial, Commercial, and Management cases). Generally, this assessment is performed primarily as a desk-based task, although some in situ visits may be required. Environmental studies are not conducted at this level. The five-case model can be briefly defined as follows:

- The Strategic Case—Whether investment is needed now or in the future;
- The Economic Case—Option impacts and value for money;
- The Financial Case—Affordability of the proposal, funding, and accounting;
- The Commercial Case—Commercial viability of the proposal and procurement strategy;
- The Management Case—Whether the proposal is deliverable.

Step 8 involves producing an Options Assessment Report (OAR) or a similar document. The OAR should align with the preceding steps outlined in this framework. It should document the first stage of identifying the need for intervention, as well as the process of developing and selecting options.

Step 9—The last step involves clarifying the modeling and appraisal methodology, as well as defining the scope for further appraisal. The methodology should be clearly outlined in an Appraisal Specification Report (ASR) or a similar document. The ASR should contain an Appraisal Specification Summary Table, which outlines the proposed methodology for the appraisal and identifies the challenges or sub-impacts in the Appraisal Specification Table (AST).

4.2. Derivation of Road DT Planning Processes (Scope) Based on the Road Planning Requirements

4.2.1. Mapping of Relevant Steps from Road Planning to Road DT Planning

Based on the concepts reviewed and defined in the previous sections, DT planning refers to a process designed for digital assets, meaning it occurs before a real asset becomes built. In other words, it is the pre-planning stage of road planning, similar to the feasibility study—the pre-project stage.

From this sense, it can be said that DT planning enhances the whole process by introducing an additional step—virtual/digital planning of roads based on real road planning steps/processes. However, it is important to note that DT planning neither improves the current processes (planning steps) nor substitutes them. Rather, it attempts to derive its parallel from the road planning process.

In the next section, we discuss what is applicable or not for creating road DT planning based on road PT planning. To determine what road planning steps would apply to DT planning, mapping existing processes to the corresponding DT functions or layers is essential. Mapping is the most commonly used technique in database engineering, including DT ontology modeling [56], and in various conceptual approaches to road infrastructure modeling [57,58].

The core assumption of this framework is that it will be implemented by relevant stakeholders/decision-makers who might need to make early decisions regarding whether to construct a road and start planning. Therefore, this framework is supposed to be used before any real project steps, even pre-planning stages (feasibility study). Road DT planning is based on applicable (pre-)planning steps in the road PT planning process.

Existing road planning processes can be broadly divided into two main phases: before and after option generation, as options are the most important results/part of solving the transportation problem. Prior to generating options, there are four basic steps, as reviewed in the previous section (Steps 1 to 4): understanding the current situation, understanding the future situation, demonstrating the need for intervention, and clarifying objectives.

Understanding the current situation is mainly based on collecting existing data and evidence and analyzing them. Thus, currently available data sources can serve as input data for DT planning as well as the modeling and analysis part. Understanding the future situation involves studying future changes, incorporating analysis and various forecasts (with or without schemes), and developing transport models. From this perspective, this step is related to the input, modeling, and analysis (parts) of DT planning.

The next step of road planning, establishing the necessity for intervention, deals with refining transportation-related problems (both current and future) and their underlying causes. Various existing data, along with supplementary survey data and models, are utilized to identify these problems and their causes. The result significantly influences the generation of options, which is a step in the planning process. Similar to the previous step, this step is possibly related to the input, modeling, and analysis (parts) of DT planning.

The next step, identifying objectives, is performed based on the findings of transportation-related problems from the prior step of the road planning process. Objectives are classified into three levels: strategic, specific, and operational. Additionally, identifying the geographical area of impact is vital for the scope of options generated in the next step. From the DT perspective, this step can be the input and modeling stage.

On the other hand, the option generation process comprises three constructive/fundamental steps, as reviewed in the previous section (Steps 5 to 7): generating options, undertaking initial sift, and developing and accessing potential options. These steps are based on the earlier stages (Steps 1 to 4) and are all commonly mapped to the analysis part of DT planning.

Generating options is a key step in the road planning process. In this step, a comprehensive range of options should be generated, including all transportation modes, infrastructure, regulations, costs, and other factors affecting travel behavior. Relevant stakeholders, such as consultants, authorities, and the public, should be involved to provide ideas, suggestions, and feedback. However, in the context of DT planning, users/decision-makers should be able to generate options independently. In this case, decision-makers may take advantage of reviewing previous proposals and benchmarking similar projects to address the transportation problem based on the input data of DT planning.

The next step is undertaking an initial sift, which involves eliminating unpromising options from those generated in the previous step. Therefore, only feasible or potential options are retained for further assessment. When assessing options, the EAST decision support tool, discussed earlier, is employed, primarily qualitative evaluation based on the knowledge and practice of transport appraisers. In the case of DTs, the analysis part will focus on assessing potential options based on appropriate input data and modeling. It may involve a multi-objective optimization (MOO) problem, as several options will be appraised for advancement to the next step. More detailed content is provided in the next section—the proposed modeling of the DT planning.

The last relevant step in road planning involves developing and assessing potential options identified in the previous step. This step takes a few possible or better-performing options and investigates them in greater detail, employing quantitative aspects where applicable. The Options Assessment Framework of the Transport Business Case criteria is applied with evidence against each option. The evaluation five-case models—strategic, economic, financial, commercial, and management—are integrated into the DT planning process, particularly during the development of objectives or measures.

Steps 8 and 9 are not considered or discussed in the mapping process due to their lesser relevance to the initial option generation and assessment. Figure 10 illustrates mapping applicable road (pre-)planning steps to road DT planning layers.

4.2.2. Key Differences of Road (Pre-)Planning and Road DT Planning

As mentioned in the previous section, road DT planning is not intended to substitute or improve the existing road (pre-)planning processes. Rather, it attempts to improve the whole process by incorporating DT technology that adds value to the entire chain.

The main idea is that DT planning takes place before even starting real road pre-planning or feasibility study, which takes time, money, and effort. When there is a required or obvious need to do something and solve the problem, a stakeholder and/or decision-maker should be able to use the road DT planning tool to decide on their own in the first place. They need to decide whether to plan (start a feasibility study) a road, perhaps widen/build a road, or use the DT technology on their systems at the early stage of the product lifecycle to solve the problem. They may need to deliver their requirements, whether related to planning, building the road, or using DTs at certain specifications, to the designers of the DTs.

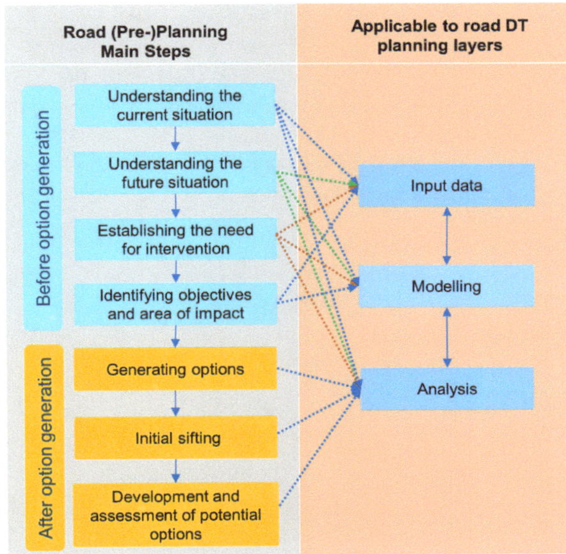

Figure 10. Mapping relevant road (pre-)planning steps to digital twin planning.

In other words, for the systematic development and adoption of DTs, DT planning is expected to hand over user requirements to the DT designers to construct DTs for the next step (see Figure 3). The key assumption is that road DT planning should be based on and derived from actual road planning steps, making it accountable and practical.

Road DT planning allows for the creation of digital roads in advance, ensuring they can be realized in the real world as PT. As discussed in the previous sections, there are applicable and non-applicable steps in the road planning processes. Table 2 compares the key differences between road (pre-)planning and road DT planning in terms of data collection, problem identification, data analysis, and option generation processes.

Table 2. Key differences between road planning (feasibility stage) and road DT planning.

	Road (Pre-)Planning	Road DT Planning
Data collection	A group, company, or team works on a data collection process that takes certain times (e.g., a few months)	A stakeholder or user can collect and input data by oneself when using DTs
Problem definition	Problems need to be identified by the stakeholders based on the data and evidence-collection processes	A stakeholder or user can define the problem by oneself based on the data, including expert knowledge
Data analysis	Experts perform relevant qualitative and quantitative analyses by themselves using related tools and forms	A stakeholder or user can use DT to analyze before and during the option generation process
Option generation and Assessment	Multi-stakeholders get involved in generating and assessing options by using various methods	DT generates options and provides optimal solutions to the problem using its functions (e.g., models, algorithms)

4.3. Framework Proposal for Planning Road Digital Twins

Based on the findings from the previous sections, we propose a road DT conceptual framework/architecture (Figure 11). This framework consists of five layers: data acquisition, data processing, data modeling and algorithms, data analysis and control, and the service layer plus users. However, it is of note that existing road planning processes must be followed, as regulations and statutory processes, and particularly some of these processes, such as public consultation during the options development phase, are essential to making relevant decisions.

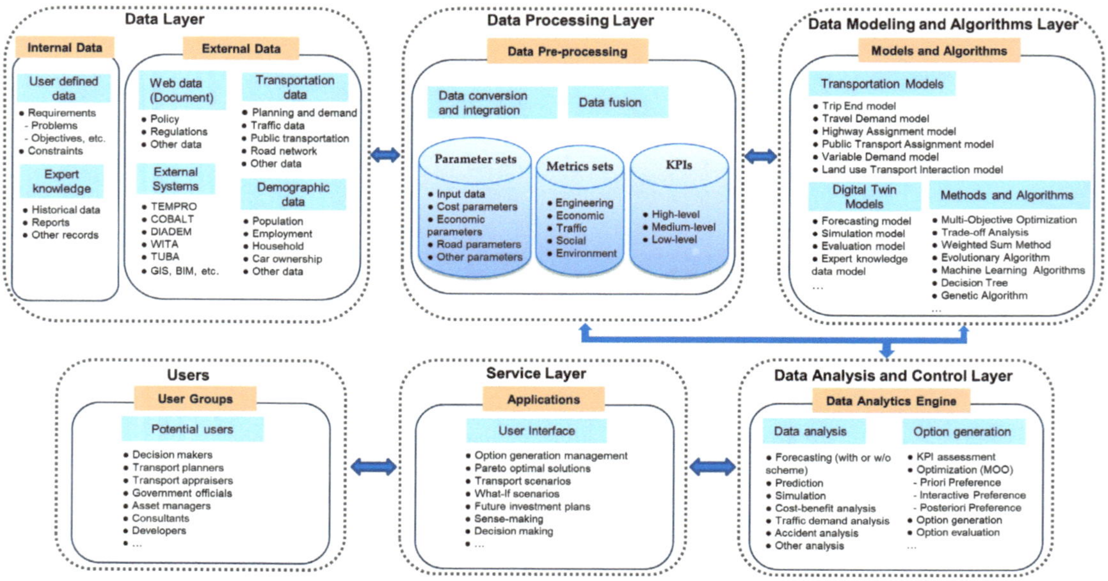

Figure 11. Proposed framework for road digital twin planning.

4.3.1. Data Acquisition Layer

This layer obtains data that are similar to the road planning steps. Most input data pertain to Steps 1 to 4, which are part of the pre-option generation phase, specifically Steps 1 and 2. Thus, the main input data for road DT planning can be generally grouped into two categories: internal and external data/storage. Internal data contain user-defined data and expert knowledge. User-defined data imply that the data inputted by users, such as general requirements for planning roads, for example, in the form of problem definitions that may be described in the text, factors, numbers, etc. Expert knowledge comprises historical data and records that might have resulted from the previous events in the DT. On the other hand, external data consist of four categories, including web data, transportation data, demographic data, and external databases/systems. Below is the more elaboration on the main parts of the data layer.

(1) Policy and regulations

Steps 1 and 2 involve gathering internet data, such as local, regional, and national policies, relevant to the site and the transportation sector, including physical, legal, and institutional constraints. The national planning policy framework and local and regional development plans should also be considered. Additionally, current travel demand, levels of service, and capacity information are important. These data can be found on the relevant authorities' websites and through UK or regional transport statistics. Moreover, local planning data and other work outputs, such as strategic environmental assessments, are valuable resources.

(2) Transportation data

Transportation data encompass various existing data sources, including planning and demand data, highway traffic data, public transportation data, road network data, and other datasets. Planning and demand data can be obtained from the NTEM dataset through TEMPRO.

Various sources provide highway traffic data. Road traffic statistics offer statistical information on the vehicle miles traveled, categorized by vehicle type, road classification, and location. Additionally, traffic data for every link between junctions on the "A" road and motorway network in Great Britain include the annual average daily flow [59]. Moreover, the government portal (data.gov.uk) provides data on road traffic volume, local authority traffic volume, traffic speed, and congestion levels. Traffic flow and journey time data can be acquired from the Highways England Traffic Information System (WebTRIS) webpage.

Public transportation data include information on buses and railways. LENNON data, a rail ticket information database, provides passengers with information based on station-to-station trips, including rail passenger counts. Data from the electronic ticket machine encompass all journey information, such as trip records covering fare stages where passengers board or alight.

Road network data are available in two-dimensional formats, including detailed representations of road center lines showing all highways and curvature details and skeletal representations of major roads for "journey planner" packages. Practitioners should preserve geometric data necessary for environmental analyses, particularly when the network is stored in a geographic information system (GIS) format. Another useful source for GIS-based modeled networks is the Integrated Transport Network from the Ordnance Survey, which aids in understanding the modeled representation of networks and conducting environmental analyses.

(3) Demographic data

A measure of travel demand is determined by demographic assumptions, including population and employment. In a broader sense, demographic data refer to population, households, car ownership, and employment. The most robust and relevant demographic data typically pertain to the resident population. Other data, such as GDP growth and fuel price trends, can also be considered. At a more detailed level, land use data provided by OS AddressBase can be used.

(4) Other transportation data

Other sources of transportation data include census data, National Travel Survey (NTS) data, and aviation data. Census data offer further information on trips between home, i.e., usual journey-to-work information. The government portal provides census journey data. The NTS, the first source of data on personal travel patterns in Great Britain, is based on household surveys conducted since 1988. The Civil Aviation Authority (CAA) provides data on airport usage in the UK. This information summarizes the results from the CAA Passenger Survey, providing details about the origin and destination of air passengers, their travel purpose, and passenger characteristics.

(5) External database/systems

External databases/systems include, but are not limited to, TEMPRO; COBALT, which assesses accident impact as part of economic appraisals; DIADEM, which deals with variable demand modeling that predicts and quantifies these changes in transport conditions;

WITA, which evaluates wider economic impacts, including the welfare impacts of employment, investment, and productivity effects, excluding those assessed by the Transport User Benefit Appraisal (TUBA), which is used for cost-benefit analysis, employing a "willingness to pay" approach to economic appraisals.

(6) Expert knowledge

Expert knowledge is a crucial part of the DT. It stores the previous knowledge in the planning process, such as option generation and decision-making. Specifically, previous DT planning tasks or decisions can be stored in the expert knowledge database as historical data and reports for retrieval and reference in current projects/tasks. Additionally, the ability to store and use DT information of its counterpart PT, whose lifecycle ended, will be a valuable source for future designs and help save time and expenses [8].

4.3.2. Data Processing Layer

The data processing layer processes the data collected through the data acquisition layer from various sources. These data must be converted into a standardized format to produce useful information for further data manipulation. Multiple data sources can be integrated into the DT data structure. As a result, based on this layer, standardized planning parameter sets, predefined metric sets, and key performance indicators (KPIs) are readily defined for data analysis and model development.

The DT data processing layer comprises planning parameter sets that are essential for data modeling and analysis. Road planning parameter sets should be formed in this layer, including cost parameters (e.g., trip, travel time, and financial cost), economic parameters (e.g., value of time, fuel costs, and income), road geometry parameters (e.g., bendiness, hilliness, and road width), model uncertainty parameters (e.g., input data such as the size of a new housing development), and other relevant parameters.

On the other hand, the planning steps need to be assessed. KPIs are the most commonly used measure for assessing the performance of any target object. KPIs serve two main purposes: evaluating the performance (past or present) of a target system and forecasting its future behavior [60]. Additionally, KPIs are categorized into different levels depending on their inclusiveness and scale.

In practice, KPIs are derived from broader performance indicators. In this framework, DT planning steps can be assessed based on the derived KPIs and quantify their outcomes. Figure 12 presents an example of a road planning objectives tree, illustrating the performance goals (main objectives) and indicators (sub-objectives) in terms of the objectives. At a more detailed level, the main features of the performance indicators, i.e., KPIs, can be determined. From this perspective, KPIs can be derived from three basic levels: high-level (evaluation goals), medium-level (evaluation indicators), and low-level (evaluation KPIs).

In general, indicators refer to higher-level (strategic) decision-making, while metrics are lower-level (operational and tactical) measurement aspects [61]. However, in this study, we applied indicators and metrics at the same level, with KPIs representing their key or main features for evaluating a planning process or business goal.

The low-level or evaluation KPIs can be based on the categories of indicators or metrics such as engineering, economic, traffic, social, and environmental [62]. In addition, metrics can be classified into strategic, economic, financial, commercial, and management, as these categories are used to evaluate road planning processes (e.g., options assessment framework).

KPIs should be developed based on a computable framework. KPIs are generated using the KPI framework, model, and specification [60]. In our case, this framework refers to DT, particularly its planning aspect. KPI conceptual modeling can be performed in modeling environments that may include profiles, metamodels, and ontologies. By leveraging these computable frameworks and models, KPIs can be specified for evaluating planning objectives.

The SPARQL is considered more advantageous over other modeling commands due to its expressiveness and compliance with semantic rules [63]. Consequently, it is used to assess and query semantic KPI models, particularly for business KPI selection and elicitation processes [64] and querying metrics within the semantic knowledge base [63]. A possible approach for validating KPI properties is illustrated in Figure 13, as reported by [65].

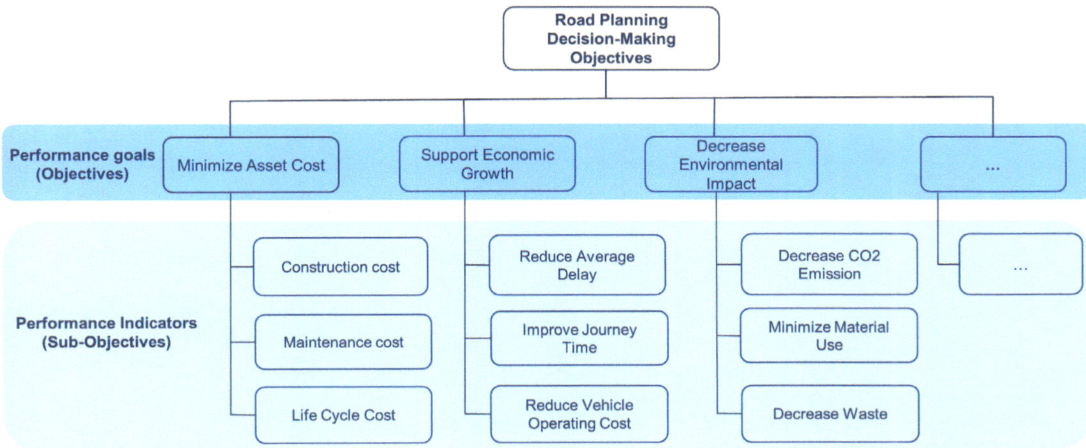

Figure 12. Planning objectives tree example.

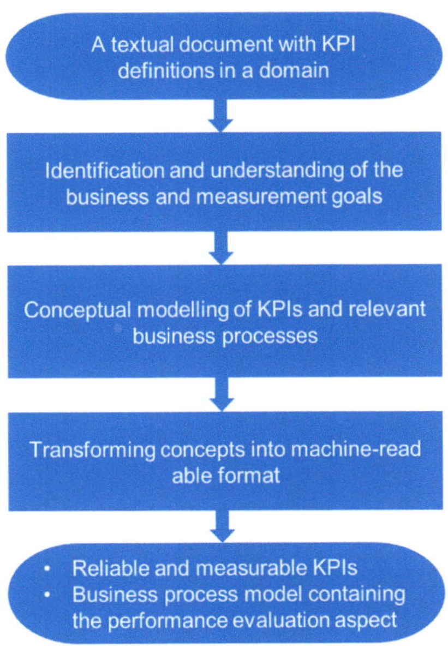

Figure 13. General steps of KPI development.

4.3.3. Data Modeling and Algorithms Layer

The third layer, the data modeling and algorithms layer, comprises several models integral to the road PT planning process. These models can be divided into transportation-related models and DT-related models. Transportation models include the trip end model, travel demand model, highway and public transport assignment models, variable demand model, and land use transport interaction models, all of which facilitate various analyses and functions in the DT. Additionally, DT-related models include the evaluation model, expert knowledge data model, simulation model, forecasting model, and other relevant models built in the DT. These models support DTs in efficiently and effectively generating viable solutions for road transportation planning problems.

As the planning process involves multi-factors, which require multi-criteria decision-making, DT should implement MOO functions that involve relevant methods, such as MOO evolutionary algorithms (MOEAs), trade-off analysis, the weighted sum method, and other machine learning algorithms, including decision trees and genetic algorithms.

The incorporation of artificial intelligence (AI) and machine learning algorithms for performing data analysis in road DT planning would be beneficial. The data modeling and algorithms layer contains machine learning algorithms that will be effectively used and power the capabilities of the data analytics in terms of forecasting, prediction, and simulation in road DT planning. Supervised machine learning algorithms such as support vector machines, K-nearest neighbors, logistic regression, and neural networks can be exploited to improve predictability for better decision-making.

The information generated in the analysis layer is deposited back to the knowledge data model within the data modeling and algorithms layer. The evaluation model supports the assessment of the generated options through the MOO process by utilizing the expert knowledge data model.

4.3.4. Data Analysis and Control Layer

The fourth layer, the analysis layer, serves as the core of the DT model, whose main purpose is to provide an analytical environment based on the DT input and models. The layer maintains a bidirectional relationship with the data processing layer and the data models and algorithms layer. The data processing layer provides standardized planning parameter sets, predefined metric sets, and KPIs, while the data modeling layer offers various planning and DT-related models and methods.

The analysis layer functions as a data analytics platform/engine with two primary roles. First, it provides an analytical environment before the option generation. Second, it facilitates platform functions, such as MOO, during the option generation process.

In its initial role, the layer includes forecasting, prediction, simulation, cost-benefit analysis, traffic demand, etc., to support option generation. By integrating planning parameter sets, KPIs from the data processing layer, and relevant models and algorithms from the data modeling and algorithms layer, the analysis layer performs DT analysis for optimal generation of options.

Regarding the second role, the layer incorporates some relevant steps of option generation and the subsequent steps from road PT planning (Steps 5 to 7). Unlike the road PT planning process, where options are generated mainly based on human activities, such as workshops, meetings, and discussions, the DT model should be capable of autonomously generating options toward (semi)automating planning decisions.

Option generation involves compromising with different road planning objectives and constraints toward optimal solutions provision. As such, DT must consider different objectives and constraints, requiring a quantitative evaluation of trade-offs. To address such conflicting objectives or KPIs in our case, a MOO approach that considers the main planning parameters with multiple objectives should be developed [66]. The MOO method provides the optimal solutions—the Pareto-optimal solutions—while taking into account competitive objective parameters [66,67].

As noted by [67], several approaches have been developed to allow decision-makers to express further preferences for the Pareto-optimal solutions. Therefore, in this paper, with the DT analysis part, we propose three options/modules to generate Pareto-optimal solutions when decision-makers interact with the DT planning system, as follows:

- Decision-makers set preferences before option generation (Priori Preference);
- Decision-makers set preferences while options are being generated (Interactive Preference);
- Decision-makers select options after relevant solutions are generated (Posteriori Preference).

In the first method, a priori preference, the decision-makers must provide preferences before optimization, i.e., preference weights (relative importance) should be given to each objective. Consequently, only Pareto optimal solutions relevant to the decision-makers are generated. Although a weighted sum method can be employed in this context, decision-makers may face difficulties in setting their preferences before any solutions are provided.

The second method, the interactive preference, enables the decision-maker to control the search by repeatedly altering optimization and preferences. This method allows them to know the interdependencies between the competing objectives based on the acquired solutions, which can help refine their preferences. However, a disadvantage of this method is that it requires substantial time and endeavor from decision-makers in the solution process.

The last method, a posteriori preference, is where the decision-maker sets preferences after optimization. In this approach, decision-makers do not need to participate actively in the option generation process. When implementing this method, a diverse set of Pareto-optimal solutions on the Pareto fronts are presented to decision-makers, who then make a final decision by exploring and negotiating among viable options. MOEAs are commonly used for this task. However, this method has two main drawbacks: developing an efficient method to generate Pareto fronts is challenging, and selecting optimal solutions from a large number of alternatives (non-dominated solutions) may impose a significant burden on decision-makers.

4.3.5. Service/Application Layer

The service/application layer is the topmost layer that interacts with users through a user interface. Users can manage and control the whole process of planning with the help of a user interface. This layer presents the results to the users, such as decision-makers, planners, and experts, who need to engage with the road DT planning system to solve their road transportation planning problems.

The primary purpose of the service layer is to provide viable solutions to road transportation planning issues. As discussed in the previous layer, they can be obtained based on three modules: a priori, interactive, and posterior approaches. Additionally, the service layer facilitates various transportation use case scenarios, what-if scenarios, and future investment plans required for the decision-making processes of transport planners and decision-makers based on all proposed DT layers.

Furthermore, all processed results and information from the system and the feedback from people should be stored in an external knowledge database to improve the DT planning process and enhance customer satisfaction.

4.4. *Road DT Planning Case Study*

This section aims to demonstrate how the proposed framework for road DT planning in this paper functions in terms of road planning. We will use the A1 North of Newcastle Feasibility Study [68] as an example to demonstrate the types of data that could serve as input for the DT framework.

The A1 route north of Newcastle, extending towards Scotland, is approximately 60 miles long. This route includes 23 miles of dual carriageway and 36 miles of single carriageway, divided into 11 sections. Stakeholders identified several issues that prompted the planning of this road as follows:

- Lack of alternative routes;
- Inconsistent carriageway standards on the route;
- Insufficient junction standards and layout;
- Large number of at-grade junctions/Private Means of Access;
- Different average speeds through the single-carriageway sections of the route and upgraded dual-carriageway;
- Relatively high proportion of heavy goods and agricultural vehicles that cause lowered speeds
- Lack of overtaking opportunities;
- Peak-hour traffic speeds significantly lower than average off-peak speeds.

Figure 14 illustrates the geographic scope of the feasibility study A1 North of Newcastle. Table 3 presents an example of data input and potential sources using the feasibility study.

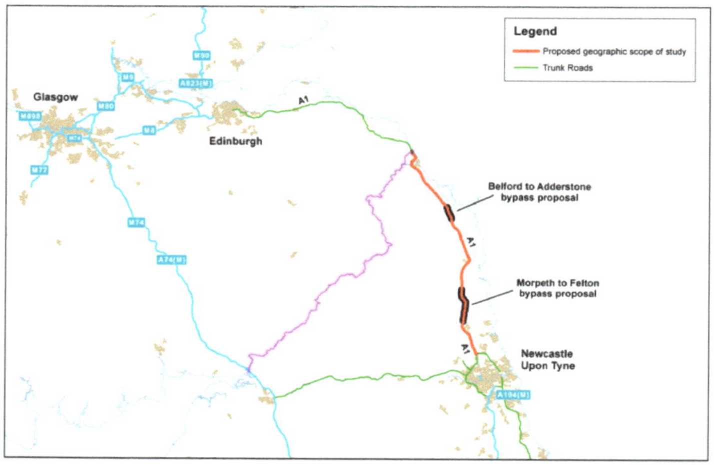

Figure 14. Geographic extent of the feasibility study A1 North of Newcastle [69].

Table 3. Relevant example input from the A1 North of Newcastle Feasibility Study.

DT Layer	Category	Data and Usage Example	Source and Measure
Data layer	Requirements (Objectives)	Improve journey times and reliability; Improve network resilience; Improve safety; Improve the conditions for strategic traffic while keeping access to local traffic Foster future economic growth; Reduce and balance any potential impacts on both the built and natural environment	Web data (documents, reports), previous studies (A1 North of Newcastle, Morpeth to Felton Dualling, etc.), and current issues, etc.
	Constraints (Environmental)	Special protection areas; Heritage coasts; Local wildlife sites; National parks; Green belt; Noise-sensitive areas; Flood zones; Main rivers; Public rights of way; Residential properties; Schools, hospitals, care homes, and worship places, etc.	Online maps, County maps, Local plans

Table 3. *Cont.*

DT Layer	Category	Data and Usage Example	Source and Measure
	Web survey data	DfT Strategic Vision and Business Plan; National Planning Policy Framework; National Infrastructure Plan, etc. Northeast Local Transportation Body, Strategic Economic Plan, Independent Economic Review, etc. Northumberland Core Strategy, Local Plan, Local Transport Plan, etc.	National policy Regional policy Local policy
	Transportation data	Traffic Volumes Heavy Goods Vehicles (HGV) Average Route Speeds Congestion Accident Rail Network Usage Future Highways Improvements Future Traffic Growth Future Rail Network Proposals	Annual Average Daily Traffic (AADT) Percentage HGV on Route Average Vehicle Speeds Free Flow Speed Accident numbers and rates Station entries and exits Planned schemes Traffic forecast by percentage Planned schemes
	Demographic data	Future Housing Future Employment	Numbers of additional houses Numbers of additional jobs
	External database/systems	TRAffic Database System (TRADS) Trafficmaster COBA-LT	Traffic volume Vehicle type and speed data Historical journey time (individual vehicle speed) Calculation of link and junction accident rates
	Expert knowledge	Historical data Previous cases	Previous DT planning cases (analysis results and data) Recycled DTs information
Data processing layer	Data conversion and integration Parameter sets Metrics sets	Internal and external data are transformed and integrated into a uniform base—an ontology or metamodel covering parameter sets, metrics/indicators, and KPIs Parameter sets, such as trips and travel times, include parameters that are imported from input data through data conversion Metrics/Indicators sets are distilled and grouped by their similarities, such as engineering metrics (e.g., length, width, junction, interchange, profile, road type); traffic metrics (e.g., traffic volume, traffic assignment, congestion, accessibility, connectivity); social metrics (e.g., traffic noise, safety, heritage, land taking, population); environmental metrics (e.g., environmental impact, carbon emission, pollution); and economic metrics (e.g., construction cost, operation and maintenance cost, asset cost, other costs) from input data based on the data conversion	
	KPIs	KPIs are key features derived from metrics established at three potential levels: high, medium, and low. They can be used for data analysis and generating options.	
Data modeling and algorithms layer	Transportation and DT models, and methods	Transportation models and DT models, as well as methods and algorithms, are built into the road DT planning platform These models and algorithms use the processed input data from the data processing layer to enable data analytics DT models, along with appropriate methods, enable data analysis that supports generating options Machine learning algorithms such as decision trees, genetic algorithms, and multi-objective evolutionary algorithms can be used to optimize the solutions	

Table 3. *Cont.*

DT Layer	Category	Data and Usage Example	Source and Measure
Data analysis and control layer	Data analysis for option generation	Users can import externally analyzed data into the DT platform. However, if a user prefers to analyze the raw data independently and generate optimal options based on their findings, they can utilize the data analytics function within the DT platform Data analysis can be performed on a quantitative and qualitative basis • Quantitative analysis can be traffic speed, flow, and volume analysis; accident analysis; journey time; vehicle operating cost analysis; etc. • Qualitative analysis can be objectives, constraints, policies, historical works, highway standards/layout, and other relevant Prediction and simulation functions analyze processed input data regarding current performance, trends, patterns, and statistical models to forecast future scenarios and suggest alternatives.	
	Option generation	Options are generated based on the results of the data analysis that support option generation KPIs are a key input to the option generation, and they are assessed toward viable solution provision Options are possibly generated as interventions related to highways, public transportation, and demand management • Highway-related options can be in the form of dualling a single carriageway, road widening, junction improvement, etc. • Public transportation-related options are possibly bus service extensions, park-and-ride schemes, improved facilities, etc. • Demand management relevant are possibly variable message signing, business travel plans, car sharing, etc. • Possibly a combination of them at certain road sections or routes	
Service/Application layer	User Interface	Users can effectively manage and control the entire planning process using the user interface Management of the option generation process is aided by the user interface Users can interact with the twin by using modules such as a priori, interactive, and posteriori Road planning-related Pareto optimal solutions considering every factor are presented Users can interact with data analytics engines to perform predictions and various scenarios, e.g., with or without schemes All these activities facilitate sense-making, which aids decision-making, and all processed results and information from this layer are recorded in expert knowledge	

4.5. Proposed Steps for Lightweight Implementation of Road DT Planning

There are several ways to implement the proposed road development planning framework. In Table 3, we presented an example case where data sources can be identified from the actual planning process (feasibility study) based on the proposed methods related to data or user requirements. Additionally, in Section 4.3.2, we discussed a KPI-based approach for evaluating the planning steps and elements.

In this section, we provide a more general and comprehensive step-by-step implementation guide that outlines the main steps for planning road DTs. This guide aims to assist practitioners and stakeholders in developing and adopting the proposed DT framework. The implementation methods presented here are just one of many possible approaches to applying the framework.

- Step 1: Understand the requirements. This step may involve a range of general requirements, such as national development plans and Sustainable Development Goals (SDGs), as well as problem-specific requirements that include the purpose of the intervention, identified issues, constraints, and data sources along with their quality. Documentation from the internet can be collected using web crawler technology.

While road planning parameters may vary slightly from country to country, we believe that the general procedure remains consistent across different contexts. Therefore, stakeholders and practitioners should identify specific steps tailored to their region's planning practices to initiate road DT planning.
Policies related to national cyber-infrastructure and digital transformation will be vital in this process. Additionally, addressing cybersecurity concerns and ensuring network security will be crucial. It is also important to define planning parameters and indicators and to derive KPIs from them, as this will facilitate modeling and implementation in subsequent steps. These processes will ultimately help define the scope of the framework.

- Step 2: Develop the modeling framework. This step begins with selecting the modeling language and the appropriate environment. During this stage, all the requirements identified in the first step need to be mapped and modeled within the selected environment (e.g., ontology). It is essential to model the planning parameters along with relevant metrics and KPIs while also identifying their relationships and dependencies. Additionally, considerations regarding data storage and database interfaces should be made in this stage, along with the selection and utilization of suitable modeling languages and tools. In this context, we introduced the KPI-based implementation method using tools such as SPARQL (see Section 4.3.2 and Figure 13).
- Step 3: Implement the framework. This step begins with the construction of a physical database (e.g., semantic database creation). This database stores all the pre-modeled parameters, metrics, and KPIs, as well as transportation and digital twin models and algorithms for analyzing data and generating options. Developed and integrated AI/ML models are utilized to measure these KPIs and quantify the results, aiming to generate optimal solutions that meet stakeholders' needs. The models and algorithms include multi-criteria decision analysis, multi-objective matching problems, and KPI optimization, among others. This process may facilitate partial or even full automation of planning and decision-making.
- Step 4: Report and monitor. This step will showcase the user interface that stakeholders will interact with. It may include results generated in the selected modeling environment or a web-based tool designed for user-friendly access. In this context, options for managing and modifying KPIs will be available, encompassing methods such as a priori, interactive, and a posteriori approaches. Stakeholders' feedback and responses can be utilized to validate both the model and the outcomes.

5. Discussion and Conclusions

This study aimed to contribute to the conceptualization and application of DT planning, focusing on the road transportation sector within the AEC industry, taking the UK as a case study. As DT planning is referred to as the early stage of the DT lifecycle, we have focused on the early-stage development of DTs. Therefore, we attempted to contribute to the research gap by clarifying certain concerns in the research and application of DT planning. Specifically, we formulated the DT planning concept and its constituents, providing implementation guidelines, including a case study as part of a DT planning framework that will support not only the adoption of DT planning but also digital transformation in the AEC industry, particularly the road transportation sector toward realizing digital roads in the future.

We defined the DT planning concept based on the relevant literature on the conceptualization of road DT planning. Concepts related to DT planning were identified as the pre-DT, early DT, fetal DT, and DT prototype. Furthermore, we explored the elements and capabilities embedded in DT planning. The main capabilities identified were strategy and planning, planning and prediction, virtual prototype, and decision-making, and their specific (sub)capabilities were reflected in conceiving road DT planning. Moreover, we examined stakeholders' relevant guidelines and documents related to the road PT planning process to identify knowledge gaps and better understand the scope of DT planning, particularly for roads.

We derived the scope and processes for road DT planning based on insights from road PT (pre)planning. First, a mapping process was performed to determine what is applicable or not for creating road DT planning based on road PT planning. Based on the mapping, we proposed a DT planning framework consisting of five layers: data acquisition, data processing, data modeling and algorithms, data analysis and control, and service layer plus users. The core of our proposed framework/architecture model is that it generates viable options for road transportation planning problems for better decision-making. It includes three modules—a priori, interactive, and posteriori approaches—which one to choose for better decision-making depends on the stakeholders or users.

DT planning occurs before PT planning and is intended for new road construction rather than existing roads. As a pre-planning stage (feasibility study) of the road PT planning process, DT planning improves the whole process by introducing a virtual planning step for digital road assets before real planning begins. However, it is important to note that the proposed framework for road planning is not designed to replace existing planning procedures as they should be followed in the process but to complement them, thus adding value and saving costs.

Moreover, road DT planning is not developed for a specific platform; therefore, it can be further used in any DT platform, i.e., the road DT design, which is the next phase of its lifecycle. Thus, DT planning is more to capture the requirements and hand them over to the DT design so that the design part can be accountable and reliable. It would be inefficient to start designing DTs without understanding of the user requirements that DT planning delivers. When implemented at the right time, DTs would ensure that the client's information requirements are met and that the right information is provided to the appropriate people, facilitating informed decision-making [25].

5.1. Research Implications and Potential Benefits

The study expanded the one potential use case of DTs in the road transportation sector, which would support the decision-making process of the road planning. Unlike the conventional road planning process, stakeholders can plan DTs early in the lifecycle for further adoption while having pre-planning stage thoughts before starting the actual planning process, including feasibility by using the road DT planning. Specifically, it will be valuable to deal with the decision-making related to the major infrastructure project development where large amounts of budgets and efforts are required.

In addition, in general, the proposed framework contributes to the advancement of the smart city and/or urban DT concept and framework by introducing elemental technology such as road DT planning. Particularly, if the smart city framework is implemented and extended at the national level, the proposed framework will be an essential part of it.

Furthermore, road DT planning will help stakeholders start planning and constructing real roads by following established road planning steps. It can also support relevant stakeholders, particularly National Highways or other clients, in initiating road planning and whether to apply DTs to their projects or businesses. Moreover, when implemented, our proposed road DT planning can be used as a reference model before starting any formal stage of the PCF. In addition, road DT planning may be implemented using a graphic-based environment such as GIS technology, where road alignments can be visualized for effective stakeholder engagement. In this case, if the road DT design is built on the GIS environment, then the handover between DTs would be easier, and consistency of systems would be retained.

Digital roads would significantly transform the UK transportation infrastructure sector. They form a crucial component of the long-term vision for the SRN, i.e., for the longer-term 2050 vision, and lay the foundation for future road developments [45]. This work contributes to the National Highways' effort to lay the groundwork for digital roads toward a fully digital age. Specifically, it emphasizes how current and future practices can be improved by harnessing such digital technology. It also aligns to integrate digital technology across every stage of the project lifecycle—from planning and design, to construction and operation [70].

Moreover, from a broader perspective, the proposed framework is expected to significantly contribute to national decarbonization and net-zero efforts. Decarbonization and transportation improvements are intended to be essential components of local transport planning [71]. Our proposed framework aims to support sustainable road transport planning activities by streamlining the necessary steps for planning major road infrastructure once it is adopted and implemented. Specifically, it is anticipated that corporate emissions—such as those from lighting, power, work travel, and energy consumption [72]—will be reduced by facilitating a preliminary phase in the planning process, allowing stakeholders to begin planning independently.

5.2. Research Limitations, Future Challenges, and Directions

This study contributes to the body of knowledge by conceptualizing a novel framework for road DT planning and proposing a system architecture that can serve as a roadmap for future development in this area. This study also offers valuable insights that reveal future research opportunities and directions for academics, researchers, and practitioners toward the successful adoption of DTs early in the project development lifecycle.

The study is limited to introducing a conceptual framework that identified some high-level requirements based on existing road planning practices. Future research should focus on compiling a comprehensive list of data or information requirements for a study about road DT planning. Additionally, a proof of concept is essential for implementing the framework presented in this paper. The proposed lightweight implementation framework could be one potential solution, but it may not encompass all the capabilities and functions suggested. Web-based applications can be designed with user interfaces, incorporating embedded models and algorithms that process input data and requirements. This enables analysis and generates optimal options. DT planning can also be created for other new infrastructure assets, such as railroads, and other relevant facilities, as this study only used a road planning approach from practice as an example to conceptualize DT planning.

Furthermore, considerations regarding costs and the overall feasibility of DTs (e.g., digital assets) should extend beyond the road planning aspect. In addition, scalability issues that relate to handling a data volume that affects system performance may be addressed. In this regard, DTs can be developed with modular unit form where scalable extension is possible in the cloud environment [73]. In DT planning, for example, it can be referred to the level of requirements such as high, medium, and low levels where decision-makers can make appropriate decisions about planning based on the corresponding input data.

Cybersecurity should also be considered, as a road DT planning system or platform should be able to interact with external databases and systems to acquire and store input data. The network security of the system should be robust enough for potential harm and sensitive information leaking. To make the DT planning system operationally secure, security mechanisms and those security best practices should be applied. Thus, the DT network system with such mechanisms can be aligned with the data verification and model validation, operation, and interaction between the external systems and internal DT systems by only verified and authorized users [74].

In addition, blockchain technology is extensively being discussed for securing digital twins based on decentralized ledgers that provide data integrity and stability. It provides transparency for all records and modifications to maintain data integrity and avoid unauthorized changes [75]. For reliable DT planning, these decentralized ledgers may be effectively used to ensure the quality and consistency of data. Ultimately, these efforts are expected to contribute to the standardization of the planning process in compliance with the National DT program [76], where DTs should be safe, secure, trustworthy, and ethical.

Author Contributions: Conceptualization, M.B. and I.B.; methodology, investigation, writing—original draft preparation, M.B.; writing—review and editing, M.B., I.B., M.P., G.E. and W.L.; visualization, M.B.; supervision, I.B., M.P., G.E. and W.L.; project administration, M.B.; funding acquisition, I.B. and W.L. All authors have read and agreed to the published version of the manuscript.

Funding: This work was supported by the Basic Science Research Program through the National Research Foundation of Korea (NRF) funded by the Ministry of Education (NRF-2020R1I1A3061750), a National Research Foundation of Korea (NRF) grant funded by the Korean government (MSIT) (NRF-2021R1A5A8033165), and the European Union's Horizon 2020 research and innovation programme under the Marie Skłodowska-Curie grant agreement No. 101034337.

Data Availability Statement: Data are contained within the article.

Conflicts of Interest: Author Matt Peck was employed by the company AtkinsRealis. Author George Economides was employed by the company Department for Transport. The remaining authors declare that the research was conducted in the absence of any commercial or financial relationships that could be construed as a potential conflict of interest.

References

1. Chen, Z.S.; Chen, K.D.; Xu, Y.Q.; Pedrycz, W.; Skibniewski, M.J. Multiobjective optimization-based decision support for building digital twin maturity measurement. *Adv. Eng. Inform.* **2023**, *59*, 102245. [CrossRef]
2. Research and Markets. Digital Twins Market by Technology, Twinning Type, Cyber-to-Physical Solutions, Use Cases and Applications in Industry Verticals 2024–2029. Available online: https://www.researchandmarkets.com/reports/5308850/digital-twins-market-by-technology-twinning?utm_source=dynamic&utm_medium=CI&utm_code=6q68tb&utm_campaign=1366076+-The+Future+of+the+Digital+Twins+Industry+to+2025+in+Manufacturing,+Smart+Cities,+Auto (accessed on 2 September 2024).
3. Grand View Research. Digital Twin Market Size & Trends. Available online: https://www.grandviewresearch.com/industry-analysis/digital-twin-market (accessed on 2 September 2024).
4. Markets and Markets. Market Research Report. Available online: https://www.marketsandmarkets.com/Market-Reports/digital-twin-market-225269522.html?gclid=CjwKCAjwoMWBhAdEwAVJ2ndn2zsFCn00pvhe8FbqHp8V5avZkh3hmDbnCwMw8Z4bQY8cU3dvAAhoCQWUQAvDBwE (accessed on 2 September 2024).
5. Gartner, Top Strategic Technology Trends 2023 Bringing Clarity to Your Tech Priorities. Available online: https://enterprise.press/wp-content/uploads/2022/10/Gartner.pdf (accessed on 2 September 2024).
6. Nour El-Din, M.; Pereira, P.F.; Poças Martins, J.; Ramos, N.M.M. Digital Twins for Construction Assets Using BIM Standard Specification. *Buildings* **2020**, *12*, 2155. [CrossRef]
7. Pregnolato, M.; Gunner, S.; Voyagaki, E.; De Risi, R.; Carhart, N.; Gavriel, G.; Tully, P.; Tryfonas, T.; Macdonald, J.; Taylor, C. Towards Civil Engineering 4.0: Concept, workflow and application of Digital Twins for existing infrastructure. *Autom. Constr.* **2022**, *141*, 104421. [CrossRef]
8. Grieves, M. Intelligent digital twins and the development and management of complex systems. *Digit. Twin* **2022**, *2*, 8. [CrossRef]
9. Davila Delgado, J.M.; Oyedele, L. Digital Twins for the built environment: Learning from conceptual and process models in manufacturing. *Adv. Eng. Inform.* **2021**, *49*, 101332. [CrossRef]
10. Bazaz, S.M.; Lohtander, M.; Varis, J. 5-Dimensional Definition for a Manufacturing Digital Twin. In Proceedings of the 29th International Conference on Flexible Automation and Intelligent Manufacturing (FAIM2019), Limerick, Ireland, 24–28 June 2019. [CrossRef]
11. Lu, Q.; Parlikad, A.K.; Woodall, P.; Don Ranasinghe, G.; Xie, X.; Liang, Z.; Konstantinou, E.; Heaton, J.; Schooling, J. Developing a Digital Twin at Building and City Levels: Case Study of West Cambridge Campus. *J. Manag. Eng.* **2020**, *36*, 1–19. [CrossRef]

12. Jiang, F.; Ma, L.; Broyd, T.; Chen, K. Digital twin and its implementations in the civil engineering sector. *Autom. Constr.* **2021**, *130*, 103838. [CrossRef]
13. Honghong, S.; Gang, Y.; Haijiang, L.; Tian, Z.; Annan, J. Digital twin enhanced BIM to shape full life cycle digital transformation for bridge engineering. *Autom. Constr.* **2023**, *147*, 104736. [CrossRef]
14. Errandonea, I.; Beltrán, S.; Arrizabalaga, S. Digital Twin for maintenance: A literature review. *Comput. Ind.* **2020**, *123*, 103316. [CrossRef]
15. Consilvio, A.; Hernández, J.S.; Chen, W.; Brilakis, I.; Bartoccini, L.; Di Gennaro, F.; van Welie, M. Towards a digital twin-based intelligent decision support for road maintenance. In Proceedings of the AIIT 3rd International Conference on Transport Infrastructure and Systems (TIS ROMA 2022), Rome, Italy, 15–16 September 2022. [CrossRef]
16. Tanne, Y.A.; Zultaqawa, Z.; Aulia, M.D.; Falderika; Farhani, S.; Rivana, D. Integrated System for Urban Road Asset Management: Conceptual Framework. In Proceedings of the 2023 9th International Conference on Signal Processing and Intelligent Systems (ICSPIS), Bali, Indonesia, 14–15 December 2023; IEEE: Piscataway, NJ, USA, 2023. [CrossRef]
17. Jiang, F.; Ma, L.; Broyd, T.; Chen, W.; Luo, H. Digital twin enabled sustainable urban road planning. *Sustain. Cities Soc.* **2022**, *78*, 103645. [CrossRef]
18. Jiang, F.; Ma, L.; Broyd, T.; Chen, K.; Luo, H. Underpass clearance checking in highway widening projects using digital twins. *Autom. Constr.* **2022**, *141*, 104406. [CrossRef]
19. Nochta, T.; Wan, L.; Schooling, J.M.; Parlikad, A.K. A Socio-Technical Perspective on Urban Analytics: The Case of City-Scale Digital Twins. *J. Urban Technol.* **2021**, *28*, 263–287. [CrossRef]
20. Machl, T.; Donaubauer, A.; Kolbe, T.H. Planning agricultural core road networks based on a digital twin of the cultivated landscape. *J. Digit. Landsc. Archit.* **2019**, *2019*, 316–327. [CrossRef]
21. Malé, C.; Lagier, T. Simulating the Interactions of Environmental and Socioeconomic Dynamics at the Scale of an Ecodistrict: Urban Modeling of Gerland. In *Ecosystem and Territorial Resilience*; Garbolino, E., Voiron-Canicio, C., Eds.; Elsevier: Amsterdam, The Netherlands, 2021; Chapter 11; pp. 299–321, ISBN 978-0-12-818215-4.
22. Schrotter, G.; Hürzeler, C. The Digital Twin of the City of Zurich for Urban Planning. *PFG-J. Photogramm. Remote Sens. Geoinf. Sci.* **2020**, *88*, 99–112. [CrossRef]
23. Dembski, F.; Wössner, U.; Letzgus, M.; Ruddat, M.; Yamu, C. Urban Digital Twins for Smart Cities and Citizens: The Case Study of Herrenberg, Germany. *Sustainability* **2020**, *12*, 2307. [CrossRef]
24. Akroyd, J.; Mosbach, S.; Bhave, A.; Kraft, M. Universal Digital Twin—A Dynamic Knowledge Graph. *Data-Centric Eng.* **2021**, *2*, e14. [CrossRef]
25. Seaton, H.; Sepasgozar, S.M.E.; Sawhney, A. Digital Twins from Design to Handover of Constructed Assets. 2022. Available online: https://www.rics.org/content/dam/ricsglobal/documents/research/digital-twins-from-design-to-handover-of-constructed-assets.pdf (accessed on 22 June 2024).
26. Fuller, A.; Fan, Z.; Day, C.; Barlow, C. Digital Twin: Enabling Technologies, Challenges and Open Research. *IEEE Access* **2020**, *8*, 1–21. [CrossRef]
27. Ariansyaha, D.; Fernàndez del Amo, I.; Erkoyuncu, J.A.; Agha, M.; Bulka, D.; De La Puante, J.; Langlois, M.; M'khinini, Y.; Sibson, J.; Penver, S. Digital Twin Development: A Step by Step Guideline. In Proceedings of the 9th International Conference on Through-Life Engineering Service, Cranfield, UK, 3–4 November 2020.
28. Madni, A.M.; Madni, C.C.; Lucero, S.D. Leveraging Digital Twin Technology in Model-Based Systems Engineering. *Systems* **2021**, *7*, 7. [CrossRef]
29. Haraguchi, M.; Funahashi, T.; Biljecki, F. Assessing governance implications of city digital twin technology: A maturity model approach. *Technol. Forecast. Soc. Chang.* **2024**, *204*, 123409. [CrossRef]
30. Sacks, R.; Brilakis, I.; Pikas, E.; Xie, H.S.; Girolami, M. Construction with digital twin information systems. *Data-Centric Eng.* **2020**, *1*, 6. [CrossRef]
31. Emir Isik, G.; Hubertus Achten, H. Operationalising Concepts of Digital Twins on Different Maturity Levels (Foetal, Child, Adult) for the Architectural Design Process. In Proceedings of the the International Conference on Engineering Design (ICED23), Bordeaux, France, 24–28 July 2023. [CrossRef]
32. Jones, D.E.; Snider, C.; Kent, L.; Hicks, B. Early stage digital twins for early stage engineering design. In Proceedings of the 22nd International Conference on Engineering Design (ICED19), Delft, The Netherlands, 5–8 August 2019. [CrossRef]
33. University of Cambridge. Digital Twin Toolkit. 2021. Available online: https://digitaltwinhub.co.uk/download/digital-twin-toolkit/ (accessed on 22 February 2024).
34. Singh, M.; Srivastava, R.; Fuenmayor, E.; Kuts, V.; Qiao, Y.; Murray, N.; Devine, D. Applications of Digital Twin across Industries: A Review. *Appl. Sci.* **2022**, *12*, 5727. [CrossRef]
35. Shahat, E.; Hyun, C.T.; Yeom, C. City Digital Twin Potentials: A Review and Research Agenda. *Sustainability* **2021**, *13*, 3386. [CrossRef]
36. Parrott, A.; Warshaw, L. *Industry 4.0 and the Digital Twin*; Deloitte Univ. Press: New York, NY, USA, 2017.

37. Singh, M.; Fuenmayor, E.; Hinchy, E.P.; Qiao, Y.; Murray, N.; Devine, D. Digital Twin: Origin to Future. *Appl. Syst. Innov.* **2021**, *4*, 36. [CrossRef]
38. Zuhr, P.; Mühlbauer, K.; Bäuml, S.; Meißner, S. An approach for target-oriented process analysis for the implementation of Digital Process Optimization Twins in the field of intralogistics. In Proceedings of the European Modeling and Simulation Symposium, EMSS, Online, 15–17 September 2021; pp. 183–191. [CrossRef]
39. Cho, N.G. Prefeasibility Study: Methods and Implementation. Available online: https://www.krihs.re.kr/boardDownload.es?bid=0014&list_no=354829&seq=2 (accessed on 25 December 2024).
40. Texas Department of Transportation. Feasibility Study Keep It Moving Dallas. US 380 Collin County Feasibility Study. Available online: https://www.keepitmovingdallas.com/projects/us-highways/us-380-collin-county-feasibility-study (accessed on 25 December 2024).
41. Texas Department of Transportation. Feasibility Study Keep It Moving Dallas. SH 34 Feasibility Study. Available online: https://www.keepitmovingdallas.com/projects/state-highways/sh-34-feasibility-study (accessed on 25 December 2024).
42. U.S. Department of Transportation, Federal Highway Administration (FHWA). Feasibility Study Checklist. Available online: https://ops.fhwa.dot.gov/publications/psechecklists/checklist_2.htm (accessed on 26 December 2024).
43. National Highways. *Strategic Road Network Initial Report (2025–2030)*; National Highways: Birmingham, UK, 2022.
44. Department for Transport. Road Investment Strategy: For the 2015/16–2019/20 Road Period. 2015. Available online: https://assets.publishing.service.gov.uk/media/5a80ef47e5274a2e8ab52fed/ris-for-2015-16-road-period-web-version.pdf (accessed on 16 February 2024).
45. National Highways. Digital Roads. Available online: https://nationalhighways.co.uk/our-work/digital-data-and-technology/digital-roads/ (accessed on 20 March 2024).
46. Department for Transport. Road Investment Strategy 2: 2020–2025. 2020. Available online: https://assets.publishing.service.gov.uk/government/uploads/system/uploads/attachment_data/file/951100/road-investment-strategy-2-2020-2025.pdf (accessed on 16 February 2024).
47. Highways Agency. The Project Control Framework Handbook. 2018. Available online: https://assets.highwaysengland.co.uk/roads/road-projects/A46+Coventry+Junctions+Upgrade/Proofs+of+evidence/J.01+PROJECT+CONTROL+FRAMEWORK+HANDBOOK++V4-NOVEMBER+2018_.pdf (accessed on 2 December 2023).
48. Department for Transport. Transport Analysis Guidance (TAG): The Appraisal Transport Process. 2018. Available online: https://assets.publishing.service.gov.uk/media/5fbfb271d3bf7f572b163fb9/tag-transport-appraisal-process.pdf (accessed on 14 December 2023).
49. Department for Transport. TAG Unit A1.1—Cost-Benefit Analysis. 2023. Available online: https://www.gov.uk/government/publications/webtag-tag-unit-a1-1-cost-benefit-analysis-may-2018 (accessed on 28 December 2023).
50. Department for Transport. TAG Unit M3.1—Highway Assignment Modelling. 2020. Available online: https://www.gov.uk/government/publications/webtag-tag-unit-m3-1-highway-assignment-modelling (accessed on 22 December 2023).
51. Department for Transport. TAG Unit M3.2—Public Transport Assignment. 2020. Available online: https://www.gov.uk/government/publications/webtag-tag-unit-m3-2-public-transport-assignment-modelling (accessed on 22 December 2023).
52. Department for Transport. TAG Unit M2.1—Variable Demand Modelling. 2020. Available online: https://www.gov.uk/government/publications/tag-unit-m2-1-variable-demand-modelling (accessed on 22 December 2023).
53. Department for Transport. TAG Unit M1.2—Data Sources and Surveys. 2020. Available online: https://www.gov.uk/government/publications/webtag-tag-unit-m1-2-data-sources-and-surveys (accessed on 26 December 2023).
54. Department for Transport. Early Assessment and Sifting Tool (EAST) Guidance. Available online: https://assets.publishing.service.gov.uk/media/5f6227d3e90e072bb82412a1/east-tool-guidance.pdf (accessed on 22 December 2023).
55. Department for Transport. TAG Unit M1—Principles of Modelling and Forecasting. 2014. Available online: https://www.gov.uk/government/publications/webtag-tag-unit-m1-1-principles-of-modelling-and-forecasting (accessed on 6 March 2024).
56. Singh, S.; Shehab, E.; Higgins, N.; Fowler, K.; Reynolds, D.; Erkoyuncu, J.A.; Gadd, P. Data management for developing digital twin ontology model. *Proc. Inst. Mech. Eng. Part B J. Eng. Manuf.* **2021**, *235*, 2323–2337. [CrossRef]
57. Buuveibaatar, M.; Lee, K.; Lee, W. Development of a Conceptual Data Model for 3D Geospatial Road Management Based on LandInfra Standard: A Case Study of Korea. *ISPRS Int. J. Geo-Inf.* **2022**, *11*, 316. [CrossRef]
58. Buuveibaatar, M.; Kim, M.G.; Shin, S.P. Towards application of landinfra standard for highway management in Korea. *Int. Arch. Photogramm. Remote Sens. Spat. Inf. Sci.* **2020**, *43*, 435–439. [CrossRef]
59. Department for Transport. Road Traffic Statistics. Available online: https://www.gov.uk/government/collections/road-traffic-statistics (accessed on 22 August 2024).
60. Domínguez, E.; Pérez, B.; Rubio, Á.L.; Zapata, M.A. A taxonomy for key performance indicators management. *Comput. Stand. Interfaces* **2018**, *64*, 24–40. [CrossRef]
61. RIB. KPIs vs. Metrics: Understanding the Differences with Tips & Example. Available online: https://www.rib-software.com/en/blogs/kpis-vs-metrics-differences (accessed on 28 August 2024).

62. Jiang, F.; Ma, L.; Broyd, T.; Li, J.; Jia, J.; Luo, H. Systematic framework for sustainable urban road alignment planning. *Transp. Res. Part D* **2023**, *120*, 103796. [CrossRef]
63. Kritikos, K.; Plexousakis, D.; Woitch, R. Towards semantic KPI measurement. In Proceedings of the 7th International Conference on Cloud Computing and Services Science (CLOSER 2017), Porto, Portugal, 24–26 April 2017; pp. 63–74. [CrossRef]
64. del Mar Roldán-García, M.; García-Nieto, J.; Maté, A.; Trujillo, J.; Aldana-Montes, J.F. Ontology-driven approach for KPI meta-modelling, selection and reasoning. *Int. J. Inf. Manag.* **2021**, *58*, 102018. [CrossRef]
65. Roubtsova, E.; Michell, V. KPIs and their properties defined with the EXTREME method. In *Proceedings of the Business Modeling and Software Design, Third International Symposium, BMSD 2013, Noordwijkerhout, The Netherlands, 8–10 July 2013*; Lecture Notes in Business Information Processing; Springer: Cham, Switzerland, 2014; Volume 173, pp. 128–149. [CrossRef]
66. Magnanini, M.C.; Melnychuk, O.; Yemane, A.; Strandberg, H.; Ricondo, I.; Borzi, G.; Colledani, M. A digital twin-based approach for multi-objective optimization of short-term production planning. *IFAC-PapersOnLine* **2021**, *54*, 140–145. [CrossRef]
67. Nagy, M.; Mansour, Y.; Abdelmohsen, S. Multi-Objective Optimization Methods as a Decision Making Strategy. *Int. J. Eng. Res.* **2020**, *9*, 3. [CrossRef]
68. Jacobs. A1 North of Newcastle Feasibility Study (Stage 1, 2, 3 Reports). 2015. Available online: https://www.gov.uk/government/publications/a1-north-of-newcastle-feasibility-study-technical-reports (accessed on 12 December 2023).
69. Department for Transport. A1 North of Newcastle Feasibility Study Summary. 2015. Available online: https://assets.publishing.service.gov.uk/media/5a757eac40f0b6397f35ede8/a1-north-of-newcastle-summary.pdf (accessed on 22 December 2023).
70. Highways England. Strategic Business Plan 2020–2025. 2020. Available online: https://assets.publishing.service.gov.uk/media/5f477ecad3bf7f5d839b64a6/strategic-business-plan-2020-25.pdf (accessed on 25 June 2024).
71. Department for Transport. Decarbonising Transport—A Better, Greener Britain. 2021. Available online: https://assets.publishing.service.gov.uk/media/610d63ffe90e0706d92fa282/decarbonising-transport-a-better-greener-britain.pdf (accessed on 26 December 2024).
72. National Highways. *Net Zero Highways: Our 2030/2040/2050 Plan*; National Highways: Birmingham, UK, 2021.
73. Noeikham, P.; Buakum, D.; Sirivongpaisal, N. Architecture designing of digital twin in a healthcare unit. *J. Health Inform.* **2024**, *30*, 1–27. [CrossRef]
74. Zhou, C.; Yang, H.; Duan, X.; Lopez, D.; Pastor, A.; Wu, Q.; Boucadair, M.; Jacquenet, C. Digital Twin Network: Concepts and Reference Architecture. 2024. Available online: https://www.ietf.org/archive/id/draft-irtf-nmrg-network-digital-twin-arch-04.html (accessed on 28 December 2024).
75. ISC2. Managing Cybersecurity in the Age of Digital Twins. Available online: https://www.isc2.org/Insights/2024/09/Cybersecurity-in-the-Age-of-Digital-Twins (accessed on 10 January 2024).
76. National Digital Twin Programme (NDTP). Available online: https://www.gov.uk/government/collections/the-national-digital-twin-programme-ndtp (accessed on 25 September 2024).

buildings

MDPI

Article

Centralized or Decentralized? Communication Network and Collective Effectiveness of PBOs—A Task Urgency Perspective

Xue Ding [1], Wenxin Shen [2,*] and Shiai Wang [1]

1 School of Financial Technology, Shanghai Lixin University of Accounting and Finance, Shanghai 201209, China; dingxue@lixin.edu.cn (X.D.)
2 School of Economics and Management, Beijing Jiaotong University, Beijing 100044, China
* Correspondence: shwenxin@bjtu.edu.cn

Abstract: In the construction industry, there are a large number of project-based organizations (PBOs), where the efficiency of communication and collaboration among organizational members greatly impacts the success of projects. For PBOs employing both centralized and decentralized communication networks, it is worth delving into the question of under what circumstances which type of network will yield better results. Based on the IMO model and organizational learning theory, this paper conducts a grouped communication experiment involving 598 engineering management personnel to explore the differences in collective effectiveness of varying communication networks from the perspective of task urgency. Beyond task performance assessments, we have included organizational member perception to form evaluation criteria for collective effectiveness. Our research results show that under conditions of weak task urgency, decentralized networks yield higher collective effectiveness. Conversely, under conditions of strong task urgency, centralized networks demonstrate superior collective effectiveness. Furthermore, this study also verifies the mediating role of knowledge sharing behavior when task urgency is strong. This research provides significant managerial insights for the establishment of appropriate communication networks for PBOs in the construction industry.

Keywords: project-based organizations (PBOs); communication network; SNA (social network analysis); task urgency; knowledge sharing; collective effectiveness

Citation: Ding, X.; Shen, W.; Wang, S. Centralized or Decentralized? Communication Network and Collective Effectiveness of PBOs—A Task Urgency Perspective. *Buildings* **2024**, *14*, 448. https://doi.org/10.3390/buildings14020448

Academic Editor: Audrius Banaitis

Received: 6 December 2023
Revised: 17 January 2024
Accepted: 4 February 2024
Published: 6 February 2024

1. Introduction

In industries such as construction, high-tech manufacturing, management consulting, and professional services, a temporary legal project-based enterprise or firm is often established based on a specific output goal [1]. These are referred to as Project-Based Organizations (PBOs). Participants with different professional backgrounds and various resources are integrated through complex connections. The organizational structure is constructed based on knowledge sharing and organizational coordination to achieve the success for the project [2]. Existing research, based on the characteristics of project types, explores the formation and performance of various organizational structures for PBOs. For example, in specific situations like defense and national missions, organizational members tend to communicate or collaborate in a more centralized manner, forming a centralized structure [3,4]. In industries like software development and consulting, where high flexibility and innovation are encouraged, a more decentralized organizational cooperation structure is common [5,6]. In the construction industry, organizational structures are often built based on contractual relationships, influencing the communication relationships among organizational members.

To accurately measure the communication relationships among members of PBOs in the construction industry, Social Network Analysis (SNA) is widely used to capture and explore the structure and performance of these communication relationships, i.e., research based on communication networks. For example, indices, including those for centralization, connection strength, and network density, are used to measure the structural features of communication networks [7]. The interactions of stakeholders and collective decision-making can be predicted in a confined communication network [8,9]. Further, from the perspective of network centralization, communication networks can be divided into centralized networks and decentralized networks [10]. In decentralized networks, organizational members can directly communicate with any other member, adapting to faster unique information flows and improving performance in knowledge-based work [11,12]. In contrast, centralized networks include a core leadership or leadership team and a peripheral group capable of obtaining all critical information in a constantly changing task environment, thereby improving collective learning and problem-solving capabilities [13].

It is noteworthy that recent studies have discovered that the communication network of PBOs in the construction industry undergo adaptive evolution as the project progresses [14]. This implies that project progress, a key factor in project management, may cause the evolution and development of the communication network structure [15]. It is well-known that as a project nears completion, the urgency of tasks often increases. Previous research has confirmed that the urgency of tasks enhances the capability of knowledge sharing among members [16]. Therefore, during different stages of construction project development, what structure of communication network should we adopt to enhance management value? Moreover, what exactly signifies the management value of a PBO? Is it just project performance? Answers to these questions will help improve the management of complex projects in the VUCA era, but related studies are still somewhat lacking.

Hence, based on many observations and empirical studies of past projects, we propose hypotheses concerning the structural measurement of communication networks and their collective effectiveness for PBOs. Most PBOs we surveyed are in different spatiotemporal realms, and their collective effectiveness might be influenced by external environments. Laboratory behavioral experiments can be used to achieve the goals of this study and provide clear explanations for correlations and trends in an experimental scenario setting [17,18]. Thus, we conducted a grouped communication experiment involving 405 engineering project management personnel, comparing the differences in collective effectiveness under two levels of task urgency within two types of communication networks to provide theoretical guidance for establishing appropriate communication network structures throughout the entire lifecycle of a project. Additionally, we discuss scenarios where knowledge sharing plays a mediating role in this process. This study explores strategies for enhancing the effectiveness of PBOs from the perspective of communication networks, supplementing current research that often focuses on improving communication performance through technical means to provide new insights for complex project management.

The next section of this paper introduces the literature review and hypothesis development. Section 3 describes the experimental design, samples, procedures, and measures. Section 4 presents data analysis and test results. Section 5 mainly discusses the results, outlining theoretical contributions and managerial implications.

2. Literature Review and Hypotheses Development

2.1. PBOs and Communication Network Structure

PBOs are organizational forms engaged in temporary work to create innovative products or services. As temporary organizations, they are characterized by project duration, unique tasks, and teamwork to achieve project goals [1,19,20]. Increasingly, people are proposing methods based on SNA to address the relational issues among members of PBOs [21,22]. Quantitative analysis based on the topology of social network relationships can assess the relational ties and the overall network structure. That is, by studying the attributes of nodes and the structure of connections between them, one can explore the interdependencies among network members and how their positions in the network influence their constraints and behaviors. On this basis, individual responses and collective outputs can be analyzed and predicted [23–25].

Currently, researchers in the field of construction project management have conducted considerable theoretical and practical exploration into integrating social networks with project organization structure management. Each engineering project contains a social interaction and collaboration network, where the flow of knowledge within the network is constrained by its topological structure. The connections and the overall structure can be quantitatively assessed by SNA. Further, the complex relationships among project members at different stages of project can be displayed using social graphs [26,27] to, for example, indicate the dynamic power of stakeholders in the implementation of social responsibility issues in construction projects [28]. Moreover, in recent years, discussions on the organizational structure of large-scale PBOs such as public–private partnership projects and major engineering projects have attracted more attention. For example, by identifying the network status and relationships of project members, there are discussions on the dynamics of stakeholders in implementing social responsibility issues [29], exploring the characteristics of relationship exchange behaviors among stakeholders in mega-projects from the perspective of stakeholder value networks [30] and organizational collaboration relationships [31].

Drawing upon the existing research on communication networks of engineering project teams featured in mainstream peer-reviewed papers and case studies [32–34], this study summarizes four types of communication networks. These networks include the fully connected network, the subgroup network, the core-periphery network, and the locally clustered network. Referring to existing classification methods [10], the first two are classified as decentralized networks, while the latter two are considered centralized networks.

2.2. Organizational Effectiveness in Different Communication Networks

Currently, there is a consensus that different types of communication networks exhibit performance variations in information transferr. A classic communication experiment manipulated the communication patterns among group members by controlling who could send information to whom and measured the impact of various communication patterns on group operation and performance. It was found that the degree of centralization—i.e., the extent to which one person acts as a communication hub—has a significant impact on both individual and group performance. The complexity of the task has also been proven to be a key moderating variable: centralization is beneficial for simple tasks, but harmful for complex tasks, according to Bavelas and Barrett [35]. A decentralized structure is the best choice when information is unevenly distributed among group members or when information is unclear [36,37]. Furthermore, within a defined communication network, the ability to transfer information between network neighbors affects the quality of collective decision-making [17], which may also be due to differences in knowledge transfer capabilities [38]. Relevant research has outlined common interest in solving communication problems in different network structures.

So, how should communication network structures be adopted in PBOs? The answer depends on the unique goals of each PBO with different customary standards. Recent research suggests that in addition to basic factors such as the cost, time, and quality of project, other factors like organizational characteristics and stakeholder interests should also be carefully considered [39,40]. But the current conclusions are not yet unanimous, as they are extremely difficult to measure.

Based on the Input-Mediator-Output (IMO) model, this study—proposes that the following indicators should be used to measure the organizational effectiveness of PBOs. Specifically speaking, problem-solving performance indicators representing organizational usefulness and member perception should also be considered. In the IMO model, interaction processes at the organizational level, environmental level, and individual level impact organizational performance and other outcomes (such as member satisfaction and group cohesion) [41]. Taking PBOs in the construction industry as an example, although task performance has been used for a long time as the main criterion for measuring project management success, it has been gradually realized that team members' positive perceptions in team cooperation can affect both the completion of the current project and future project cooperation. Therefore, we believe that member perceptions including job satisfaction and participation, tacit understanding, organizational commitment, and willingness to continue cooperation should be added to constitute the category of collective effectiveness.

2.3. Urgency of the Project Process

The establishment of PBOs is aimed at harnessing creativity, collaboration, and coordination to complete a new task within an urgent timeframe. Due to the temporary nature of PBOs, members might not have prior experience working together. To succeed, they must rapidly establish a common understanding of their task and develop plans throughout the project. Project members need to utilize their time to understand the task, identify necessary steps for its accomplishment, and devise mechanisms for coordinating multiple interdependent activities [42]. For a new project, which often includes different stages like initiation, execution, and final deadlines, organizational members might suffer from uncertainty about when certain events or tasks should occur, incompatible priorities, or different pacing styles, leading to wasted efforts or inertia in the first half of the time [43]. As deadlines approach and time pressures increases, the collective tends to reorganize and adopt new methods to complete tasks [44]. Therefore, in PBOs aimed at developing complex products, organizational performance management must focus on time-based control mechanisms, including overall deadlines and the synchronicity of activities within the project [4,45]. Detailed conclusions about task urgency can also be found in existing research; for example, in high-tech-oriented project organizations, high task urgency can negatively impact inter-project communication and knowledge transfer intentions [46,47].

In the construction industry, completing projects on schedule has always been one of the standards of project performance management [48]. Therefore, research on project urgency and its impact on organizational performance has never ceased. For instance, studies on leadership discuss how project deadlines affect the objective pressure on project managers [49,50]. In recent years, debates on the impact of urgency on information sharing and knowledge transfer have not stopped. Most studies believe that under limited deadlines and tight schedules, project teams do not have enough time for communication and knowledge sharing [51] and that under centralized structures, the decision-making process becomes more concentrated, reducing the time for communication and coordination. Conversely, some studies suggest that when the task urgency is not strong, project team members are willing to seek knowledge from other project teams when needed [52], thereby enhancing the collective decision-making ability.

Thus, we can affirm that the degree of urgency in project progress is crucial for the study of collective effectiveness in PBOs. To further clarify the nature of the impact, we propose the following hypotheses:

H1a. *Under condition of weak task urgency, decentralized communication networks will yield higher collective efficacy.*

H1b. *Under condition of strong task urgency, centralized communication networks will yield higher collective efficacy.*

2.4. The Mediating Role of Knowledge Sharing

For the successful achievement of organizational goals, members of an organization endeavor to share and disseminate their unique information and knowledge to others, externalizing internal knowledge into collective knowledge that is owned, shared, and agreed upon by PBOs [53,54]. The aforementioned processes of knowledge acquisition, sharing, transfer, and integration largely depend on the communication structure. In recent years, social networks have been seen as channels for organizational learning, playing a role in information sharing and knowledge transfer [55–57]. For example, in a global PBO, we found that each member relies heavily on networks for acquiring knowledge, communication, and learning from others [58]. Thus, social networks can reveal patterns of knowledge integration [59]. In centralized networks, close ties and social cohesion among organization members who are geographically closer are more likely to facilitate knowledge transfer and learning among members [60]. On the other hand, decentralized networks, due to their heterogeneity of knowledge, can effectively enhance the effectiveness of shared knowledge [61,62].

Furthermore, organizational learning theory and group polarization theory can provide a theoretical explanation for communication networks having a key impact on the quality of collective task completion through the degree of knowledge sharing [63,64]. On the one hand, based on organizational learning theory, PBOs can benefit from integrating different types of knowledge, as people tend to use shared information as the basis for collective task completion, and a lack of non-shared information often leads to detrimental task completion outcomes [65,66]. On the other hand, according to group polarization theory, in interpersonal communication network structures where knowledge sharing is less than ideal, biased group polarization is more likely to occur, leading to lower satisfaction with collective task completion [67].

The aforementioned literature primarily demonstrates that communication network structures influence collective decision-making through their potential impact on organizational knowledge sharing and transfer. In urgent situations, quickly and effectively solving problems often requires stimulating collaboration and knowledge sharing among team members, as each individual may hold key information or expertise necessary for solving the problem. Faced with urgent tasks, team members may need to think and innovate quickly. Under such pressure, people are often more willing to share and explore new ideas, thereby promoting the flow of knowledge and the generation of new solutions, which also helps reduce knowledge hoarding behavior. Therefore, regarding the conditions and mechanisms of knowledge sharing behavior in PBOs, we propose the following hypothesis:

H2. *Under condition of strong task urgency, collective communication will stimulate knowledge sharing behavior, thereby enhancing collective effectiveness.*

The theoretical model of this study is shown in Figure 1.

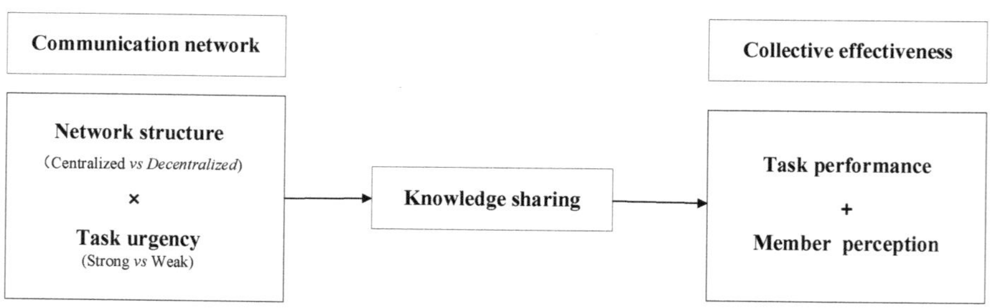

Figure 1. The theoretical model.

3. Methodology

3.1. Research Design and Experimental Treatment

Based on the Bavelas–Leavitt–Guetzkow series experiment [36,68,69] and a recent communication network experiment [13], this study presents an experimental platform to instrument the connection between communication networks and collective effectiveness for PBOs for two levels of urgency. The requirements of the platform were as follows: first, maximum verisimilitude, which means both that the presented networks had real-world analogues and that the means for accomplishing the task similar to real information delivery work in PBOs; second, maximum accessibility, which required the task to be easily understandable and implementable by subjects; third, maximum instrumentation, which required that results obtained by the task and feedback from participants be captured as richly as possible in subsequently analyzable data. Referring to the reality of PBOs in the construction industry, participants in the organization often have different information dimensions of previous cases due to their different professional backgrounds and experiences.

Therefore, this paper uses group-based experiments to explore the mechanism of how communication networks affect the collective effectiveness of PBOs. Specifically, we aim to investigate the interaction effects between communication networks (fully connected/subgroup/core-periphery/locally clustered) and two levels of task urgency on task performance and member perception (H1a and H1b) as well as the emergent scenarios of the mediating effect of knowledge sharing (H2). To complete a theater construction project, a PBO was formed consisting of owners, designers, builders, and supervisors. In this experiment, twelve participants were designated to play the roles of Owner A, Designer B, Builder C, and Supervisor D. Each role was represented by three members: the Owners (A1, A2, A3), Designers (B1, B2, B3), Builders (C1, C2, C3), and Supervisors (D1, D2, D3). Members within the same role operated independently of each other, with no formal relationship beyond their association as team members, as depicted in Figure 2.

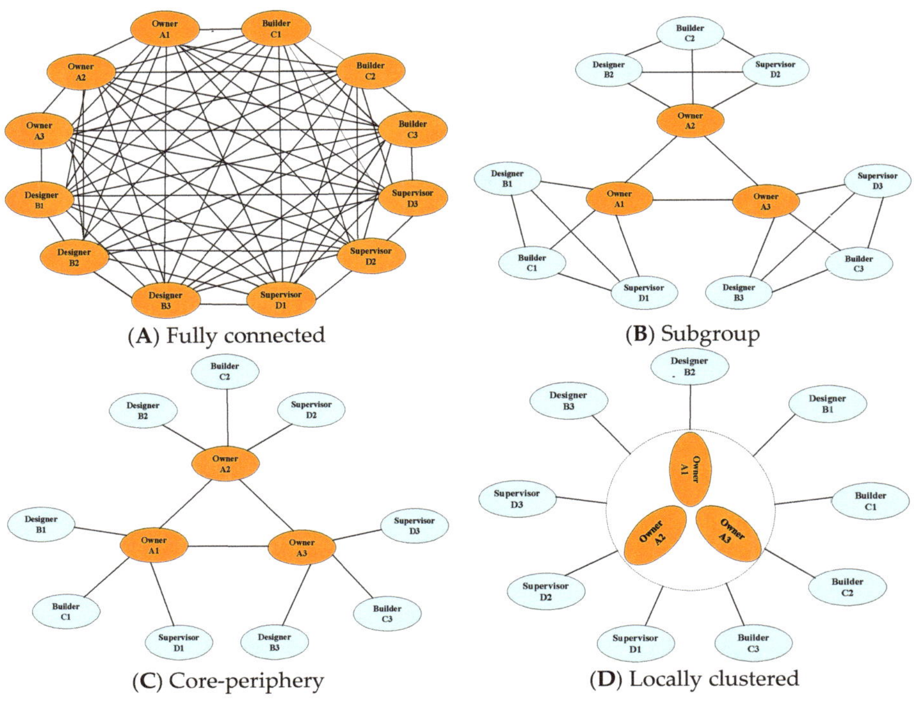

Figure 2. Visualizations of the communication networks.

Referring to the studies of Mason and Watts [70] and Enemark, et al. [71], we adopted a messaging task to measure the communication of PBOs who were in different network structures. Specifically, participants were faced with the following tasks: (a) each member initially had 33 unique messages for role category; (b) team members performed a messaging task with an optional communication partner during a designated communication time; and (c) 99 four-dimensional messages were integrated one by one to form a complete message. The overall task goal of the project was for each member to complete effective information transfer and integration within the time limit and for the individual with decision-making authority to aggregate valid four-dimensional information and make collective decisions accordingly. The main experiment was divided into two parts, each lasting ten minutes, with the results of the first part being carried over to the second part.

3.2. Participants and Procedure

Through expert interviews and recommendations, we adopted a snowball sampling method and recruited 598 individuals with over two years of experience in engineering project management from fifteen construction companies and engineering firms in Shanghai, China, to participate in this experiment. Those who successfully completed the experiment were each rewarded with a ¥100 supermarket coupon. To avoid the interference from learning effects and legacy effects due to the increasing familiarity of the experiment participants as well as the empirical summaries of the experimental steps or experimental order effects, this experiment adopted a between-group design method, thus each person could only participate in the experiment once. Before the main experiment, we played a video for all participants about the experiment rules and examples. Afterwards, participants were required to answer five questions regarding their willingness to participate and the rules of the experiment. Those who passed were then randomly assigned to groups for the main experiment. We used a random number method to group participants, striving to ensure a balanced number of participants and groups in each experimental setting. Next, participants entered their respective groups, with each sitting in front of a computer in a private booth, filling out personal information, and logging into the experimental platform. All activities were conducted through an online platform, with data being automatically collected and scanned by the system. The total duration of the experiment was 20 min. Ten minutes into the main experiment, the system paused to allow participants to fill out a questionnaire on 'perception' (including job satisfaction and communication experience). At the end of the main experiment, each participant filled out the questionnaire on perception again and provided feedback on the experiment. The feedback form was used to measure whether the subjects correctly executed the experimental procedures and rules, thus conducting an operational check of the experimental data [72]. Participants had to correctly answer all six questions. For instance, do you agree that passing information is to support decision-making activities? When information is shared, it is necessary to confirm whether the recipient had the information before. According to the rules, who can you communicate with?

The experiment was conducted from April to September 2023 in Building No. 2 of the Shanghai Lixin University of Accounting and Finance, yielding 49 sets of experimental data. After excluding one set of erroneous data, 48 sets of valid experimental data were obtained, ensuring 12 groups for each of the four networks. From a basic demographic analysis, there were 267 males (46.3%) and 309 females (53.7%), indicating a relatively balanced gender distribution. It was found that 368 persons (63.9%) had two to five years of work experience, 141 persons (24.5%) had five to ten years of work experience, and 67 persons (11.6%) had more than ten years of work experience. In addition, 62 persons (10.8%) held positions of department manager or above, 289 persons (50.2%) were project managers, and 225 persons (39.0%) were engineers and related technical personnel. Additionally, Harman's single-factor test was used to check for potential common method biases. The highest variance of all member-reported variables was 24.328%, less than half of the total explained variance. This suggests that our analysis did not suffer from severe common method bias issues.

3.3. Measures

This study measured collective effectiveness in terms of task performance and member perception. Referring to existing methods [73], task performance in this experiment was defined as the total amount of complete information obtained by the group within a set time—i.e., the cumulative number of complete pieces of information obtained by decision-makers after excluding duplicate data. Additionally, referencing the measurement methods of subjective experiences in existing communication experiments [10], this experiment measured organizational member perception through a questionnaire with five items (Appendix A). Knowledge sharing is manifested as effective private information transmission between members through individual-based behaviors. Following the methods used in behavioral experiments to measure information sharing [18], we collected the total amount of information shared and acknowledged by others within a limited time as the data source for knowledge sharing. Data was collected through records of information transmission by participants during the experiment. Considering the gradually increasing urgency of engineering construction project tasks and the results of the pilot experiment, the 10th minute of the formal experimental process was used as the critical point for dividing urgency.

To eliminate the influence of confounding factors on the experimental results, this study considered the following covariates: the participants' professional backgrounds, work experience, industry characteristics, knowledge related to organizational structure, communication experience, and previous collaboration experience. We categorized the samples into two groups based on the average scores of these six items at the collective level. We compared intergroup task performance, member perception, and knowledge sharing, with p-values of 0.456, 0.865 and 0.558, respectively, and found no significant differences ($p > 0.05$). Therefore, we disregarded these covariates in subsequent analyses.

4. Results

4.1. Manipulation Check

Following the recommendations of existing research [10], we conducted a manipulation check on the structure of communication networks and task urgency. Initially, by examining the occupancy of available communication paths among team members, there was no significant difference in occupancy between members in decentralized networks and centralized networks under two levels of task urgency ($p > 0.1$). However, in core-periphery networks and locally clustered networks, the core nodes had a slightly higher occupancy of available communication paths compared to other members ($p < 0.1$), indicating that the manipulation of communication network structures was effective. On the other hand, analysis of the frequency of transmissions among members in the four types of communication networks showed that under conditions of strong urgency, the average transmission frequency was significantly higher than under conditions of weak urgency ($p < 0.001$). This result indicates that the manipulation of urgency was effective.

4.2. Reliability and Validity Test

Firstly, regarding the reliability of the member perception scale, the Cronbach's α value was 0.965, indicating good reliability of the questionnaire. Next, the validity of the scale was tested through exploratory factor analysis (EFA) and confirmatory factor analysis (CFA). The KMO value for member perception was 0.823, with p-values at 0.001, suggesting that the structural validity meets the requirements. Furthermore, the composite reliability (CR) value exceeded 0.7, the average VVariance extracted (AVE) value was greater than 0.5, and the factor loading (FL) values were above 0.6, indicating good convergent validity of the constructs [74]. FL values of all items under their respective variables were significantly higher than those under other variables. The square root of AVE was greater than the inter-construct correlations. Additionally, IFI, TLI values were above 0.9, and the RMSEA value was 0.031, indicating good discriminant validity [75]. Therefore, this study demonstrates good reliability and acceptable validity in the measurement of member perception.

4.3. Hypothesis Testing

Table 1 shows the descriptive statistical results for task performance and member perception within four communication networks under two conditions of urgency.

Table 1. Results of descriptive statistics.

Task Urgency	Network Structure	Task Performance		Members Perception	
		Mean	**SD**	**Mean**	**SD**
Weak	Locally clustered	5.250	0.968	3.250	0.661
	Core-periphery	6.250	1.089	3.000	0.408
	Subgroup	7.750	1.198	3.375	0.857
	Fully connected	16.350	2.057	3.875	0.331
Strong	Locally clustered	78.375	5.521	4.750	0.433
	Core-periphery	36.000	2.121	3.750	0.661
	Subgroup	32.625	2.176	3.500	0.645
	Fully connected	28.875	2.571	2.750	0.433

We adopted two-factor completely randomized multi-group design to test the hypotheses usingone-way ANOVA, two-way MANOVA and Tukey-HSD post hoc test, respectively [76,77]. First, the linearity of most groups was obvious, except for perception within the fully connected network and Subgroup network under weak task urgency. There was no multicollinearity in the correlation coefficient between task performance and member perception ($|r| < 0.9$). Next, according to the Box plot and Mahalanobis test, there were no univariate outliers and no multivariate outliers. Additionally, we used the Shapiro-Wilk method to test if task performance and member perception met the requirements of multivariate normal distribution ($p > 0.05$). The Box test indicated that the assumption of equal covariance matrices is valid ($p = 0.009$).

Further, there was a significant interaction effect between the variables task performance and member perception (F-value= 4.051, $p = 0.003$, Wilk's $\lambda = 0.752$, partial $\eta^2 = 0.122$). Further, results for the one-way between-subject ANOVA test showed the interaction between network structure and task urgency had a significant effect on task performance (F-value = 8.426, $p = 0.001$, partial $\eta^2 = 0.189$), as well as member perception (F-value = 4.961, $p = 0.002$, partial $\eta^2 = 0.136$).

Next, the results of main effect test were completed (Figure 3). The fully connected network (M = 16.350, SD = 2.057) led to better task performance compared to the locally clustered network (M = 5.250, SD = 0.968, CI [9.987, 12.213], $p < 0.001$) and the core-periphery network (M = 6.250, SD = 1.089, CI [8.959, 11.240], $p < 0.001$) under the condition of weak task urgency. In addition, the fully connected network (M = 3.875, SD = 0.331) inspired more positive member perception compared to the core-periphery network (M = 3.000, SD = 0.408, CI [0.581, 1.169], $p < 0.001$) and the locally clustered network (M = 3.250, SD = 0.661, CI [0.262, 0.988], $p < 0.001$) under the condition of weak task urgency. Thus, H1a is validated.

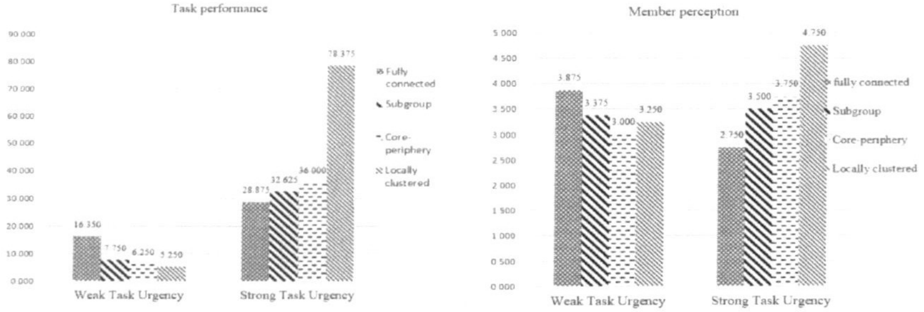

Figure 3. Results of means comparisons.

However, under conditions of strong task urgency, the locally clustered network (M = 78.375, SD = 5.521) improved task performance compared to the subgroup network

(Mean = 32.625, SD = 2.176, CI [42.841, 48.659], $p < 0.001$) and the fully connected network (M = 28.875, SD = 2.571, CI [46.497, 52.503], $p < 0.001$). Moreover, the locally clustered network (M = 4.750, SD = 0.433) also stimulated more positive member perception compared to the subgroup network (Mean = 3.500, SD = 0.645, CI [0.578, 1.922], $p < 0.001$) and the fully connected network (Mean = 2.750, SD = 0.433, CI [1.633, 2.367], $p < 0.001$). Therefore, H1b is supported.

Furthermore, we found a positive correlation between task performance and member perception (Figure 4). That is to say, as the level of member perception increases, task performance shows an upward trend under various levels of task urgency. Additionally, under higher task urgency, for any given level of member perception, task performance is highest in a decentralized structure. In a decentralized network, members with the same level of perception tend to achieve higher task performance. However, under conditions of weaker task urgency, although decentralized networks can achieve better member perception, they do not necessarily have the highest task performance; conversely, centralized networks exhibit significant advantages in task performance. Thus, H2 is verified.

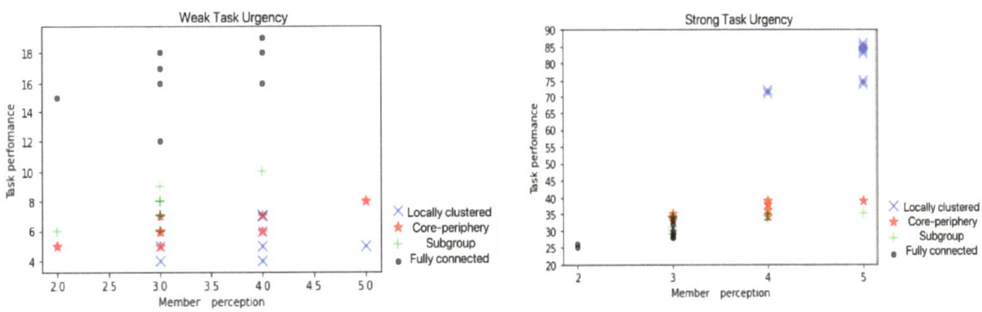

Figure 4. Scatterplot of task performance vs. member perception.

To confirm the mediating role of knowledge sharing, we further used a bootstrap mediation method with 5000 samples by process [78]. Since the independent variable is categorical and there are four centralization levels of communication networks, we designed three dummy variables. As shown in Table 2, the indirect effect of network structure and task performance via knowledge sharing was significant under conditions of strong task urgency because the 95% confidence intervals did not include zero. Similarly, the indirect effect of network structure and member perception via knowledge sharing was significant under the condition of strong task urgency. On the contrary, the mediating effect of knowledge sharing was not significant under condition of weak task urgency (all CIs include zero).

Table 2. The mediating role of knowledge sharing.

Independent Variable	Task Urgency	Network Structure	Effect	SE	95% Confidence Interval (CI)
Task performance	Weak	D1	0.186	0.088	[−0.022, 0.375]
		D2	0.310	0.144	[−0.039, 0.618]
		D3	1.120	0.515	[−2.106, 2.210]
	Strong	D1	0.407	0.443	[0.317, 1.015]
		D2	0.682	0.746	[0.647, 1.700]
		D3	2.460	2.718	[3.748, 6.205]
Member perception	Weak	D1	−0.138	0.225	[−0.606, 0.283]
		D2	−0.178	0.353	[−0.794, 0.354]
		D3	−0.215	0.447	[−0.973, 0.429]
	Strong	D1	1.469	0.756	[0.146, 3.167]
		D2	1.899	0.953	[0.183, 3.966]
		D3	2.301	1.143	[0.224, 4.755]

5. Discussion and Conclusions

5.1. Main Findings

This study involving 598 engineering project management personnel participating in group experiments based on information sharing and collective decision-making found that both centralized and decentralized communication networks have advantages in enhancing collective effectiveness under different levels of task urgency. Specifically, under condition of weak task urgency, decentralized networks contribute to improved task performance and more positive member perception. Conversely, as task urgency increases, centralized networks show advantages in enhancing task performance and also bringing about more positive member perception. The possible explanation for this phenomenon is multifaceted.

Firstly, the equal participation of all organizational members is emphasized in decentralized networks. Through effective and rapid information flow among individuals with different experiences, they integrate shared skills and knowledge to improve project outcomes and achieve strategic objectives [79–81]. Therefore, in the early stages of a project with weaker task urgency, each member or node usually has more autonomy, enhancing the sense of responsibility and involvement of organizational members. From the perspective of systems engineering, the organization is more stable at this time. That is, when individual nodes encounter problems, other nodes can continue to operate independently. This redundancy and fault-tolerance mechanism is particularly important in non-urgent tasks, as it provides additional time to correct or optimize specific parts of the system [82,83]. In contrast, centralized networks can help to quickly coordinate organizational members and production resources under urgent and coordination-challenging situations, improving the production efficiency of the project organization. The case of the 2010 World Expo shows that establishing a project command composed of experienced executives from relevant government departments can integrate various resources, accelerate project progress, and improve completion quality through efficient communication and coordination [84]. Studies of the large-scale 2012 London Olympics project also support this conclusion, suggesting that project managers play a key role in strengthening organizational coordination and team integration [31,85].

Additionally, this study measured PBOs members' perceptions with regard to work involvement and job satisfaction. The results indicate that under weaker task urgency, decentralized networks inspire more positive member perception, while under stronger task urgency, centralized networks lead to more positive member perception. In other words, as task urgency changes, organizational members' experiences of different communication networks also change. This conclusion also confirms the complementary relationship between task performance and member perception. The reason is that enhanced motivation and participation can promote better cooperation and work output. The achievement of task performance, in turn, feeds back into the perception of the work process, thus forming a virtuous cycle. Our research results are consistent with the literature emphasizing that job satisfaction enhances work performance [86,87].

Furthermore, this study also explored the internal mechanism by which the interaction of communication network and task urgency affects collective effectiveness. Under different level of task urgencies, the mediating effect of knowledge sharing in the communication network and task urgency is evolutionary. That is, when project processes are under weak urgency, the mediating effect of knowledge sharing is not significant. However, when tasks become more urgent, the transformation of collective behavior based on communication and decision-making will enhance task performance and member perception through knowledge contribution. A possible explanation is that knowledge sharing is a fundamental organizational capability needed to promote the integration of expertise, whose implementation relies on stronger linkage relationships [88,89]. Consistent with related literature, the process of knowledge sharing aims to break down barriers between knowledge owners, achieving a certain degree of free knowledge flow within a certain range, which plays a crucial role in enhancing organizational learning, knowledge creation, and organizational performance [64,90]. In situations of close relationships and high cohesion, effective knowledge sharing will help organizational members gain access to others' unique knowledge, avoid resource waste due to repetitive knowledge production, thereby strengthening the collective belief in completing tasks and improving the perception of collaboration [91–93].

5.2. Theoretical Contributions

First, although there is an increasing number of studies on PBOs management focusing on inter-member relationships from a social network analysis perspective [31,94], discussions from the task urgency viewpoint are still somewhat lacking. This study advances our understanding of organizational structure relationships of PBOs by revealing how the impact of communication network structure on collective effectiveness is moderated by task urgency. Specifically, for PBOs in the construction industry in which long construction cycles are prone to turbulence, complexity, uncertainty, and ambiguity, centralized networks can provide stronger resilience with increasing task urgency, thus enhancing collective effectiveness [95].

Second, this study expands the emergent mechanisms of knowledge sharing in PBOs. The results confirm that under certain communication networks, knowledge sharing simultaneously drives task performance and member perception, reemphasizing the importance of organizational network structure indicators in this study. Moreover, from the perspective of organizational learning theory, the important mediating role of knowledge sharing between communication networks and collective effectiveness depends on the increased strength of connections among organizational members, with task urgency also enhancing these linkages. The results of this study reveal the dynamic nature of knowledge sharing channels in PBOs [14].

Third, this paper supports the current expansion of project management value beyond just focusing on expected outcomes and benefits from a reductionist viewpoint [96–98]. Combining task performance and member perception, this study sets an expanded definition and measurement standards for the management success of PBOs. This aligns with the more complex, multidimensional technical project management characteristics of PBOs, reflecting the contextualized, strategic nature of a wider range of projects across various sectors and those undertaken for competitive advantage [99]. Additionally, our study also found an interactive relationship between task performance and member perception, which has rarely been explored in previous research. This finding highlights the positive correlation between organizational structure and member job satisfaction in non-structured tasks as suggested by motivation theories and self-efficacy theories [100–102]. Therefore, future research can enhance the management value of PBOs on a more solid theoretical and methodological basis.

5.3. Managerial and Practical Implications

This study also provides important insights for decision-makers and leaders in the construction industry. Specifically, it is necessary to design organizations that are adaptable to the environment through the coordination between member behavior and interactions [1,22]. In other words, the communication structure has a crucial impact on organizational management performance. Specific suggestions are as follows.

Communication and coordination among organizational members are significant in enhancing project outcomes. When executing projects, decision-makers should gather human resources with different knowledge backgrounds and establish appropriately centralized organizational structures according to the type of and environmental factors relevant to the project. Therefore, leaders should recognize that even if formal task interdependence channels have been formed among team members within a certain relationship structure, interactions may not effectively occur. Project organization leaders should investigate and recognize the actual needs and difficulties of knowledge senders and receivers and promote cohesion between them through network interventions. For example, the appointment of a central coordinator as a hub is crucial for large PBOs like the 2020 World Expo. It is advisable to implement coordination among sub-projects by building management teams and coordinating resources to mitigate the negative impacts of potential daily changes and disruptions, thus improving collective outcomes [31]. This is a solution from a central collaboration perspective. On the other hand, in situations with relatively relaxed schedules or high cohesion among organizational members, members with different professional expertise in a decentralized structure have more defined responsibilities and rights in the collective, promoting the exchange and sharing of knowledge, forming efficient task decisions based on lean construction concepts. This organizational structure also provides adaptive collaboration for frequently occurring work coordination issues in project progress. Additionally, decision-makers and leaders of public organization projects should avoid exclusively using economic or financial performance indicators and adopt more diversified evaluation standard. Of course, especially for large-scale projects like public facilities that require significant investment, have extensive coverage, and large social impact, assessing task performance is essential. However, paying attention to PBO members' perceptions of work (participation, satisfaction, etc.) is very important. This not only helps to improve project task performance but also enhances the well-being of project members, contributing to the formation of an experiential knowledge base. This aligns with the long-term development needs of the construction industry.

6. Conclusions

This study, based on the IMO model and organizational learning theory and from the perspective of social network analysis, explores the relationship between communication networks, knowledge sharing, and collective effectiveness in PBOs. Utilizing a classic communication experiment design and the practice of contract management in construction engineering projects, we confirmed through implementation using an online platform that the interaction between communication networks and task urgency is an important way to enhance the collective effectiveness of PBOs. The research results show that under stronger task urgency, it is recommended to adopt a centralized communication network, which leads to higher task performance and member perception. Conversely, a decentralized communication network is recommended under weaker task urgency. The measurement indicators of collective effectiveness should include task performance and member perception.

However, due to the diversity of PBOs, this study cannot precisely simulate and measure all structures in reality. For example, some PBOs are combinations of multiple structures or large project plans, which could bring noticeable changes to the results. To improve the stability of the research, future studies can combine field experiments and expand the types of projects. Additionally, our study is applicable to the construction industry in China and does not consider the impact of regional cultural differences on the behavior of organizational members. Furthermore, whether its results are applicable to developed markets or even emerging markets still needs further verification. In the future, experiments can be conducted for different industrial backgrounds and from the perspectives of participants.

Author Contributions: Conceptualization, methodology, investigation and data curation, X.D.; writing—original draft, X.D. and S.W.; writing—review and editing, W.S. and S.W.; funding acquisition, X.D. and W.S. All authors have read and agreed to the published version of the manuscript.

Funding: This research was funded by Shanghai Universities Young Faculty Training and Support Program (grant number ZZ202220047) and National Natural Science Foundation of China (grant number 72201027).

Data Availability Statement: The raw data supporting the conclusions of this article will be made available by the corresponding author on request.

Acknowledgments: The authors would like to express their sincere gratitude to the editor and the anonymous reviewers for their insightful and constructive comments.

Conflicts of Interest: The authors declare no conflicts of interest.

Appendix A

Table A1. Questionnaire for member perception.

Item	Measure
1	The team task is interesting and it can engage you.
2	Other members of the team are also highly engaged in the coordination and cooperation for the task.
3	Due to the importance of your information transfer for the task, you take it more seriously.
4	You are very satisfied with your own performance.
5	You are very satisfied with the team's performance.
6	You have gained professional knowledge and experience.

References

1. Miterev, M.; Mancini, M.; Turner, R. Towards a design for the project-based organization. *Int. J. Proj. Manag.* **2016**, *35*, 479–491. [CrossRef]
2. Cai, X.; Huang, J.; Peng, C. Research on Construction Workers' Safety Risk Sharing in Tunneling Projects Based on a Two-Mode Network: A Case Study of the Shangwu Tunnel. *Buildings* **2023**, *13*, 2689. [CrossRef]
3. Tang, W. Holistic hydropower scheme for China. *Nature* **2016**, *532*, 37. [CrossRef]
4. Melkonian, T.; Picq, T. Building project capabilities in PBOs: Lessons from the french special forces. *Int. J. Proj. Manag.* **2011**, *29*, 455–467. [CrossRef]
5. Brunetta, F.; Boccardelli, P.; Lipparini, A. *The Role of Networks for Innovation in Temporary and Project-Based Organizations*; Palgrave Macmillan: Cham, Switzerland, 2018.
6. Gann, D.M.; Salter, A.J. Innovation in project-based, service-enhanced firms: The construction of complex products and systems. *Res. Policy* **2000**, *29*, 955–972. [CrossRef]
7. Wang, H.; Lu, W.; Soderlund, J.; Chen, K. The interplay between formal and informal institutions in projects: A social network analysis. *Proj. Manag. J.* **2018**, *49*, 20–35. [CrossRef]
8. Zhang, D.; Guo, P.; Zhao, J. The motives system for developing project-based inter-organizational cooperation. *Int. J. Proj. Manag.* **2021**, *40*, 167–180. [CrossRef]
9. Zhao, D.; Duva, M.; Mollaoglu, S.; Frank, K.A.; Garcia, A.J.; Tait, J.R. Integrative collaboration in fragmented project organizations: Network perspective. *J. Constr. Eng. Manag.* **2021**, *147*, 04021115. [CrossRef]
10. Argote, L.; Aven, B.L.; Kush, J. The effects of communication networks and turnover on transactive memory and group performance. *Organ. Sci.* **2018**, *29*, 191–206. [CrossRef]
11. Lee, M.Y.; Edmondson, A. Self-managing organizations: Exploring the limits of less-hierarchical organizing. *Res. Organ. Behav.* **2017**, *37*, 35–58. [CrossRef]
12. Bernstein, E.; Bunch, J.; Canner, N.; Lee, M. Beyond the holacracy hype. *Harv. Bus. Rev.* **2016**, *94*, 38–49.
13. Shore, J.; Bernstein, E.; Jang, A.J. Network centralization and collective adaptability to a shifting environment. *Organ. Sci.* **2023**, *34*, 2064–2096. [CrossRef]
14. Garcia, A.J.; Mollaoglu, S.; Frank, K.A.; Duva, M.; Zhao, D. Emergence and evolution of network structures in complex interorganizational project teams. *J. Manag. Eng.* **2021**, *37*, 04021056. [CrossRef]
15. Atkinson, R. Project management: Cost, time and quality, two best guesses and a phenomenon, its time to accept other success criteria. *Int. J. Proj. Manag.* **1999**, *17*, 337–342. [CrossRef]
16. Ren, X.; Yan, Z.; Wang, Z.; He, J. Inter-project knowledge transfer in project-based organizations: An organizational context perspective. *Manag. Decis.* **2019**, *58*, 844–863. [CrossRef]
17. Kearns, M.; Suri, S.; Montfort, N. An experimental study of the coloring problem on human subject networks. *Science* **2006**, *313*, 824–827. [CrossRef] [PubMed]
18. Xiao, Y.; Zhang, H.; Basadur, T.M. Does information sharing always improve team decision making? An examination of the hidden profile condition in new product development. *J. Bus. Res.* **2016**, *69*, 587–595. [CrossRef]
19. Thiry, M.; Deguire, M. Recent developments in project-based organisations. *Int. J. Proj. Manag.* **2007**, *25*, 649–658. [CrossRef]
20. Choi, S.; Cho, I.; Han, S.H.; Kwak, Y.H.; Chih, Y.-Y. Dynamic capabilities of project-based organization in global operations. *J. Manag. Eng.* **2018**, *34*, 04018027. [CrossRef]
21. Burke, C.M.; Morley, M.J. Developed Mind-Undeveloped Group: Structural and Functional Mechanisms in Temporary Organizations. *Acad. Manag. Proc.* **2017**, *2017*, 16187. [CrossRef]
22. Leiringer, R.; Zhang, S. Organisational capabilities and project organising research. *Int. J. Proj. Manag.* **2021**, *39*, 422–436. [CrossRef]
23. Bartsch, V.; Ebers, M.; Maurer, I. Learning in project-based organizations: The role of project teams' social capital for overcoming barriers to learning. *Int. J. Proj. Manag.* **2013**, *31*, 239–251. [CrossRef]

24. Lingard, H.; Pirzadeh, P.; Blismas, N.; Wakefield, R.; Kleiner, B. Exploring the link between early constructor involvement in project decision-making and the efficacy of health and safety risk control. *Constr. Manag. Econ.* **2014**, *32*, 918–931. [CrossRef]
25. Zhou, Q.; Deng, X.; Hwang, B.-G.; Ji, W. Integrated framework of horizontal and vertical cross-project knowledge transfer mechanism within project-based organizations. *J. Manag. Eng.* **2020**, *36*, 04020062. [CrossRef]
26. Chinowsky, P.; Diekmann, J.; Galotti, V. Social network model of construction. *J. Constr. Eng. Manag.* **2008**, *134*, 804–812. [CrossRef]
27. Lienert, J.; Schnetzer, F.; Ingold, K. Stakeholder analysis combined with social network analysis provides fine-grained insights into water infrastructure planning processes. *J. Environ. Manag.* **2013**, *125*, 134–148. [CrossRef]
28. Lin, X.; Ho, C.M.; Shen, G.Q. Who should take the responsibility? Stakeholders' power over social responsibility issues in construction projects. *J. Clean. Prod.* **2017**, *154*, 318–329. [CrossRef]
29. Mok, K.Y.; Shen, G.Q.; Yang, J. Stakeholder management studies in mega construction projects: A review and future directions. *Int. J. Proj. Manag.* **2015**, *33*, 446–457. [CrossRef]
30. Zheng, X.; Chen, J.; Han, Y.; Ren, L.; Shi, Q. Unveiling complex relational behavior in megaprojects: A qualitative-quantitative network approach. *Int. J. Proj. Manag.* **2021**, *39*, 738–749. [CrossRef]
31. Shen, W.; Ying, W. Large-scale construction programme resilience against creeping disruptions: Towards inter-project coordination. *Int. J. Proj. Manag.* **2022**, *40*, 671–684. [CrossRef]
32. Du, J.; Zhao, D.; Zhang, O. Impacts of human communication network topology on group optimism bias in Capital Project Planning: A human-subject experiment. *Construct. Manag. Econ.* **2019**, *37*, 44–60. [CrossRef]
33. Ding, X.; Feng, L.; Huang, Y.; Li, W. The interactive effects of communication network structure and organizational size on task performance in project-based organizations: The mediating role of bootleg innovation behavior. *Buildings* **2024**, *14*, 98. [CrossRef]
34. Du, J.; Zhao, D.; Issa RR, A.; Singh, N. Bim for improved project communication networks: Empirical evidence from email logs. *J. Comput. Civil. Eng.* **2020**, *34*, 04020027. [CrossRef]
35. Bavelas, A.; Barrett, D. *An Experimental Approach to Organizational Communication*; American Management Association: New York, NY, USA, 1951.
36. Leavitt, H.J. Some effects of certain communication patterns on group performance. *J. Abnorm. Soc. Psychol.* **1951**, *46*, 38. [CrossRef]
37. Shaw, M.E. *Group Dynamics: The Psychology of Small Group Behavior*; McGraw-Hill: New York, NY, USA, 1971.
38. Gao, S.; Song, X.; Ding, R. Promoting information transfer in collaborative projects through network structure adjustment. *J. Constr. Eng. Manag.* **2020**, *146*, 04019108. [CrossRef]
39. Reagans, R.E.; McEvily, B. Network structure and knowledge transfer: The effects of cohesion and range. *Adm. Sci. Q.* **2003**, *48*, 240–267. [CrossRef]
40. Senaratne, S.; Jin, X.; Balasuriya, K. Exploring the role of networks in disseminating construction project knowledge through case studies. *Eng. Constr. Archit. Manag.* **2017**, *24*, 1281–1293. [CrossRef]
41. Mathieu, J.; Maynard, M.T.; Rapp, T.; Gilson, L. Team effectiveness 1997-2007: A review of recent advancements and a glimpse into the future. *J. Manag.* **2008**, *34*, 410–476. [CrossRef]
42. Siddiquei, A.N.; Fisher, C.D.; Hrivnak, G.A. Temporal leadership, team processes, and project team task performance. *Int. J. Proj. Manag.* **2022**, *40*, 715–724. [CrossRef]
43. Van Hooft, E.A.; Van Mierlo, H. When teams fail to self-regulate: Predictors and outcomes of team procrastination among debating teams. *Front. Psychol.* **2018**, *9*, 464. [CrossRef] [PubMed]
44. McGrath, J.E. Time, interaction, and performance (TIP) A Theory of Groups. *Small Group Res.* **1991**, *22*, 147–174. [CrossRef]
45. Klein, K.J.; Ziegert, J.C.; Knight, A.P.; Xiao, Y. Dynamic delegation: Shared, hierarchical, and deindividualized leadership in extreme action teams. *Adm. Sci. Q.* **2006**, *51*, 590–621. [CrossRef]
46. Zheng, W.; Yang, B.; McLean, G.N. Linking organizational culture, structure, strategy, and organizational effectiveness: Mediating role of knowledge management. *J. Bus. Res.* **2010**, *63*, 763–771. [CrossRef]
47. Zhao, D.; Zuo, M.; Deng, X.N. Examining the factors influencing cross-project knowledge transfer: An empirical study of IT services firms in China. *Int. J. Proj. Manag.* **2015**, *33*, 325–340. [CrossRef]
48. Schieg, M. Risk management in construction project management. *J. Bus. Econ. Manag.* **2006**, *7*, 77–83. [CrossRef]
49. White, D.; Fortune, J. Current practice in project management—An empirical study. *Int. J. Proj. Manag.* **2002**, *20*, 1–11. [CrossRef]
50. Leung, M.-Y.; Chan, Y.-S.; Olomolaiye, P. Impact of stress on the performance of construction project managers. *J. Constr. Eng. Manag.* **2008**, *134*, 644–652. [CrossRef]
51. Moud, H.I.; Abbasnejad, B. *Factors Affecting Knowledge Transfer in Project Based Organizations (PBOs)*; Department of Construction Management, Chalmers University of Technology: Gothenburg, Sweden, 2012.
52. Newell, S.; Bresnen, M.; Edelman, L.; Scarbrough, H.; Swan, J. Sharing knowledge across projects: Limits to ICT-led project review practices. *Manag. Learn.* **2006**, *37*, 167–185. [CrossRef]
53. Fiore, S.M.; Rosen, M.A.; Smith-Jentsch, K.A.; Salas, E.; Letsky, M.; Warner, N. Toward an understanding of macrocognition in teams: Predicting processes in complex collaborative contexts. *Hum. Factors* **2010**, *52*, 203–224. [CrossRef]
54. Grand, J.A.; Braun, M.T.; Kuljanin, G.; Kozlowski, S.W.J.; Chao, G.T. The dynamics of team cognition: A process-oriented theory of knowledge emergence in teams. *J. Appl. Psychol.* **2016**, *101*, 1353. [CrossRef]
55. Pryke, S. *Projects as Networks of Relationships*; Blackwell Publishing: Hoboken, NJ, USA, 2006; pp. 213–235.

56. Ruuska, I.; Ahola, T.; Artto, K.; Locatelli, G.; Mancini, M. A new governance approach for multi-firm projects: Lessons from Olkiluoto 3 and Flamanville 3 nuclear power plant projects. *Int. J. Proj. Manag.* **2011**, *29*, 647–660. [CrossRef]
57. Müller, R.; Andersen, E.S.; Kvalnes, Ø.; Shao, J.; Sankaran, S.; Turner, J.R.; Biesenthal, C.; Walker, D.; Gudergan, S. The interrelationship of governance, trust, and ethics in temporary organizations. *Proj. Manag. J.* **2013**, *44*, 26–44. [CrossRef]
58. Wiewiora, A.; Chang, A.; Smidt, M. Individual, project and organizational learning flows within a global project-based organization: Exploring what, how and who. *Int. J. Proj. Manag.* **2020**, *38*, 201–214. [CrossRef]
59. Ruan, X.; Ochieng, E.G.; Price, A.D.F.; Egbu, C. Knowledge integration process in construction projects: A social network analysis approach to compare competitive and collaborative working. *Constr. Manag. Econ.* **2011**, *30*, 5–19. [CrossRef]
60. Phelps, C.C.; Heidl, R.A.; Wadhwa, A. Knowledge, networks, and knowledge networks: A review and research agenda. *Soc. Sci. Res. Netw.* **2011**, *38*, 1115–1166. [CrossRef]
61. Anderson, E.W.; Sullivan, M.W. The antecedents and consequences of customer satisfaction for firms. *Mark. Sci.* **1993**, *12*, 125–143. [CrossRef]
62. Lipshitz, R.; Levy, D.L.; Orchen, K. Is this problem likely to be solved? A cognitive schema of effective problem solving. *Think. Reason.* **2006**, *12*, 413–430. [CrossRef]
63. Bresman, H.; Zellmer-Bruhn, M.E. The structural context of team learning: Effects of organizational and team structure on internal and external learning. *Organ. Sci.* **2013**, *24*, 1120–1139. [CrossRef]
64. Jia, J.; Ma, G.; Wu, Z.; Wu, M.; Jiang, S. Unveiling the impact of task conflict on construction project performance: Mediating role of knowledge integration. *J. Manag. Eng.* **2021**, *37*, 04021060. [CrossRef]
65. Baker, D.F. Enhancing Group Decision Making: An Exercise to Reduce Shared Information Bias. *J. Manag. Educ.* **2010**, *34*, 249–279. [CrossRef]
66. Argote, L.; Miron-Spektor, E. Organizational learning: From experience to knowledge. *Organ. Sci.* **2011**, *22*, 1123–1137. [CrossRef]
67. Luhan, W.J.; Kocher, M.G.; Sutter, M. Group polarization in the team dictator game reconsidered. *Exp. Econ.* **2009**, *12*, 26–41. [CrossRef]
68. Guetzkow, H.S.; Simon, H.A. The impact of certain communication nets Upon organization and performance in task-oriented groups. *Manag. Sci.* **1955**, *1*, 233–250. [CrossRef]
69. Bavelas, A. Communication patterns in task-oriented groups. *J. Acoust. Soc. Am.* **1950**, *22*, 725–730. [CrossRef]
70. Mason, W.; Watts, D.J. Collaborative learning in networks. *Proc. Natl. Acad. Sci. USA* **2012**, *109*, 764–769. [CrossRef] [PubMed]
71. Enemark, D.; McCubbins, M.D.; Weller, N. Knowledge and networks: An experimental test of how network knowledge affects coordination. *Soc. Netw.* **2014**, *36*, 122–133. [CrossRef]
72. Perdue, B.C.; Summers, J.O. Checking the success of manipulations in marketing experiments. *J. Mark. Res.* **1986**, *23*, 317–326. [CrossRef]
73. Lu, S.; Wang, L.; Ni, D.; Shapiro, D.L.; Zheng, X. Mitigating the harms of abusive supervision on employee thriving: The buffering effects of employees' social-network centrality. *Hum. Relat.* **2023**, *76*, 1441–1473. [CrossRef]
74. Hair, J.F.; Black, W.C.; Babin, B.J.; Anderson, R.E. *Multivariate Data Analysis: A Global Perspective*, 7th ed.; Pearson Prentice Hall: Upper Saddle River, NJ, USA, 2010.
75. Ren, X.; Deng, X.; Liang, L. Knowledge transfer between projects within project-based organizations: The project nature perspective. *J. Knowl. Manag.* **2018**, *22*, 1082–1103. [CrossRef]
76. Navidi, W.C. *Principles of Statistics for Engineers and Scientists*; McGraw-Hill, Higher Education: Boston, MA, USA, 2010.
77. Weitzl, W.J.; Hutzinger, C. The effects of marketer- and advocate-initiated online service recovery responses on silent bystanders. *J. Bus. Res.* **2017**, *80*, 164–175. [CrossRef]
78. Hayes, A.F.; Preacher, K.J. Statistical mediation analysis with a multicategorical independent variable. *Br. J. Math. Stat. Psychol.* **2014**, *67*, 451–470. [CrossRef] [PubMed]
79. Ahmad, I.; Azhar, N.; Chowdhury, A. Enhancement of IPD Characteristics as Impelled by Information and Communication Technology. *J. Manag. Eng.* **2019**, *35*, 04018055. [CrossRef]
80. Paik, J.E.; Miller, V.D.; Mollaoglu, S.; Sun, W. Interorganizational projects: Reexamining innovation implementation via IPD cases. *J. Manag. Eng.* **2017**, *33*, 04017017. [CrossRef]
81. Su, G.; Hastak, M.; Deng, X.; Khallaf, R. Risk Sharing Strategies for IPD Projects: Interactional Analysis of Participants' Decision-Making. *J. Manag. Eng.* **2021**, *37*, 04020101. [CrossRef]
82. de Rooij, M.M.G.; Janowicz-Panjaitan, M.; Mannak, R.S. A configurational explanation for performance management systems' design in project-based organizations. *Int. J. Proj. Manag.* **2019**, *37*, 616–630. [CrossRef]
83. Shi, Q.; Chen, X.; Xiao, C.; Han, Y. Network Perspective in Megaproject Management: A Systematic Review. *J. Constr. Eng. Manag.* **2022**, *148*, 03122002. [CrossRef]
84. Zhai, Z.; Ahola, T.; Le, Y.; Xie, J. Governmental governance of megaprojects: The case of EXPO 2010 Shanghai. *Proj. Manag. J.* **2017**, *48*, 37–50. [CrossRef]
85. Davies, A.; Mackenzie, I. Project Complexity and Systems Integration: Constructing the London 2012 Olympics and Paralympics Games. *Int. J. Proj. Manag.* **2014**, *32*, 773–790. [CrossRef]
86. Sy, T.; Tram, S.; O'Hara, L.A. Relation of employee and manager emotional intelligence to job satisfaction and performance. *J. Vocat. Behav.* **2006**, *68*, 461–473. [CrossRef]

87. Torlak, N.G.; Kuzey, C. Leadership, job satisfaction and performance links in private education institutes of Pakistan. *Int. J. Product. Perform. Manag.* **2019**, *68*, 276–295. [CrossRef]
88. Javernick-Will, A. Motivating Knowledge Sharing in Engineering and Construction Organizations: Power of Social Motivations. *J. Manag. Eng.* **2012**, *28*, 193–202. [CrossRef]
89. Wen, Q.; Qiang, M. Coordination and knowledge sharing in construction project-based organization: A longitudinal structural equation model analysis. *Autom. Constr.* **2016**, *72*, 309–320. [CrossRef]
90. Bartol, K.M.; Srivastava, A. Encouraging knowledge sharing: The role of organizational reward systems. *J. Leadersh. Organ. Stud.* **2002**, *9*, 64–76. [CrossRef]
91. Schaubroeck, J.; Lam, S.S.; Xie, J.L. Collective efficacy versus self-efficacy in coping responses to stressors and control: A cross-cultural study. *J. Appl. Psychol.* **2000**, *85*, 512–525. [CrossRef]
92. Walumbwa, F.O.; Lawler, J.J.; Avolio, B.J.; Peng, W.; Kan, S. Transformational Leadership and Work-Related Attitudes: The Moderating Effects of Collective and Self-Efficacy Across Cultures. *J. Leadersh. Organ. Stud.* **2005**, *11*, 2–16. [CrossRef]
93. Ding, X.; Li, W.; Huang, D.; Qin, X. Does Innovation Climate Help to Effectiveness of Green Finance Product R&D Team? The Mediating Role of Knowledge Sharing and Moderating Effect of Knowledge Heterogeneity. *Sustainability* **2022**, *14*, 3926.
94. Liu, L.; Zhao, M.; Fu, L.; Cao, J. Unraveling local relationship patterns in project networks: A network motif approach. *Int. J. Proj. Manag.* **2021**, *39*, 437–448. [CrossRef]
95. Naderpajouh, N.; Matinheikki, J.; Keeys, L.A.; Aldrich, D.P.; Linkov, I. Resilience and projects: An interdisciplinary crossroad. *Proj. Leadersh. Soc.* **2020**, *1*, 100001. [CrossRef]
96. Turner, J.R.; Xue, Y. On the success of megaprojects. *Int. J. Manag. Proj. Bus.* **2018**, *11*, 783–805. [CrossRef]
97. Shenhar, A.J.; Holzmann, V. The Three Secrets of Megaproject Success: Clear Strategic Vision, Total Alignment, and Adapting to Complexity. *Proj. Manag. J.* **2017**, *48*, 29–46. [CrossRef]
98. Flyvbjerg, B. What you Should Know about Megaprojects and Why: An Overview. *Proj. Manag. J.* **2014**, *45*, 6–19. [CrossRef]
99. Müller, R.; Jugdev, K. Critical success factors in projects: Pinto, Slevin, and Prescott the elucidation of project success. *Int. J. Manag. Proj. Bus.* **2012**, *5*, 757–775. [CrossRef]
100. Blomquist, T.; Farashah, A.D.; Thomas, J.L. Project management self-efficacy as a predictor of project performance: Constructing and validating a domain-specific scale. *Int. J. Proj. Manag.* **2016**, *34*, 1417–1432. [CrossRef]
101. Osterloh, M.; Frey, B.S. Motivation, knowledge transfer, and organizational forms. *Organ. Sci.* **2000**, *11*, 538–550. [CrossRef]
102. Quigley, N.R.; Tesluk, P.E.; Locke, E.A.; Bartol, K.M. A multilevel investigation of the motivational mechanisms underlying knowledge sharing and performance. *Organ. Sci.* **2007**, *18*, 71–88. [CrossRef]

buildings

MDPI

Systematic Review

A Bibliometric Analysis of Supply Chain Management within Modular Integrated Construction in Complex Project Management

Yuhang Zhang [1], Geoffrey Qiping Shen [2] and Jin Xue [3,*]

[1] School of Management Science & Real Estate, Chongqing University, Chongqing 400044, China; yhzhang@stu.cqu.edu.cn
[2] Department of Building and Real Estate, The Hong Kong Polytechnic University, Hong Kong, China; bsqpshen@polyu.edu.hk
[3] School of Project Management, The University of Sydney, Sydney, NSW 2037, Australia
* Correspondence: jin.xue@sydney.edu.au

Abstract: As construction projects become increasingly complex, modular integrated construction (MiC) has emerged as a pivotal solution, driving integrated development in complex projects. However, the reliance on prefabricated modules underscores the crucial role of supply chain management (SCM) in MiC, necessitating strategic planning and operational control. This study aimed to use bibliometric analysis to map the SCM knowledge domain within MiC. Through the use of keywords related to "supply chain" and "MiC", 196 relevant papers were extracted from the Web of Science database. These papers were subjected to co-citation analysis, keyword co-occurrence analysis, and time span analysis to elucidate the historical evolution, multidisciplinary domains, and future directions in planning and control within SCM-MiC. The research identified two milestones in SCM-MiC's historical trajectory, enhancing our understanding of its foundations. Moreover, 11 clusters were identified, illustrating the multidisciplinary nature of SCM-MiC. Dividing the literature into seven stages of the supply chain, the research outlined four research directions aligned with project complexity and technological development, highlighting current hotspots and gaps of the strategic planning and control. These directions bridge the construction management and information technology domains, guiding future SCM-MiC research within complex project management.

Keywords: modular integrated construction; supply chain management; complex project management; bibliometric analysis

Citation: Zhang, Y.; Shen, G.Q.; Xue, J. A Bibliometric Analysis of Supply Chain Management within Modular Integrated Construction in Complex Project Management. *Buildings* **2024**, *14*, 1667. https://doi.org/10.3390/buildings14061667

Academic Editor: Jurgita Antucheviciene

Received: 29 April 2024
Revised: 21 May 2024
Accepted: 23 May 2024
Published: 5 June 2024

1. Introduction

The construction industry is currently experiencing significant evolution, fueled by the integration of pioneering methodologies aimed at addressing the escalating complexity and requirements of contemporary construction projects. Within this context, modular integrated construction (MiC) has emerged as an innovative construction methodology, attracting increasing interest for its potential to revolutionize conventional building practices [1]. MiC resembles prefabricated construction, in which a considerable part of the construction work is carried out offsite in a controlled factory setting and then transported to the building site for assembly [2]. This approach has been increasingly recognized for its potential to fundamentally change how projects are conceptualized, procured, designed, constructed, and managed [3]. The transition towards MiC reflects a broader shift in the construction sector towards more integrated and efficient processes, aligning closely with the principles of industrialization and sustainability [4].

However, the shift to MiC also imposes new demands on the planning and control mechanisms within the construction industry. Specifically, it underscores the crucial role of supply chain management (SCM) in the successful implementation of MiC strategies [5].

The prefabrication of modules at offsite facilities means that construction is no longer a predominantly site-based activity but is rather one that is spread across multiple locations and phases, integrating suppliers, manufacturers, and assembly processes seamlessly and efficiently [6]. Effective SCM is pivotal in maximizing the efficiency of MiC projects, as it addresses the complexities of logistics and management of materials, which are crucial for timely project execution and controlling costs. Therefore, a key factor that profoundly affects the success of MiC projects is the management of supply chains [3].

Existing research has recognized that the integral role of SCM within MiC (SCM-MiC) necessitates a thorough examination of how traditional supply chain strategies adapt to the modular construction environment [3,7]. Despite the burgeoning interest and increasing volume of literature on MiC, a significant gap persists: a comprehensive scholarly review that integrates these diverse insights into a unified framework is notably absent. Most current studies have focused on specific elements of SCM-MiC, such as logistical challenges or technological integration, failing to provide a holistic perspective on how these elements are interconnected and collectively influence overall delivery of the project [3,7]. This approach often results in an incomplete understanding of the evolution of the research, the existing gaps, and the broader implications of the topic. Therefore, the research question is as follows. What are the key trends, pivotal research contributions, and existing gaps in the field of supply chain management within modular integrated construction from a perspective of supply chain lifecycle?

To overcome the labor-intensive and potentially biased challenges inherent in traditional manual review methods [8], the use of bibliometric analysis has been proposed as an effective alternative [9]. Bibliometric techniques offer a quantitative approach to systematically identify the bibliographic relationships within a large volume of literature [10]. Tools such as Citespace and VOSviewer are instrumental in this regard, enabling researchers to visualize the knowledge network accurately and comprehensively [11,12]. These tools have the advantage of minimizing subjective interpretation by providing objective, data-driven insights into the literature's structure [13]. By using bibliometric analysis, researchers can achieve a more nuanced and interconnected view of the SCM-MiC field, facilitating a deeper understanding of its foundational and emerging themes.

In this research, we undertook a detailed bibliometric analysis to dissect the landscape of SCM in every step of the supply chain from design to demolition within MiC. The research aimed to reveal the evolving trends, seminal research, and notable gaps within the existing research in SCM-MiC, suggesting directions for future research. The article is structured as follows. First, the research background is described. The second part describes the development of the research methodology based on the perspectives of bibliometric analysis. On this basis, the filtered papers were all analyzed. Finally, the discussion and conclusion are presented.

2. Research Background

2.1. Modular Integrated Construction in Complex Projects

Traditional complex construction projects often encounter issues such as inefficiencies, cost overruns, and delays [14–16]. For instance, empirical studies have indicated that the average cost overrun for the development of infrastructure across Asian nations has surpassed 25% [17], while a survey of 130 public projects revealed that 106 projects (81.5%) experienced delays averaging over 289 days [18]. Modular integrated construction (MiC) is increasingly recognized for its innovation in construction, offering substantial improvements over traditional construction techniques [3]. MiC involves the prefabrication of building modules at an offsite location, which are then transported to the construction site for assembly. These modules can include structural elements, architectural finishes, and mechanical, electrical, and plumbing systems that are pre-installed in controlled factory settings [19]. Unlike the traditional approach, which depends heavily on on-site assembly, MiC involves the offsite prefabrication of standardized modules that are later assembled at the construction site [20]. This method effectively addresses many challenges of the construction industry in hospital construction during pandemics [16], construction of high-rise buildings [21], and so forth.

Numerous studies have highlighted the efficiency of MiC, primarily through reduced construction time. Gibb and Isack [22] provided evidence that MiC could cut projects' overall timelines by up to 50% due to the parallel progression of site work and production of the modules. Deep et al. [23] also noted significant time savings as being a critical factor in a project's delivery and crucial for complex projects where delays are costly. MiC facilitates enhanced quality control, as the components are manufactured in a controlled environment, reducing the variability seen in traditional construction methods [24]. Additionally, MiC demonstrates potential environmental benefits, including reduced waste and lower carbon emissions, emphasizing the sustainable nature of offsite construction practices [25]. While the initial costs can be higher due to the need for specialized designs and transportation, the project's overall costs may be reduced through decreased on-site labor, less waste, and a shorter project duration [26].

Despite its advantages, MiC faces challenges, particularly in logistical complexities and integration with traditional construction methods [8,27]. Transporting large prefabricated modules can present significant logistical challenges, especially in urban areas with restricted site access [28]. The complexity increases with the size and design intricacies of the modules. Integrating MiC with traditional construction methods poses challenges, especially in terms of coordination and management of the interface between modular and non-modular components [29]. These challenges of integration are heightened in complex projects that may require bespoke solutions. Regulatory frameworks often lag behind innovations in construction methods, presenting a barrier to the adoption of MiC. The lack of standardized codes specific to modular construction can hinder its implementation [30]. Moreover, market acceptance remains variable, influenced by conservatism in the industry and the novelty of the approach.

2.2. Supply Chain Management in Modular Integrated Construction

MiC offers significant benefits but encounters challenges that could limit its widespread adoption [31]. These challenges include the high initial investment costs, issues with scalability, regulatory obstacles, and the stakeholders' resistance [32]. Addressing these challenges effectively requires a comprehensive approach to supply chain management for modular integrated construction (SCM-MiC), which encompasses procurement, transportation, inventory management, and coordination among stakeholders, ensuring efficient flows of materials, components, and information [33]. Initially developed in manufacturing, SCM now encompasses procurement, transportation, and inventory management, and has proved essential for increasing productivity [34], minimizing waste [35], and creating value within construction projects [36]. The supply chain of construction is defined as the sequence of stages that the resources of construction (materials, equipment, and personnel) pass through in their entirety, from the points of supply to the construction site [37].

Critical to the effectiveness of SCM in construction are integration and collaboration among stakeholders [38]. The success of SCM in construction hinges on the stakeholders' integration and collaboration. The sector's supply chains necessitate immediate sharing of data and strong communication for effective coordination. Such cooperation, rooted in solid stakeholder relationships, markedly improves a project's outcomes by reducing lead times [39], shortening delivery periods [40], and enhancing efficiency [41]. Effective SCM-MiC is essential for overcoming obstacles, as it facilitates the on-time delivery of modules [42], optimizes inventory levels [43], minimizes transportation costs [44], etc. Moreover, leveraging technological innovations such as robotic automation and digital tools can greatly enhance the framework of SCM-MiC. Implementing sustainable SCM practices can also contribute to the environmental sustainability of MiC projects over the long term [15,45].

However, the existing research has identified specific SCM-related challenges within MiC projects. Poor management of MiC-SCM stands out as a significant barrier hindering the broader adoption of MiC [46]. This deficiency in management stems from ineffective planning of the supply chain that fails to integrate the stages of MiC-SCM [47], as well as from a lack of collaboration, trust, and real-time sharing of information among the stakeholders of MiC-SCM [48]. Consequently, this results in a weakly integrated, unresponsive, and suboptimal supply chain that undermines the advantages of MiC. These challenges are pivotal to the effective execution of MiC. For instance, transporting the modular components from the factories to the construction sites is often hindered by logistical issues such as traffic congestion and inadequate infrastructure, affecting the project's timelines and efficiency [49]. Effective coordination of delivery schedules and optimization of transportation routes are vital to circumvent these obstacles [42]. Moreover, achieving successful stakeholder collaboration requires detailed management and robust communication strategies [50]. Despite the recognized advantages of MiC and its logistic and SCM challenges, there is an evident need for further investigation in this area. Future research should aim to develop strategies that address the transportation challenges, improve coordination of the supply chain, and tailor MiC approaches to accommodate specific site conditions.

2.3. The Utilization of Bibliometric Analysis

Several reviews have utilized bibliometric analysis to conduct literature reviews within the field of MiC. For instance, Li et al. [51] conducted a bibliometric review focusing on the management issues in prefabricated construction, analyzing 100 articles. Similarly, Jin et al. [52] conducted a more extensive bibliometric analysis, examining 349 articles to decipher the interconnections among the keywords, as well as the collaborative networks among the research sources, scholars, and nations. Additionally, Han et al. [53] executed a bibliometric evaluation of 131 articles, engaging in qualitative discussions to delineate the prevailing trends and gaps within the SCM-MiC research area.

However, many reviews have often addressed the logistics and supply chain aspects of MiC projects in a fragmented manner. They may have narrowly focused on one aspect or concentrated excessively on specific topics such as such as policies [54], environmental and economic performance [55], stakeholders [56], and critical factors [57], losing sight of the broader context and failing to provide a comprehensive overview of the field. This fragmented approach has hindered the development of a coherent understanding of MiC and its associated challenges of supply chain management. Therefore, despite their contributions, these reviews fall short of providing insights into the distinct operational problems at each stage of the supply chain and how these challenges have been resolved. There remains a lack of in-depth systematic analysis of the modeling approaches and problem formulations across the stages of SCM-MiC, spanning design, manufacturing, procurement, transportation, inventory, and installation. Addressing these gaps would contribute to a more comprehensive understanding of SCM-MiC and guide future research in the field. In essence, bibliometric analysis offers a robust methodology for synthesizing and analyzing vast amounts of literature, providing valuable insights into the landscape of research into complex project management. By mapping out citation networks; identifying the key authors, institutions, and research themes; and uncovering emerging trends, bibliometric analysis serves as a powerful tool for understanding the evolution and current state of the field.

3. Research Methodology

3.1. Method of Bibliometric Analysis

Advancements in information technology and scientific visualization have established bibliometrics as essential for researchers studying trends in academic literature. This approach examines bibliographic data to reveal complex scholarly relationships. For example, co-citation analysis groups scientific articles based on their citations' connections and thematic similarities, revealing the main research communities and foundational studies [58]. Keyword co-occurrence analysis highlights the dominant research topics and methods, offering insights into current and emerging scholarly discussions [59]. Timespan analysis tracks the evolution of research themes, marking critical developments in academic discourse [60].

Tools such as Citespace, Citnet, Bibexcel, Sci2, and VantagePoint are crucial for conducting bibliometric analyses [61]. These tools help researchers visualize and statistically analyze the knowledge landscape. We chose Citespace (Version 6.1.R6, Manufactured by Chaomei Chen, Philadelphia, PA, USA) for several reasons: it offers comprehensive visualization capabilities and it excels in dynamic analysis for tracking the evolution of research topics [62]. Additionally, it integrates seamlessly with major bibliographic database such as Web of Science, ensuring efficient data retrieval. In construction management, bibliometric methods have proven effective in synthesizing and exploring extensive research landscapes [63,64].

3.2. Data Collection

3.2.1. The Database

This research leveraged the Web of Science (WoS) database for collecting high-quality bibliometric data. Recognized for its inclusion of over 12,000 reputable and influential academic journals, WoS aggregates content from the Science Citation Index (SCI) and the Social Science Citation Index (SSCI), offering a broad spectrum of high-quality data necessary for the scope and depth of bibliometric analysis envisioned in this study [12]. Its established use in previous studies highlights its effectiveness in providing detailed bibliographic records for statistical analysis [9]. The study specifically targeted peer-reviewed articles and reviews for their acknowledged academic rigor. This approach aligned with the methodologies of prior reviews by Xue et al. [63] and Derakhshan et al. [65], ensuring a focus on literature with recognized scholarly integrity.

3.2.2. The Search Terms

The search terms utilized in this study were structured in two components to comprehensively capture literature that was relevant to logistics, supply chain, and modular integrated construction. The first component included terms such as "logistics" and "supply chain", drawn from previous review studies, to ensure coverage of the relevant literature in these domains. The second component comprised project-related vocabulary, expanded to encompass various terms associated with modular construction, as detailed earlier in Section 2.2. A preliminary search was conducted for articles published between 2000 and 2022 to capture relevant literature over a significant timeframe.

The building and construction industry uses various alternative terms to describe concepts related to prefabrication. These terms, such as "industrialized buildings", "offsite construction", and "modular building", were included in the search strategy to ensure comprehensive coverage of the relevant literature. The formulation for retrieval of the literature was established as follows: ("logistics" OR "supply chain") AND ("modular integrated construction" OR "modular construction" OR "prefabricated building" OR "prefabricated construction" OR "prefabricated prefinished volumetric construction" OR "industrialized building" OR "industrialized construction" OR "offsite building" OR "offsite construction" OR "precast construction" OR "precast building"). The initial search yielded 271 articles.

The WoS core database, renowned for its credibility, served as the primary source for collecting the relevant literature. Review articles and other irrelevant publications were filtered out according to predetermined criteria. Only articles exclusively linked to the WoS core were retained after this process. Next, publications lacking specific keywords in their titles or abstracts were excluded, followed by a manual review of the titles and abstracts to remove less relevant literature swiftly. The remaining records were then organized for subsequent bibliometric analysis. As a result, 196 articles were deemed eligible for further analysis in the subsequent phase. The diagram of the PRISMA process is shown in Figure 1.

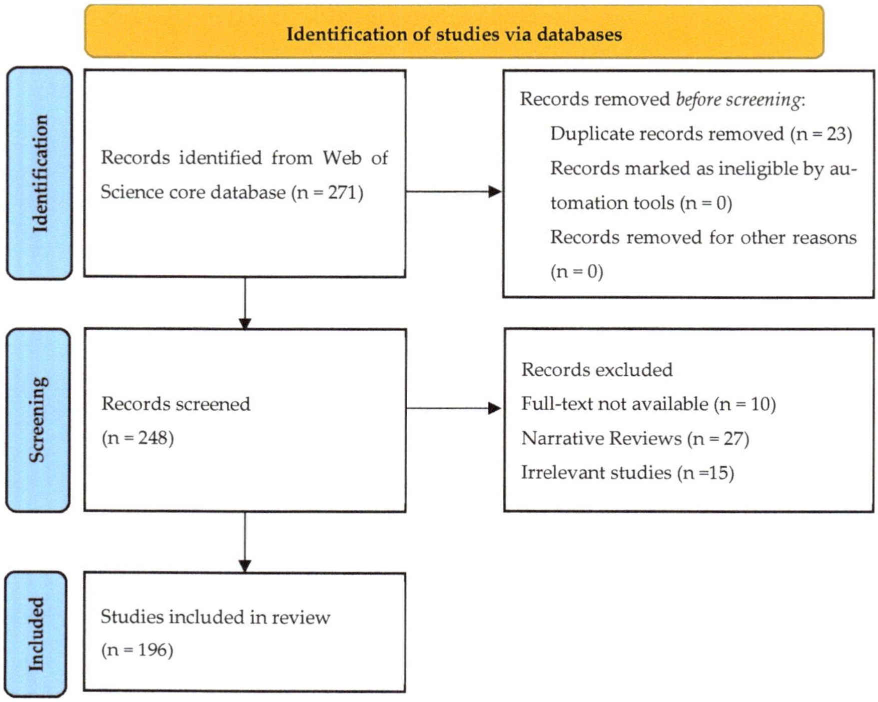

Figure 1. Diagram of the PRISMA process.

3.2.3. Bibliometric Records

Each bibliographic record contains the metadata of a published article, including the authors' names, the articles' titles, abstracts, keywords, volume numbers, DOI references, and all references cited within the articles.

3.3. *Data Analysis*

3.3.1. Co-Citation Analysis

Document co-citation analysis involves tracking articles that are often cited together, signaling shared research interests [66]. This method connects articles in a co-citation network, treating each as a node. The connections or links between nodes reveal the relationships and group articles into clusters, identifying distinct knowledge domains and specific research topics. Visualizing these networks helps elucidate the structure of the knowledge domains and the interconnectedness of articles.

3.3.2. Keyword Co-Occurrence Analysis

Co-occurrence analysis determines the prevalence of specific keywords within the dataset, indicating their relevance in the research area [13]. This analysis reveals the core concepts and themes within the research domain, showing what topics are the most frequently explored [67]. By identifying these patterns, researchers can discern the primary focus areas and conceptual underpinnings of the literature.

3.3.3. Time-Span Citation Analysis

Time-span citation analysis identifies pivotal articles across different periods, reflecting the evolution of a research topic [63]. It involves counting citations to highlight frequently cited works within specific timeframes. Tracking these citations over time reveals shifts in influential articles, indicating how research trends and seminal works have developed, thus offering insights into the dynamic nature of the research domains.

4. Results

4.1. Results of Data Collection

The literature review conducted on research into logistics and supply chain management within the construction industry, particularly modular integrated construction (MiC), yielded valuable insights into the developmental trends of the related research. Figure 2 illustrates the progression of SCM-MiC research from 2000 to 2023, based on data obtained from the sample in this study. In general, the quantity of published articles demonstrated a pattern of gradual then accelerated expansion. Consequently, the timeframe of publication for the studies can be divided into three distinct phases: an initial stage, a phase of fluctuating growth, and a phase of rapid development.

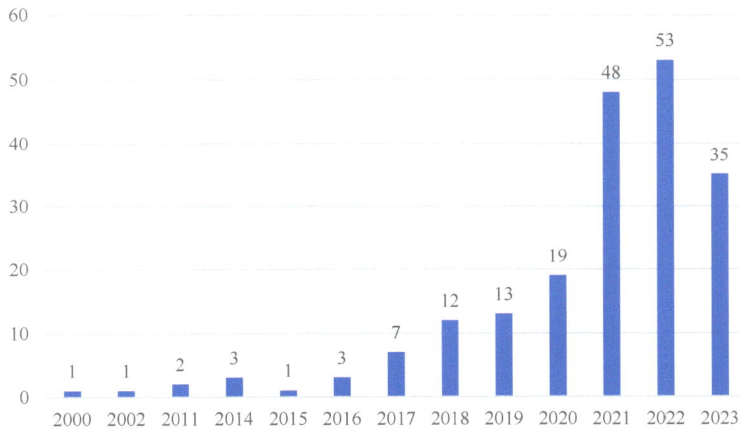

Figure 2. Year of publication.

(1) Initial stage (2000–2010): Before 2010, there was a relatively modest number of published articles. Much of the literature on MiC comprised policy recommendations for program and project management, with limited academic exploration of supply chain management in MiC.
(2) Fluctuating growth phase (2011–2020): During this period, there was a gradual increase in the number of publications. As construction standards evolved and international competition intensified, MiC began to have greater significance. Many researchers recognized the need to integrate supply chain activities to optimize MiC activities.
(3) Rapid development phase (2021–present): The number of published articles experienced rapid growth, peaking in 2022. Compared with 2021, the volume of literature in 2022 surged by over 150%, indicating a growing emphasis on the study of SCM-MiC worldwide.

4.2. Knowledge Domains

The provided text explains how the literature's co-citation network was organized into clusters based on the interconnections between the nodes. These clusters were then labeled using a statistical technique called the LLR (log-likelihood ratio) test. This method is a statistical technique utilized in the analysis of co-citation networks to determine the significance of words or terms associated with clusters [58]. A higher LLR value indicates that the observed frequency of a term within a cluster is significantly different from what would be expected by chance alone, suggesting that the term is meaningful or characteristic of that cluster [68]. By applying the LLR method, clusters in co-citation networks can be labeled with terms that are statistically significant and representative of the content within each cluster.

Here, we discuss the outcomes of clustering analysis conducted in the field of logistics in modular integrated construction (MiC). Table 1 presents a comprehensive list of 10 significant clusters, each containing at least five group members. Notably, the silhouette value for each cluster exceeded 0.7, indicating the reliability of the results for further interpretation.

Table 1. Top-ranked clusters.

Cluster ID	Size	Silhouette Value	Mean Year	Cluster Name	Category
0	49	0.837	2018	Social network analysis	Methods
1	49	0.712	2018	Optimization	Directions
2	48	1	2012	Prospects and challenges	Challenges
3	42	0.76	2017	Game theory	Methods
4	26	1	2011	Factors of critical assessment	Challenges
5	19	1	2012	Management of the products' lifecycle	Directions
6	14	0.913	2014	Mitigation strategies	Methods
7	8	0.935	2013	Decision support systems	Methods
8	6	1	2010	Building an information model	Directions
9	5	0.977	2020	Supply chain's capabilities	Directions

The results highlighted the most substantial cluster, focusing on stakeholders (Cluster 0), which included 49 publications. This emphasizes the importance of collaboration within logistics in construction, with social network analysis being the most prevalent method studied. Additionally, the results revealed other significant clusters such as Cluster 3, focusing on game theory and supply chain management, and Cluster 2, which included theoretical studies on prospects and challenges of the process of logistics. Furthermore, clusters such as Cluster 4 (factors of critical assessment) and Cluster 9 (the supply chain's capability) address challenges that may arise in the process of logistics, highlighting the importance of management within logistics in MiC (Figure 3).

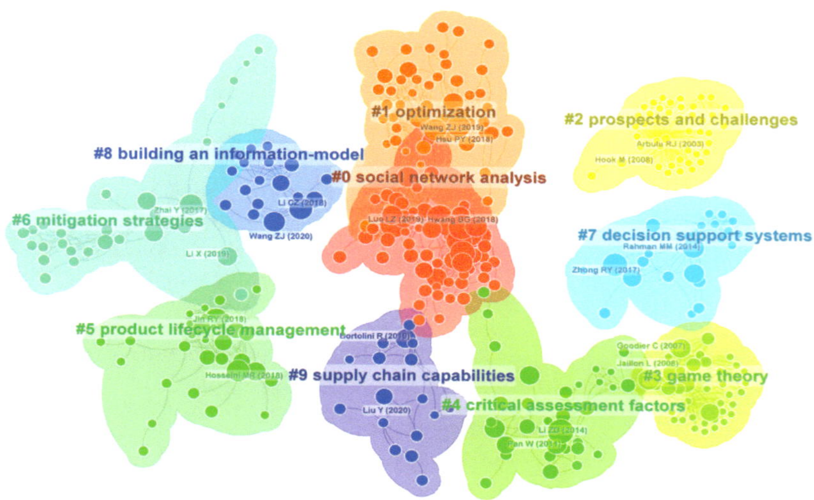

Figure 3. Clusters of knowledge domains within the related research. Luo L.Z. (2019) [50]; Hwang B.G. (2018) [69]; Hsu P.Y. (2018) [70]; Wang Z.J. (2019) [71]; Arbulu R.J. (2003) [72]; Hook M. (2008) [73]; Jaillon L. (2008) [74]; Goodier C (2007) [75]; Li Z.D. (2014) [76]; Pan W. (2011) [77]; Jin R.Y. (2018) [52]; Hosseini M.R. (2018) [78]; Zhai Y. (2017) [39]; Li X. (2019) [79]; Rahman M.M. (2014) [80]; Zhong R.Y. (2017) [81]; Li C.Z. (2018) [82]; Wang Z.J. (2020) [48]; Bortolini R. (2019) [83]; Liu Y. (2020) [84].

4.3. Knowledge Evolution

Here, we present the results of the time-span analysis, depicted in Figure 4, which revealed a distinct division of the evolution of knowledge into two stages on the basis of the timeline. In the initial stage, predating 2015, research and scholarly contributions regarding logistics in modular integrated construction (MiC) were emerging but limited. Attention and investigation in this field were relatively modest, with fewer publications and scholarly discussions dedicated to logistics-specific issues within the context of construction.

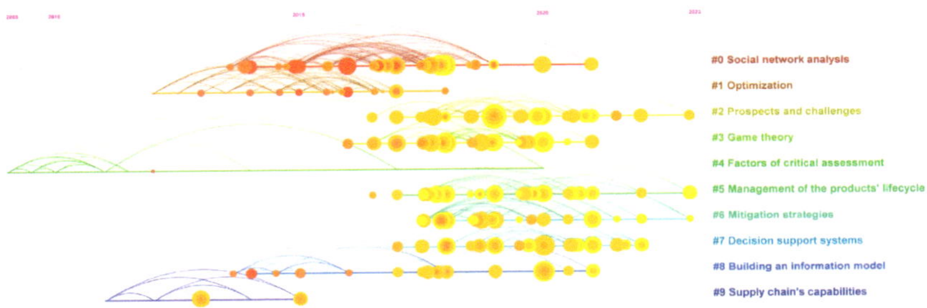

Figure 4. Timeline of research domains.

However, the second stage, spanning from 2016 to the present, witnessed significant growth and development in the knowledge base concerning logistics in MiC. Researchers and industry practitioners increasingly recognized the significance of logistics in project delivery, supply chain management, and handling of materials. Consequently, there was a surge in the number of publications, research studies, and academic contributions during this period.

Furthermore, the results highlighted a shift in the focus of research between the two stages. In the first stage, there was more emphasis on the technical methods of logistics in MiC, whereas in the second stage, the focus of research shifted towards the capability and resilience of the supply chain. This change underscores the evolution of priorities and interests within the domain of research, suggesting that relevant research in this field will continue to gain attention.

4.4. Knowledge Frontiers

This section provides a summary of the findings derived from the keyword co-occurrence analysis spanning from 2010 to 2023, which is visually represented in Figure 5. In this analysis, the most frequently occurring keywords are depicted with larger characters, revealing three primary research aspects. First, the research delves into supply chain management, decision-making processes, blockchain technology, management of demolition waste, and optimization of the supply chain. Second, it focuses on research targets such as prefabricated buildings, modular construction, and offsite construction methodologies. Third, it explores research methods including model research, social network analysis, and case studies.

Additionally, the results revealed recent strong citation bursts depicted in Figure 5, indicating emerging trends in research into supply chains and logistics within MiC. These trends encompass the integration of artificial intelligence and machine learning techniques, the adoption of building-model technology for enhanced traceability, the process of reverse logistics, and exploration of the principles of the circular economy in construction supply chains. The literature review underscored the importance of future research endeavors centered on evaluating the practical implementation of these innovations and assessing their impact on the supply chain's performance within the context of MiC.

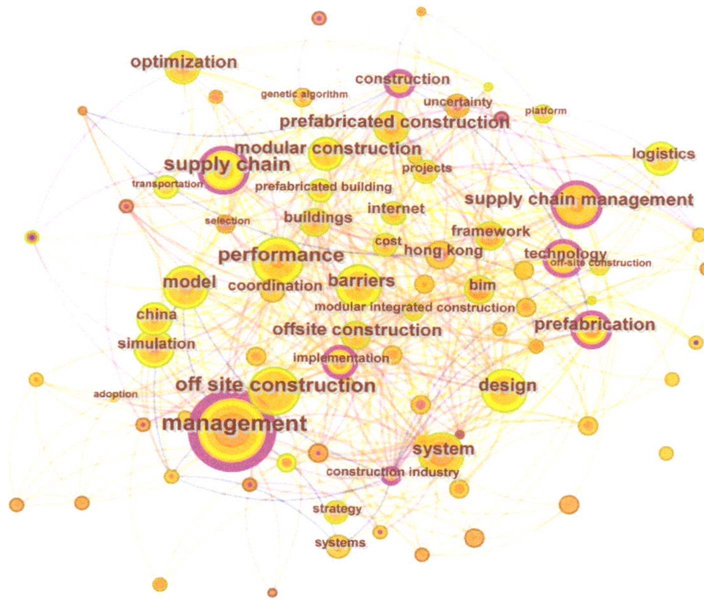

Figure 5. Network of keywords' co-occurrence.

5. Discussion and Implications

5.1. Milestones in the Development of SCM-MiC

The chronological markers identified through the timespan analysis, as depicted in Figure 4, serve as crucial indicators of the progress made in SCM-MiC projects. Initially, the focus of research was on fundamental principles and methodologies of supply chain management [85,86]. This foundational work established the basic framework for understanding and managing supply chains in modular integrated construction (MiC). Over time, there has been a clear progression towards the application of advanced technologies and theories across various sub-domains, such as digitalization, automation, and sustainable practices [4,8]. This trend showed the evolution of SCM-MiC from the exploration of basic theories to the integration of multidisciplinary fields within the construction industry. The adoption of technologies such as building information modeling (BIM) and the internet of things (IoT) has been particularly noteworthy, marking significant milestones in enhancing efficiency and collaboration in SCM-MiC projects [87,88]. These chronological markers not only highlight the key developments in SCM-MiC but also reflect the dynamic nature of this field, driven by continuous innovation and interdisciplinary integration.

5.1.1. Birth Stage: Concept and Method of SCM-MiC

The birth stage of SCM-MiC marked the initial forays into fundamental applications and methodologies within the field. During this phase, researchers focused on exploring the potential of leveraging the supply chain to optimize the efficiency of construction, although their understanding of SCM in modular building processes was still developing. A significant milestone during this period is the work of Doran and Giannakis [86], titled "An examination of a modular supply chain: a construction sector perspective". This seminal article delved into modular practices and principles within the construction sector's supply chain, aiming to elucidate the alignment between supply chain practices and principles of modular construction.

Additionally, research during this stage delved into the overarching prospects and challenges inherent in SCM-MiC, such as the supply chain's transparency [89], deeper explorations of the challenges [90], and analyses of the major barriers [91]. Moreover, this stage witnessed an exploration of management methodologies and emerging technologies in supply-chain-related research, aiming to enhance the efficiency of construction. Projects focused on management of the projects' lifecycle (Cluster 5) and building an information model (Cluster 8) aligned closely with the temporal aspects of the timespan analysis. For instance, the study by O'Connor et al. [92] provided insights into the characteristics and benefits of modular projects with standard designs, while the research by Arashpour et al. [93] optimized decision-making processes in offsite construction networks, ultimately aiming to eliminate schedule delays and increase efficiency in modular construction practices. Arashpour et al. [94] built an autonomous production tracking method to eliminate schedule delays and the increased costs of deficiencies in offsite construction.

5.1.2. Development Stage: The Application of Technology and the Focus of SCM-MiC

After the year 2015, the landscape of research within SCM-MiC experienced a significant expansion, marking the onset of a growth phase. This period was characterized by a concerted effort to address the complex challenges posed by the multifaceted aspects of supply chain dynamics within the MiC domain. While previous research predominantly focused on the risks encountered during the construction stage of MiC projects, a comprehensive assessment revealed that schedule-related risks permeated modular integrated construction across the entire supply chain. Hsu et al. [8] highlighted this observation, emphasizing that risks manifested at every stage, from design and manufacturing to logistics and on-site assembly. In response to this understanding, Li et al. [95] delved into stakeholder-related risks, exploring their intricate cause-and-effect relationships. Their work aimed to uncover the network of potential risk factors associated with stakeholders, which significantly impact the complex framework of MiC projects. This perspective was further supported by Luo et al. [50], who meticulously identified and analyzed these risk factors, underscoring their pivotal role within the MiC landscape. Similarly, the contribution of Arshad and Zayed [7] offered valuable insights into the challenges arising from the complex supply chain and the sensitive logistical processes inherent in MiC projects.

Following the exploration of the supply chain's risks within the context of the MiC approach, researchers shifted their focus towards examining the supply chain's resilience, vulnerability, and capability, which represent critical dimensions contributing significantly to the robustness and effectiveness of SCM-MiC [96,97]. In light of the complex risks inherent in MiC projects, there was an increasing emphasis on the supply chain's resilience, which involves equipping supply chains with the ability to anticipate, withstand, and recover from disruptions such as manufacturing delays, transportation issues, or unforeseen design changes [98,99]. In addition to the supply chain's resilience, assessments of vulnerability played a vital role. Researchers analyzed potential weak points in MiC supply chains that were susceptible to disruptions such as material shortages, breakdowns of transportation, or regulatory shifts to develop strategies to mitigate these vulnerabilities [100]. Furthermore, the supply chain's capability, referring to the inherent ability of a supply chain to efficiently deliver products or services, garnered attention from researchers. They sought to identify factors that enhance the supply chain's capabilities, including accurate predictions of demand, streamlined stakeholder coordination, and optimized inventory management strategies [40].

The prevailing themes of this research direction, which included optimization, simulation, scheduling, and decision-making, collectively underscored the commitment to enhancing the efficiency, effectiveness, and overall performance of projects within the subject's domain. Researchers incorporated diverse approaches such as social network analysis (SNA), fuzzy logic techniques, and game theory, among others. SNA highlights stakeholder-associated risks, interactions, and challenges, aiming to provide a structured framework for effectively managing these risks within MiC projects [101,102]. Additionally, researchers leveraged fuzzy logic approaches to discern a comprehensive array of the critical factors [90,103,104]. Moreover, game theory is widely utilized in optimization and simulation scenarios for modeling strategic decision-making among interdependent participants [53,105,106]. In this context, the research methodology centers around the identification and examination of these pivotal factors, constituting a holistic approach to understanding and enhancing the performance of the supply chain within the realm of MiC projects.

5.2. *Major Knowledge Domains of SCM-MiC*

The co-citation analysis identified three key dimensions within SCM-MiC: strategies, topics, and challenges. SCM strategies encompass methodologies such as social network analysis, game theory, and decision support systems, which are crucial for understanding the stakeholders' interactions and navigating complex decision-making processes. In today's global and risk-prone supply chains, these strategies facilitate agile responses to disruptions [107,108]. Additionally, focusing on the refinement of processes, leveraging information models, and coordinating the stages of the products' lifecycle underscore the new demands on the planning and control mechanisms of SCM. Integration of technological advancements such as AI and big data analytics is imperative for optimizing supply chains [109,110]. Addressing the challenges of SCM involves grasping the industry's trends, overcoming obstacles, and aligning SCM-MiC initiatives with organizational goals through the factors of critical assessment.

5.2.1. Directions of Supply Chain Management

In the evolving landscape of SCM-MiC, the importance of sophisticated planning and control mechanisms cannot be overstated. These complexities are reflected in the analytical methodologies, which are distributed across four significant clusters, highlighting critical areas such as optimization, management of the products' lifecycle, modeling information, and enhancement of the supply chain's capability. However, uncertainties and risk factors inherent in any link can significantly impact the overall operation of the supply chain [111]. Analytical methodologies in SCM-MiC, represented by four clusters, underscore the importance of improvements in the supply chain [112,113].

Firstly, optimization (Cluster 2 in Table 1) of MiC projects is crucial. Optimization of the supply chain aims to streamline the processes to achieve the best possible outcomes, balancing cost, speed, quality, and sustainability. Utilizing data analysis, mathematical modeling, and advanced algorithms, optimization techniques enhance forecasts of demand, management of the inventory, scheduling production, routing transportation, and resource allocation [32,83,114].

Secondly, management of the products' lifecycle (Cluster 5 in Table 1) plays a significant role. Studying the lifecycle of the production system provides valuable insights and benefits for organizations involved in manufacturing and business operations. In supply chain management, the lifecycle of the production system ensures that products are designed with considerations for manufacturability, sustainability, and end of life [115]. By facilitating collaboration and information-sharing, it reduces the time to market, minimizes defects in the products, and enhances the supply chain's overall efficiency [116,117].

Thirdly, establishing a comprehensive information model (Cluster 8 in Table 1) is critical for effective supply chain management. A robust information model provides transparency in the supply chain, enabling better decision-making, faster resolution of issues, and improved collaboration among stakeholders [32,83]. And finally, the supply chain's capability, introduced in Section 5.1.2, is a vital area of research. Enhancing the supply chain's capabilities involves investment in technology, talent, and processes to meet customers' demands and strategic objectives. A resilient and agile supply chain with strong capabilities can navigate disruptions, seize opportunities, and consistently deliver value to customers.

5.2.2. Methods of Supply Chain Management

In the realm of SCM-MiC, the evolving landscape demands a profound re-evaluation of the planning and control mechanisms to navigate the intricacies of the stakeholders' dynamics, strategic decision-making, and optimization of the supply chain. Among these, social network analysis (Cluster 0 in Table 1) stands out as a major knowledge domain. Recognizing stakeholders' participation as a critical driving force stimulating the demand for SCM-MiC studies (as stated in Section 5.1), numerous discussions have focused on effectively managing a project's stakeholders [50,101]. Given that construction projects often attract significant public attention due to their use of public funds and profound social impact, it is imperative to identify the stakeholders' concerns clearly to detect the key challenges and associated risks of the stakeholders [63,114].

In addition to social network analysis, game theory (Cluster 3 in Table 1) has emerged as another frequently utilized simulation method. Game theory has proven invaluable for identifying suitable suppliers of various modular components and facilitating contract negotiations that align the goals of different suppliers, contractors, and stakeholders [32]. Moreover, the social equation model (SEM) has also been used to demonstrate that integrating different aspects of the supply chain positively impacts the operational performance, thereby improving the financial performance of the companies involved in the process of integration [118].

In advanced MiC, making robust supply decisions is vital. As the construction industry becomes more industrialized and digital, a decision support system (Clusters 7 in Table 1) utilizing ontology and multiagent methods is recommended for SCM-MiC. Through mathematical analysis and evaluations of real projects, this approach enhances decision-making in the manufacturing of prefabricated product [119,120].

5.2.3. Topics of Supply Chain Management

Supply chain management in modular integrated construction presents several unique challenges due to the combination of the attributes of manufacturing and construction, involving the offsite manufacturing of building components and their subsequent on-site assembly. These challenges span every aspect of the supply chain, with numerous studies discussing potential solutions. The overarching goal is to find effective strategies for addressing the challenges related to capability, resilience, vulnerability, and efficiency.

Identifying critical factors (Cluster 4 in Table 1) is paramount, including the optimization of scheduling, decision-making, and transportation [119,121,122]. Additionally, one study modeled the dynamic impact of the supply chains' vulnerabilities and capabilities on the supply chains [98]. Wang et al. [123] systematically analyzed risks to the supply chain and calculated the risks' balance point, concluding that risks can persist in the supply chain for extended periods.

Delving into the prospects and challenges (Cluster 2 in Table 1), delays in the schedule of construction have emerged as a fundamental challenge. These delays can stem from various factors and have far-reaching consequences for all stakeholders involved [8,124]. Additionally, accurately predicting the lead times for the manufacturing and transportation of modular components poses a challenge, given the coordination required among various processes and the potential disruptions impacting delivery schedules [20]. While modular construction offers potential cost savings, effectively managing the costs across the supply chain necessitates meticulous monitoring and control, especially considering potential variations in the costs of materials and the efficiency of production [125]. Implementing circular economy principles, such as reusing and recycling materials, within the supply chain of modular integrated construction can be complex. Designing components with future repurposing or recycling in mind can challenge the traditional thinking of a linear supply chain [126]. Moreover, concrete operational challenges persist. Adapting modular components to specific on-site conditions can be operationally challenging. Ensuring a seamless fit with the existing infrastructure and making the necessary adjustments efficiently are essential for smooth assembly [127].

5.3. *Trends in the Supply Chain Management in MiC*

Utilizing co-occurrence network analysis from 2011 to 2023 as a foundational framework, this study delved into the focal points at the cutting edge of SCM-MiC, as illustrated in Figure 5. The forthcoming changes in SCM-MiC align with the lifecycle processes and new technologies commonly found in modern construction projects. We compiled the essential co-occurrence keywords into a two-dimensional classification matrix, distinguishing between the SCM-MiC process and the main research domains of SCM-MiC summarized above. By leveraging such analytical tools, researchers and practitioners can pinpoint the vital intersections between emerging technologies and the lifecycle processes of complex construction projects.

Regarding the lifecycle of construction processes, Figure 6 depicts the distribution of studies concerning the stages of construction and their research emphases. The stages of the supply chain encompass the overarching stages of the supply chains' research, design, production, procurement, transportation, inventory, installation, and demolition. Predominant research directions center around optimization, simulation, scheduling, and decision-making. These investigations are often underpinned by specific methods, including building information modeling (BIM), artificial intelligence (AI), blockchain, the internet of things (IoT), and radio frequency identification (RFID). A color map is utilized to better visualize these matrices. Each graph is supplemented with a two-dimensional polynomial trend line to demonstrate the accumulation of knowledge in each of these research areas. Therefore, these figures provide an overview of mainstream research on the modeling of SCM-MiC and can guide researchers in identifying areas that need further study.

Research focus \ Stages of the supply chain		Design	Production	Procurement	Transportation	Inventory	installation	Demolition and reuse	Overall counts
Directions	Optimization	27	35	3	18	5	5	1	94
	Simulation	15	19	1	14	4	6	1	60
	Scheduling	6	18	2	13	0	2	1	42
	Decision-making	20	20	2	6	3	2	0	53
Methods	BIM	28	18	3	6	2	4	0	61
	RFID	1	1	0	3	2	0	0	7
	Blockchain	6	3	4	0	0	0	0	13
	IoT	1	1	0	0	0	0	0	2
Overall counts		141	150	19	70	20	22	3	

Figure 6. Research trends and 2D classification matrix of SCM-MiC based on stages of the supply chain and solution methods. Note: Red color indicates there are 0 relevant papers; orange color indicates 1–4 relevant papers; yellow color indicates 5–9 relevant papers; light green color indicates 10–20 relevant papers; deep green color indicates more than 20 relevant papers.

5.3.1. Specific Research into the Stages of SCM-MiC

According to Figure 6, it is evident that the directions of SCM-MiC, including optimization, simulation, scheduling, and decision-making, are closely related to the field of project planning and control. These terms are commonly used to enhance processes, make efficient decisions, and optimize various aspects of systems. However, the distribution of the stages of research stages appears to be uneven, with earlier steps exhibiting higher article counts.

Examining the supply chain process depicted in Figure 6, prior researchers have primarily focused on addressing challenges at the stages of design (104 studies involved in total), production (115 studies involved in total), and transportation (60 studies involved in total). Design choices made during the initial phases of MiC projects significantly influence the effectiveness of the subsequent stages of the supply chain. Consequently, fostering close collaboration among design teams across diverse disciplines becomes imperative [20]. Furthermore, the research trend aimed to simplify uncertainties in the process of production related to goals and methods, to resolve conflicts between different industries on-site, and to address interdependencies among the supply chain's members [128]. Moreover, the intricate logistics of transporting materials and equipment procured from the suppliers to designated construction sites necessitate meticulous planning to optimize route selection, modes of transportation, and scheduling [122].

The number of research articles focusing on procurement (15 studies involved in total), inventory management (16 studies involved in total), installation procedures (19 studies involved in total), and demolition strategies (3 studies involved in total) is relatively limited. Effective procurement strategies demand a thorough evaluation of the critical factors such as quality, cost, availability, and lead times [129]. Establishing collaborative partnerships with suppliers is crucial to ensure reliable sources and negotiate favorable terms [130]. Efficient inventory management is essential for the seamless functioning of the supply chain, ensuring timely access to the necessary materials and preventing disruptions caused by shortages or excess inventory [70]. Technological advancements play a vital role in enhancing the precision of inventory management, with specialized software tailored to the construction industry and real-time tracking systems improving the efficacy of oversight [131]. Addressing the complexities of installing large components and promoting streamlined collaboration in lean construction requires a focus on installation methodologies and the integration of information technology solutions [83]. Additionally, as modular integrated buildings continue to develop, attention to the management of demolition becomes imperative, considering the entire lifecycle of the building. Scholars are increasingly exploring demolition strategies and the reuse of components for specific buildings, highlighting the importance of sustainability in construction practices [132]. As research in SCM-MiC progresses, addressing the lacunae in these understudied areas will be crucial for advancing the field's understanding and enhancing its practical applications.

5.3.2. Improvements in the Directions of SCM-MiC

In terms of the research directions, planning and optimization of the control of complex projects has received the most attention from previous researchers in SCM-MiC, followed by simulation, decision-making, and scheduling. This trend reflects a concerted effort to enhance the operations and management of the supply chain's processes within the context of modular integrated construction. This indicates a sustained interest and recognition of the importance of optimizing the supply chain's processes, simulating various scenarios, making informed decisions, and scheduling activities efficiently to overcome the challenges and complexities inherent in MiC projects.

In the realm of SCM-MiC, research into optimization is directed towards identifying optimal decisions and strategies across various aspects such as resource constraints, supply arrangements, and resource allocation. Researchers have used mathematical models, algorithms, and computational tools to optimize factors such as the cost, time, resource utilization, and quality, with the overarching goal of minimizing waste, maximizing resource utilization, and enhancing the overall performance of the supply chain. In recent years, there has been a notable emphasis on optimizing the processes of fitting and decision-making. Traditionally, managing the dimensional and geometric variability in MiC involved trial and error approaches and standardized tolerance values, which often led to the risk of reworking [46]. However, the adoption of decision-making and optimization approaches in SCM-MiC has introduced considerations of uncertainty [119]. For instance, techniques such as fuzzy set theory and the analytic hierarchy process (AHP) have been utilized to capture uncertainty effectively [133]. Moreover, uncertainty analysis, aimed at supporting optimization efforts, has been facilitated by various methods such as multi-attribute utility theory [134], media richness theory [135], discrete event simulation experiments [6], and cloud models [136]. The integration of uncertainty analysis has substantially enriched the field of SCM-MiC, highlighting the importance of addressing uncertainties in decision-making and the processes of optimization.

Based on the outcomes derived from numerical simulation, a set of managerial implications has been formulated to drive the advancement of MiC. These recommendations encompass strategies for widespread awareness, obligatory enforcement, enhanced integration of the industrial chain, and more robust guidance for the factories making the components [137]. To address environmental concerns, innovative scheduling methods for on-time production of precast components have been produced, leveraging integrated simulation–optimization algorithms, with the differential evolution–simulation approach standing out for effectively minimizing total weighted earliness and tardiness [138,139]. Numerous methods have been used to simulate scenarios of supply chain management in modular integrated construction, including approaches such as game theory and genetic algorithms. Effective coordination and timing of operations throughout the supply chain of offsite construction are crucial for project success. An orderly framework is presented to manage the supply chain, covering strategic planning, master planning and scheduling, and detailed planning and scheduling [140]. Additionally, a distinct approach involving integrated scheduling of offsite logistics for high-rise modular building projects has been suggested. This approach effectively addresses project-specific constraints and uncertainties, enhancing the project's overall efficiency and performance [47].

5.3.3. Generalization of Research Methods in SCM-MiC

The most frequent solution methods used to solve the planning and control problems of the different stages of the SCM-MiC are BIM, blockchain, the IoT, and RFID. During ongoing technological progress, BIM plays a crucial role in navigating complex issues within MiC projects [83]. Functioning as an information exchange and dissemination platform among MiC projects, building information modeling (BIM) presents a novel avenue for the accurate and instantaneous collection of stakeholders' information. A promising avenue is its integration with internet of things (IoT) technology [15,141]. In instances such as MiC projects, the combination of BIM models with RFID labels has been used to share stakeholders' information, leading to enhanced schedule performance [142,143].

Additionally, the integration of blockchain technology holds promise for enhancing transparency, traceability, and trust within construction supply chains. Mechanisms such as smart contracts and secure data sharing streamline contractual procedures and mitigate conflicts, thereby promoting operational efficiency [48].

Looking forward to the Industrial Revolution 4.0, there is an expectation of the emergence of more advanced integrated information tools that will bolster SCM-MiC's capacity to address sustainability, complexity, and uncertainty in construction projects [144]. In the current research landscape, there is a notable trend towards exploring and adopting advanced digital solutions to enhance efficiency, collaboration, and decision-making throughout the process of construction. Artificial intelligence (AI) stands out as a powerful tool with diverse capabilities, although its utilization in the construction field remains relatively limited compared with other sectors. AI holds significant potential, particularly in complex construction projects such as MiC, where it can accurately predict events, risks, and costs during the planning stage, leading to improved project outcomes and streamlined processes [145]. There is ample opportunity for further research into the applications of AI in SCM-MiC to unlock its full potential for driving innovation.

5.3.4. Research on Supply Chain Systems in MiC

After delving into the risks that supply chains face in the context of complex MiC projects, researchers have shifted their focus to explore four critical aspects: the supply chain's resilience, the supply chain's vulnerability, the supply chain's capability, and the supply chain's efficiency [146–148]. This line of research has predominantly focused on analyzing the entire system within SCM-MiC, aiming to understand its intricate workings and interactions (Figure 7).

Research methods \ Topics	Capability	Resilience	Vulnerability	Efficiency
BIM	2	4	0	7
RFID	1	0	0	2
Blockchain	0	1	0	1
loT	0	0	0	0
Game theory	1	1	1	3
Social network analysis	1	4	1	4
Genetic algorithm	0	0	0	2
Fuzzy synthetic evaluation	3	5	1	4
Overall	8	15	3	23

Figure 7. Quantitative analysis of research methods within supply chains' systems in MiC.

The most prominent area of research revolves around enhancing the supply chain's efficiency. The methodologies that have been used are diverse, with BIM being extensively utilized to amplify productivity, refine efficiency, and optimize effectiveness [149]. Additionally, there has been a growing tendency to enhance stakeholders' relationships and pinpoint the essential factors for boosting the efficiency of SCM-MiC [150,151].

The supply chain's capability and resilience have been less important research topics in research into the supply chain's system. Recognizing the intricate risks embedded in MiC projects, considerable attention is being placed on enhancing the supply chain's resilience. This entails preparing supply chains to anticipate, endure, and recover from disruptions such as delays in manufacturing, transportation glitches, or sudden design modifications. Furthermore, researchers are delving into the concept of the supply chain's capability, which pertains to its inherent capacity to efficiently provide products or services. The objective here is to uncover factors that enhance the capabilities of supply chains. This encompasses areas such as accurate predictions of demand, seamless coordination among stakeholders, and optimized management of the inventory.

Building upon an understanding of the supply chain's risks, the assessment of vulnerability has emerged as a crucial step. Researchers have turned to finding the underlying components of vulnerabilities and developing a model to explore the correlational impacts of vulnerabilities [40,104,151]. More researchers are conducting analyses to identify potential weak points within MiC supply chains that are susceptible to disturbances such as shortages of materials, breakdowns in transportation, or shifts in regulations [100].

5.4. A Knowledge Map of Supply Chain Management in MiC

The depiction of the knowledge landscape within SCM-MiC, illustrated in Figure 8, resembles a residential model, representing its foundational, structural, and visionary components. The base represents the historical evolution of knowledge of SCM-MiC, providing insights into its classical underpinnings through pivotal milestones. These milestones serve as reference points for scholars, offering a deeper understanding of the field's trajectory over time.

The base part of the graphic is composed of co-occurring keywords, which are mainly categorized as related to construction, planning and control, and methodology. The keywords in bold are the most frequent ones in each category. Anchored upon this foundation are three key aspects, forming 10 distinct clusters that shape the knowledge domain. These clusters act as foundational pillars, representing the multidisciplinary nature of SCM-MiC research and its correlation with contemporary stakeholder studies. This structural analogy provides scholars with a novel perspective on the existing research landscape within SCM-MiC, facilitating a comprehensive understanding of its multidimensional aspects.

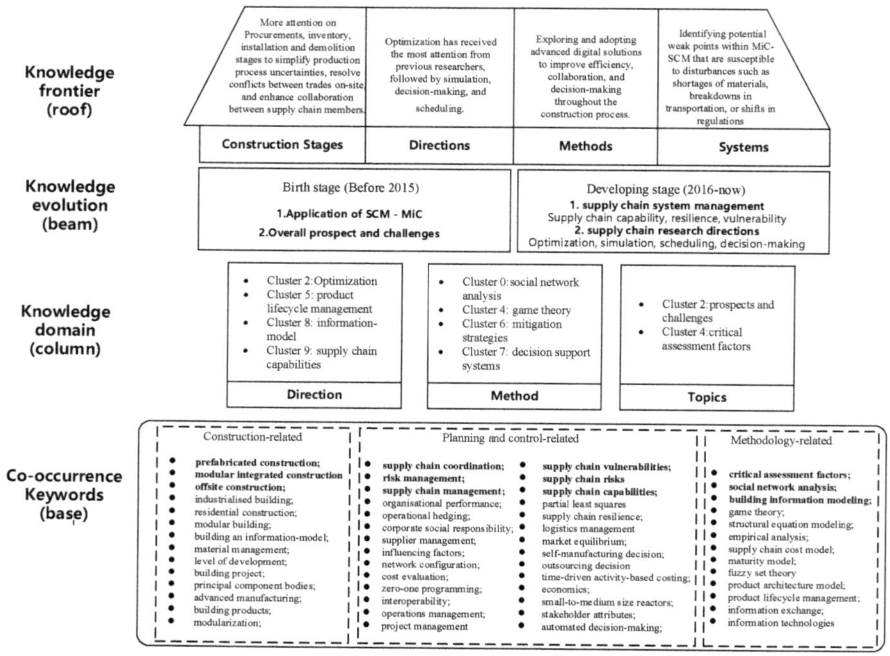

Figure 8. "Housing model" of the knowledge map of SCM-MiC.

Positioned atop this structure is the roof, symbolizing the trajectory of the knowledge frontiers across four dimensions: stages of construction, directions, methods, and systems. This visionary path outlines four synchronized research directions aligned with the complexities of construction-related project management and the evolution of construction-related information technology. These proposed research directions serve as bridges connecting stakeholder studies, construction-related project management, and domains of information technology, guiding future research endeavors in SCM-MiC.

6. Conclusions

The increasing complexity of construction projects has driven the development of MiC. The requirement for more refined MiC has made supply chain management progressively more topical. Our research used a bibliometric analysis method to delve into the evolution and frontiers of SCM-MiC research. First, we extracted data from the Web of Science (WoS) database due to its comprehensive coverage of the relevant literature. Following this, we utilized advanced bibliometric analysis tools to map and visualize the structure of knowledge within the SCM-MiC domain. In total, 196 articles centered around SCM-MiC concerns were extracted. Systematic analysis of the bibliographic records enabled the structured extraction of the evolution of knowledge, the scope of the domain, and insights into the frontier within the realm of MiC. Through a comprehensive analysis spanning multiple dimensions, including historical evolution, exploration of the domain, perspectives of the frontier, and knowledge mapping, we have gained valuable insights into the past, present, and future trajectories of SCM-MiC.

The study's results indicated that the development of SCM-MiC can be traced through two pivotal milestones, marking significant advancements and challenges faced along the way. From the early stages of exploration to the integration of emerging technologies, researchers have diligently paved the path for understanding and optimizing the SCM-MiC domain. The 10 clusters were categorized into three critical dimensions within the knowledge domain: strategies, topics, and challenges. Looking towards the research frontier, the integration of construction-related information technology, including BIM and IoT, promises to revolutionize SCM-MiC, enhancing efficiency, collaboration, and decision-making throughout the process of construction. The metaphorical "house" of SCM-MiC, constructed upon the foundation of historical evolution, offers a comprehensive tableau capturing the past, present, and prospective dimensions of the field.

This research makes a distinctive theoretical and practical contribution to the field of SCM-MiC. The knowledge map presented in this study systematically traces the evolution, domains, and frontiers of SCM-MiC, shedding light on historical milestones and their impact on the knowledge system. By identifying key research areas within the latest knowledge frontier, the study has demonstrated how SCM-MiC research has expanded into a multidisciplinary field encompassing stakeholder research, construction-related project management, and information technology. These insights can inform practitioners and policymakers about the emerging trends and areas of focus within the industry, guiding decision-making processes related to planning projects, resource allocation, and the adoption of technology. Additionally, by pinpointing hotspots and gaps in the current research landscape, the study offers valuable guidance for stakeholders involved in the management of complex projects, facilitating the development of strategies to enhance efficiency, mitigate the risks, and promote sustainable practices in construction projects.

Despite the comprehensive analysis conducted in this study, it is important to acknowledge certain limitations that may have influenced the findings and conclusions. Firstly, for software compatibility reasons, only the literature resources of the WoS database were used. Furthermore, the analysis focused on academic research and may not have incorporated insights from industry practitioners involved in MiC projects.

As we pivot towards the future of SCM-MiC, it is imperative to focus on the potential and challenges presented by cutting-edge technologies such as artificial intelligence, blockchain, and augmented reality. These innovations hold the promise of fundamentally transforming SCM-MiC by optimizing the supply chain's processes, which, in turn, can significantly enhance decision-making capabilities and operational efficiencies in complex projects. Furthermore, the integration of the principles of sustainability into SCM-MiC practices is crucial. This aligns SCM-MiC with broader global environmental objectives, prioritizing resource efficiency and reductions in waste. Incorporating these principles requires meticulous planning and control mechanisms to manage the sustainable practices within complex projects effectively. Cross-sectoral and cross-regional collaboration is another critical area that can amplify the benefits of SCM-MiC. By bridging the gap between theoretical research and practical applications, and by fostering an environment of international standardization, SCM-MiC can achieve greater uniformity and predictability in its outcomes, which are essential for global scalability and efficiency. Lastly, there is a pressing need for a thorough exploration of risk management strategies within SCM-MiC. As global challenges such as climate change and geopolitical instability increase the supply chain's vulnerabilities, enhancing resilience through robust risk management becomes paramount. This not only involves identifying the potential risks but also developing strategies to mitigate these risks proactively. Integrating risk management with strong planning and control processes in the context of managing complex projects is vital for preemptively addressing these challenges and ensuring the robustness of SCM-MiC initiatives.

Author Contributions: Conceptualization, Y.Z. and G.Q.S.; methodology, J.X.; software, Y.Z.; validation, Y.Z., G.Q.S. and J.X.; formal analysis, Y.Z.; investigation, Y.Z.; resources, G.Q.S.; data curation, J.X.; writing—original draft preparation, Y.Z.; writing—review and editing, G.Q.S. and J.X.; visualization, Y.Z.; supervision, J.X.; project administration, G.Q.S.; funding acquisition, G.Q.S. All authors have read and agreed to the published version of the manuscript.

Funding: The Research Grants Council of the Hong Kong Special Administrative Region, China (Project No. PolyU/RGC PolyU15225822).

Conflicts of Interest: The authors declare no conflict of interest.

References

1. Abdelmageed, S.; Zayed, T. A study of literature in modular integrated construction-Critical review and future directions. *J. Clean. Prod.* **2020**, *277*, 124044. [CrossRef]
2. Hussein, M.; Eltoukhy, A.E.; Karam, A.; Shaban, I.A.; Zayed, T. Modelling in off-site construction supply chain management: A review and future directions for sustainable modular integrated construction. *J. Clean. Prod.* **2021**, *310*, 127503. [CrossRef]
3. Wuni, I.Y.; Shen, G.Q. Critical success factors for modular integrated construction projects: A review. *Build. Res. Inf.* **2020**, *48*, 763–784. [CrossRef]
4. Nguyen, T.D.H.N.; Moon, H.; Ahn, Y. Critical review of trends in modular integrated construction research with a focus on sustainability. *Sustainability* **2022**, *14*, 12282. [CrossRef]
5. Wuni, I.Y.; Shen, G.Q. Exploring the critical success determinants for supply chain management in modular integrated construction projects. *Smart Sustain. Built Environ.* **2023**, *12*, 258–276. [CrossRef]
6. Arashpour, M.; Wakefield, R.; Lee, E.; Chan, R.; Hosseini, M.R. Analysis of interacting uncertainties in on-site and off-site activities: Implications for hybrid construction. *Int. J. Proj. Manag.* **2016**, *34*, 1393–1402. [CrossRef]
7. Arshad, H.; Zayed, T. Critical influencing factors of supply chain management for modular integrated construction. *Autom. Constr.* **2022**, *144*, 104612. [CrossRef]
8. Hsu, P.Y.; Aurisicchio, M.; Angeloudis, P. Risk-averse supply chain for modular construction projects. *Autom. Constr.* **2019**, *106*, 102898. [CrossRef]
9. Donthu, N.; Kumar, S.; Mukherjee, D.; Pandey, N.; Lim, W.M. How to conduct a bibliometric analysis: An overview and guidelines. *J. Bus. Res.* **2021**, *133*, 285–296. [CrossRef]
10. Merigó, J.M.; Yang, J.-B. A bibliometric analysis of operations research and management science. *Omega* **2017**, *73*, 37–48. [CrossRef]
11. Liu, X.; Zhao, S.; Tan, L.; Tan, Y.; Wang, Y.; Ye, Z.; Hou, C.; Xu, Y.; Liu, S.; Wang, G. Frontier and hot topics in electrochemiluminescence sensing technology based on CiteSpace bibliometric analysis. *Biosens. Bioelectron.* **2022**, *201*, 113932. [CrossRef]
12. Wen, S.; Tang, H.; Ying, F.; Wu, G. Exploring the global research trends of supply chain management of construction projects based on a bibliometric analysis: Current status and future prospects. *Buildings* **2023**, *13*, 373. [CrossRef]
13. Chen, C.; Hu, Z.; Liu, S.; Tseng, H. Emerging trends in regenerative medicine: A scientometric analysis in CiteSpace. *Expert Opin. Biol. Ther.* **2012**, *12*, 593–608. [CrossRef]
14. Wuni, I.Y.; Shen, G.Q.; Hwang, B.-G. Risks of modular integrated construction: A review and future research directions. *Front. Eng. Manag.* **2020**, *7*, 63–80. [CrossRef]
15. Zhou, J.X.; Shen, G.Q.; Yoon, S.H.; Jin, X. Customization of on-site assembly services by integrating the internet of things and BIM technologies in modular integrated construction. *Autom. Constr.* **2021**, *126*, 103663. [CrossRef]
16. Tsz Wai, C.; Wai Yi, P.; Ibrahim Olanrewaju, O.; Abdelmageed, S.; Hussein, M.; Tariq, S.; Zayed, T. A critical analysis of benefits and challenges of implementing modular integrated construction. *Int. J. Constr. Manag.* **2023**, *23*, 656–668. [CrossRef]
17. Andrić, J.M.; Mahamadu, A.-M.; Wang, J.; Zou, P.X.; Zhong, R. The cost performance and causes of overruns in infrastructure development projects in Asia. *J. Civ. Eng. Manag.* **2019**, *25*, 203–214. [CrossRef]
18. Al-Momani, A.H. Construction delay: A quantitative analysis. *Int. J. Proj. Manag.* **2000**, *18*, 51–59. [CrossRef]
19. Lawson, M.; Ogden, R.; Goodier, C.I. *Design in Modular Construction*; CRC Press: Boca Raton, FL, USA, 2014; Volume 476.
20. Hussein, M.; Zayed, T. Critical factors for successful implementation of just-in-time concept in modular integrated construction: A systematic review and meta-analysis. *J. Clean. Prod.* **2021**, *284*, 124716. [CrossRef]
21. Pan, W.; Yang, Y.; Pan, M. Implementing modular integrated construction in high-rise high-density cities: Perspectives in Hong Kong. *Build. Res. Inf.* **2023**, *51*, 354–368. [CrossRef]
22. Gibb, A.; Isack, F. Re-engineering through pre-assembly: Client expectations and drivers. *Build. Res. Inf. Syst. Front.* **2003**, *31*, 146–160. [CrossRef]
23. Deep, S.; Banerjee, S.; Dixit, S.; Vatin, N.I. Critical factors influencing the performance of highway projects: An empirical evaluation. *Buildings* **2022**, *12*, 849. [CrossRef]
24. Shin, J.; Choi, B.J.B. Design and implementation of quality information management system for modular construction factory. *Buildings* **2022**, *12*, 654. [CrossRef]
25. Jaillon, L.; Poon, C.S. The evolution of prefabricated residential building systems in Hong Kong: A review of the public and the private sector. *Autom. Constr.* **2009**, *18*, 239–248. [CrossRef]
26. Bertram, N.; Fuchs, S.; Mischke, J.; Palter, R.; Strube, G.; Woetzel, J. Modular construction: From projects to products. *McKinsey Co. Cap. Proj. Infrastruct.* **2019**, *1*, 1–34.
27. Tafazzoli, M.; Mousavi, E.; Kermanshachi, S. Opportunities and challenges of green-lean: An integrated system for sustainable construction. *Sustainability* **2020**, *12*, 4460. [CrossRef]
28. Choi, J.O.; Chen, X.B.; Kim, T.W. Opportunities and challenges of modular methods in dense urban environment. *Int. J. Constr. Manag.* **2019**, *19*, 93–105. [CrossRef]

29. Ivanov, D.; Dolgui, A.; Sokolov, B.; Ivanova, M. Literature review on disruption recovery in the supply chain. *Int. J. Prod. Res.* **2017**, *55*, 6158–6174. [CrossRef]
30. Thai, H.T.; Ngo, T.; Uy, B. A review on modular construction for high-rise buildings. In *Structures*; Elsevier: Amsterdam, The Netherlands, 2020; Volume 28, pp. 1265–1290.
31. Xu, Z.; Zayed, T.; Niu, Y. Comparative analysis of modular construction practices in mainland China, Hong Kong and Singapore. *J. Clean. Prod.* **2020**, *245*, 118861. [CrossRef]
32. Luo, L.; Jin, X.; Shen, G.Q.; Wang, Y.; Liang, X.; Li, X.; Li, C.Z. Supply chain management for prefabricated building projects in Hong Kong. *J. Manag. Eng.* **2020**, *36*, 05020001. [CrossRef]
33. Shen, G.Q.; Xue, J. Managing stakeholder dynamics and complexity in mega infrastructure projects. *Front. Eng. Manag.* **2021**, *8*, 148–150. [CrossRef]
34. Zhou, J.X.; Huang, L.; Shen, G.Q.; Wu, H.; Luo, L. Modeling stakeholder-associated productivity performance risks in modular integrated construction projects of Hong Kong: A social network analysis. *J. Clean. Prod.* **2023**, *423*, 138699. [CrossRef]
35. Banihashemi, S.; Tabadkani, A.; Hosseini, M.R. Integration of parametric design into modular coordination: A construction waste reduction workflow. *Autom. Constr.* **2018**, *88*, 1–12. [CrossRef]
36. Iacovidou, E.; Purnell, P.; Tsavdaridis, K.D.; Poologanathan, K. Digitally enabled modular construction for promoting modular components reuse: A UK view. *J. Build. Eng.* **2021**, *42*, 102820. [CrossRef]
37. Irizarry, J.; Karan, E.P.; Jalaei, F. Integrating BIM and GIS to improve the visual monitoring of construction supply chain management. *Autom. Constr.* **2013**, *31*, 241–254. [CrossRef]
38. Isatto, E.L.; Azambuja, M.; Formoso, C.T. The role of commitments in the management of construction make-to-order supply chains. *J. Manag. Eng.* **2015**, *31*, 04014053. [CrossRef]
39. Zhai, Y.; Zhong, R.Y.; Li, Z.; Huang, G. Production lead-time hedging and coordination in prefabricated construction supply chain management. *Int. J. Prod. Res.* **2017**, *55*, 3984–4002. [CrossRef]
40. Ekanayake, E.; Shen, G.; Kumaraswamy, M.M. Critical capabilities of improving supply chain resilience in industrialized construction in Hong Kong. *Eng. Constr. Archit. Manag.* **2021**, *28*, 3236–3260. [CrossRef]
41. RezaHoseini, A.; Noori, S.; Ghannadpour, S.F. Integrated scheduling of suppliers and multi-project activities for green construction supply chains under uncertainty. *Autom. Constr.* **2021**, *122*, 103485. [CrossRef]
42. Li, J.; Han, Y.; Duan, P.; Han, Y.; Niu, B.; Li, C.; Zheng, Z.; Liu, Y. Meta-heuristic algorithm for solving vehicle routing problems with time windows and synchronized visit constraints in prefabricated systems. *J. Clean. Prod.* **2020**, *250*, 119464. [CrossRef]
43. Yang, H.; Chung, J.K.H.; Chen, Y.; Pan, Y.; Mei, Z.; Sun, X. Ordering Strategy Analysis of Prefabricated Component Manufacturer in Construction Supply Chain. *Math. Probl. Eng.* **2018**, *2018*, 4062871. [CrossRef]
44. Ahn, S.; Han, S.; Al-Hussein, M. Improvement of transportation cost estimation for prefabricated construction using geo-fence-based large-scale GPS data feature extraction and support vector regression. *Adv. Eng. Inform.* **2020**, *43*, 101012. [CrossRef]
45. Jiang, Y.; Li, M.; Guo, D.; Wu, W.; Zhong, R.Y.; Huang, G.Q. Digital twin-enabled smart modular integrated construction system for on-site assembly. *Comput. Ind.* **2022**, *136*, 103594. [CrossRef]
46. Wuni, I.Y.; Shen, G.Q. Barriers to the adoption of modular integrated construction: Systematic review and meta-analysis, integrated conceptual framework, and strategies. *J. Clean. Prod.* **2020**, *249*, 119347. [CrossRef]
47. Yang, Y.; Pan, M.; Pan, W. Integrated offsite logistics scheduling approach for high-rise modular building projects. *J. Constr. Eng.* **2022**, *148*, 04022032. [CrossRef]
48. Wang, Z.; Wang, T.; Hu, H.; Gong, J.; Ren, X.; Xiao, Q. Blockchain-based framework for improving supply chain traceability and information sharing in precast construction. *Autom. Constr.* **2020**, *111*, 103063. [CrossRef]
49. Vrijhoef, R.; Koskela, L. The four roles of supply chain management in construction. *Eur. J. Purch. Supply Manag.* **2000**, *6*, 169–178. [CrossRef]
50. Luo, L.; Qiping Shen, G.; Xu, G.; Liu, Y.; Wang, Y. Stakeholder-associated supply chain risks and their interactions in a prefabricated building project in Hong Kong. *J. Manag. Eng.* **2019**, *35*, 05018015. [CrossRef]
51. Li, Z.; Shen, G.Q.; Xue, X. Critical review of the research on the management of prefabricated construction. *Habitat Int.* **2014**, *43*, 240–249. [CrossRef]
52. Jin, R.; Gao, S.; Cheshmehzangi, A.; Aboagye-Nimo, E. A holistic review of off-site construction literature published between 2008 and 2018. *J. Clean. Prod.* **2018**, *202*, 1202–1219. [CrossRef]
53. Han, Y.; Xu, X.; Zhao, Y.; Wang, X.; Chen, Z.; Liu, J. Impact of consumer preference on the decision-making of prefabricated building developers. *J. Civ. Eng. Manag.* **2022**, *28*, 166–176. [CrossRef]
54. Luo, T.; Xue, X.; Wang, Y.; Xue, W.; Tan, Y. A systematic overview of prefabricated construction policies in China. *J. Clean. Prod.* **2021**, *280*, 124371. [CrossRef]
55. Aghasizadeh, S.; Tabadkani, A.; Hajirasouli, A.; Banihashemi, S. Environmental and economic performance of prefabricated construction: A review. *Environ. Impact Assess. Rev.* **2022**, *97*, 106897. [CrossRef]
56. Mok, K.Y.; Shen, G.Q.; Yang, J. Stakeholder management studies in mega construction projects: A review and future directions. *Int. J. Proj. Manag.* **2015**, *33*, 446–457. [CrossRef]
57. Wuni, I.Y.; Shen, G.Q.P.; Mahmud, A.T. Critical risk factors in the application of modular integrated construction: A systematic review. *Int. J. Constr. Manag.* **2022**, *22*, 133–147. [CrossRef]

58. Chen, C.; Ibekwe-SanJuan, F.; Hou, J. The structure and dynamics of cocitation clusters: A multiple-perspective cocitation analysis. *J. Am. Soc. Inf. Sci.* **2010**, *61*, 1386–1409. [CrossRef]
59. Su, H.N.; Lee, P.C. Mapping knowledge structure by keyword co-occurrence: A first look at journal papers in Technology Foresight. *Scientometrics* **2010**, *85*, 65–79. [CrossRef]
60. Abu, K. Visualizing the citation patterns of parasitology research in India: A study using CiteNet explorer. *Int. J. Innov. Knowl. Concepts* **2016**, *2*.
61. Cobo, M.J.; López-Herrera, A.G.; Herrera-Viedma, E.; Herrera, F. Science mapping software tools: Review, analysis, and cooperative study among tools. *J. Am. Soc. Inf. Sci. Technol.* **2011**, *62*, 1382–1402. [CrossRef]
62. Chen, C. *CiteSpace: A Practical Guide for Mapping Scientific Literature*; Nova Science Publishers: Hauppauge, NY, USA, 2016.
63. Xue, J.; Shen, G.Q.; Yang, R.J.; Wu, H.; Li, X.; Lin, X.; Xue, F. Mapping the knowledge domain of stakeholder perspective studies in construction projects: A bibliometric approach. *Int. J. Proj. Manag.* **2020**, *38*, 313–326. [CrossRef]
64. Udomsap, A.D.; Hallinger, P. A bibliometric review of research on sustainable construction, 1994–2018. *J. Clean. Prod.* **2020**, *254*, 120073. [CrossRef]
65. Derakhshan, R.; Turner, R.; Mancini, M. Project governance and stakeholders: A literature review. *Int. J. Proj. Manag.* **2019**, *37*, 98–116. [CrossRef]
66. Chen, C. CiteSpace II: Detecting and visualizing emerging trends and transient patterns in scientific literature. *J. Am. Soc. Inf. Sci. Technol.* **2006**, *57*, 359–377. [CrossRef]
67. Li, X.; Wu, P.; Shen, G.Q.; Wang, X.; Teng, Y. Mapping the knowledge domains of Building Information Modeling (BIM): A bibliometric approach. *Autom. Constr.* **2017**, *84*, 195–206. [CrossRef]
68. Dunning, T. Accurate methods for the statistics of surprise and coincidence. *Comput. Linguist.* **1994**, *19*, 61–74.
69. Hwang, B.G.; Shan, M.; Looi, K.Y. Key constraints and mitigation strategies for prefabricated prefinished volumetric construction. *J. Clean. Prod.* **2018**, *183*, 183–193. [CrossRef]
70. Hsu, P.Y.; Angeloudis, P.; Aurisicchio, M. Optimal logistics planning for modular construction using two-stage stochastic programming. *Autom. Constr.* **2018**, *94*, 47–61. [CrossRef]
71. Wang, Z.; Hu, H.; Gong, J.; Ma, X.; Xiong, W. Precast supply chain management in off-site construction: A critical literature review. *J. Clean. Prod.* **2019**, *232*, 1204–1217. [CrossRef]
72. Arbulu, R.; Tommelein, I.; Walsh, K.; Hershauer, J. Value stream analysis of a reengineered construction suppply chain. *Build Res. Inf.* **2003**, *31*, 161–171. [CrossRef]
73. Höök, M.; Stehn, L. Applicability of lean principles and practices in industrialized housing production. *Build. Res. Inf.* **2008**, *26*, 1091–1100. [CrossRef]
74. Jaillon, L.; Poon, C.S. Sustainable construction aspects of using prefabrication in dense urban environment: A Hong Kong case study. *Constr. Manag. Econ.* **2008**, *26*, 953–966. [CrossRef]
75. Goodier, C.; Gibb, A. Future opportunities for offsite in the UK. *Constr. Manag. Econ.* **2007**, *25*, 585–595. [CrossRef]
76. Li, Z.; Shen, G.Q.; Alshawi, M. Measuring the impact of prefabrication on construction waste reduction: An empirical study in China. *Resour. Conserv. Recycl.* **2014**, *91*, 27–39. [CrossRef]
77. Pan, W.; Sidwell, R. Demystifying the cost barriers to offsite construction in the UK. *Constr. Manag. Econ.* **2011**, *29*, 1081–1099. [CrossRef]
78. Hosseini, M.R.; Martek, I.; Zavadskas, E.K.; Aibinu, A.A.; Arashpour, M.; Chileshe, N. Critical evaluation of off-site construction research: A Scientometric analysis. *Autom. Constr.* **2018**, *87*, 235–247. [CrossRef]
79. Li, X.; Shen, G.Q.; Wu, P.; Yue, T. Integrating Building Information Modeling and Prefabrication Housing Production. *Autom. Constr.* **2019**, *100*, 46–60. [CrossRef]
80. Rahman, M.M. Barriers of Implementing Modern Methods of Construction. *J. Manag. Eng.* **2014**, *30*, 69–77. [CrossRef]
81. Zhong, R.Y.; Peng, Y.; Xue, F.; Fang, J.; Zou, W.; Luo, H.; Ng, S.T.; Lu, W.; Shen, G.Q.; Huang, G.Q. Prefabricated construction enabled by the Internet-of-Things. *Autom. Constr.* **2017**, *76*, 59–70. [CrossRef]
82. Li, C.Z.; Xue, F.; Li, X.; Hong, J.; Shen, G.Q. An Internet of Things-enabled BIM platform for on-site assembly services in prefabricated construction. *Autom. Constr.* **2018**, *89*, 146–161. [CrossRef]
83. Bortolini, R.; Formoso, C.T.; Viana, D.D. Site logistics planning and control for engineer-to-order prefabricated building systems using BIM 4D modeling. *Autom. Constr.* **2019**, *98*, 248–264. [CrossRef]
84. Liu, Y.; Dong, J.; Shen, L. A conceptual development framework for prefabricated construction supply chain management: An integrated overview. *Sustainability* **2020**, *12*, 1878. [CrossRef]
85. London, K.; Singh, V. Integrated construction supply chain design and delivery solutions. *Archit. Eng. Des. Manag.* **2013**, *9*, 135–157. [CrossRef]
86. Doran, D.; Giannakis, M. An examination of a modular supply chain: A construction sector perspective. *Supply Chain Manag. Int. J.* **2011**, *16*, 260–270. [CrossRef]
87. Baghalzadeh Shishehgarkhaneh, M.; Keivani, A.; Moehler, R.C.; Jelodari, N.; Roshdi Laleh, S. Internet of Things (IoT), Building Information Modeling (BIM), and Digital Twin (DT) in construction industry: A review, bibliometric, and network analysis. *Buildings* **2022**, *12*, 1503. [CrossRef]
88. Tang, S.; Shelden, D.R.; Eastman, C.M.; Pishdad-Bozorgi, P.; Gao, X. A review of building information modeling (BIM) and the internet of things (IoT) devices integration: Present status and future trends. *Autom. Constr.* **2019**, *101*, 127–139. [CrossRef]

89. Čuš-Babič, N.; Rebolj, D.; Nekrep-Perc, M.; Podbreznik, P. Supply-chain transparency within industrialized construction projects. *Comput. Ind.* **2014**, *65*, 345–353. [CrossRef]
90. Zhang, X.L.; Skitmore, M.; Peng, Y. Exploring the challenges to industrialized residential building in China. *Habitat Int.* **2014**, *41*, 176–184. [CrossRef]
91. Mao, C.; Shen, Q.; Pan, W.; Ye, K. Major Barriers to Off-Site Construction: The Developer's Perspective in China. *J. Manag. Eng.* **2015**, *31*, 04014043. [CrossRef]
92. O'Connor, J.T.; O'Brien, W.J.; Choi, J.O. Standardization strategy for modular industrial plants. *J. Constr. Eng. Manag.* **2015**, *141*, 04015026. [CrossRef]
93. Arashpour, M.; Wakefield, R.; Blismas, N.; Minas, J. Optimization of process integration and multi-skilled resource utilization in off-site construction. *Autom. Constr.* **2015**, *50*, 72–80. [CrossRef]
94. Arashpour, M.; Wakefield, R.; Blismas, N.; Maqsood, T. Autonomous production tracking for augmenting output in off-site construction. *Autom. Constr.* **2015**, *53*, 13–21. [CrossRef]
95. Li, C.Z.; Hong, J.; Xue, F.; Shen, G.Q.; Xu, X.; Mok, M.K. Schedule risks in prefabrication housing production in Hong Kong: A social network analysis. *J. Clean. Prod.* **2016**, *134*, 482–494. [CrossRef]
96. Lu, J.Q.; Wang, J.W.; Song, Y.H.; Yuan, C.B.; He, J.J.; Chen, Z. Influencing Factors Analysis of Supply Chain Resilience of Prefabricated Buildings Based on PF-DEMATEL-ISM. *Buildings* **2022**, *12*, 1595. [CrossRef]
97. Zhang, S.X.; Li, Z.F.; Ma, S.B.; Li, L.; Yuan, M.Q. Critical Factors Influencing Interface Management of Prefabricated Building Projects: Evidence from China. *Sustainability* **2022**, *14*, 5418. [CrossRef]
98. Ekanayake, E.; Shen, G.Q.P.; Kumaraswamy, M.M.; Owusu, E.K.; Saka, A.B. Modeling Supply Chain Resilience in Industrialized Construction: A Hong Kong Case. *J. Constr. Eng. Manag.* **2021**, *147*, 05021009. [CrossRef]
99. Zhang, M.J.; Liu, Y.M.; Ji, B.Y. Influencing Factors of Resilience of PBSC Based on Empirical Analysis. *Buildings* **2021**, *11*, 467. [CrossRef]
100. Luo, L.Z.; Li, Y.K.; Wang, X.; Jin, X.; Qin, Z.Y. Supply Chain Vulnerability in Prefabricated Building Projects and Digital Mitigation Technologies. *IEEE Trans. Eng. Manag.* **2023**, 1–13. [CrossRef]
101. Yuan, M.Q.; Li, Z.F.; Li, X.D.; Luo, X.W. Managing stakeholder-associated risks and their interactions in the life cycle of prefabricated building projects: A social network analysis approach. *J. Clean. Prod.* **2021**, *323*, 129102. [CrossRef]
102. Wang, Q.Y.; Gong, Z.Q.; Liu, C.K. Risk Network Evaluation of Prefabricated Building Projects in Underdeveloped Areas: A Case Study in Qinghai. *Sustainability* **2022**, *14*, 6335. [CrossRef]
103. Bui, T.D.; Tsai, F.M.; Tseng, M.L.; Tan, R.D.R.; Yu, K.D.S.; Lim, M.K. Sustainable supply chain management towards disruption and organizational ambidexterity: A data driven analysis. *Sustain. Prod. Consum.* **2021**, *26*, 373–410. [CrossRef]
104. Ekanayake, E.; Shen, G.Q.P.; Kumaraswamy, M.M. A fuzzy synthetic evaluation of capabilities for improving supply chain resilience of industrialised construction: A Hong Kong case study. *Prod. Plan. Control* **2023**, *34*, 623–640. [CrossRef]
105. Jiang, W.; Liu, M.L.; Gan, L.; Wang, C. Optimal Pricing, Ordering, and Coordination for Prefabricated Building Supply Chain with Power Structure and Flexible Cap-and-Trade. *Mathematics* **2021**, *9*, 2426. [CrossRef]
106. Zhai, Y.; Zhong, R.Y.; Huang, G.Q. Buffer space hedging and coordination in prefabricated construction supply chain management. *Int. J. Prod. Econ.* **2018**, *200*, 192–206. [CrossRef]
107. Gao, J.; You, F. A stochastic game theoretic framework for decentralized optimization of multi-stakeholder supply chains under uncertainty. *Comput. Chem. Eng.* **2019**, *122*, 31–46. [CrossRef]
108. Vazifeh, Z.; Mafakheri, F.; An, C. Biomass supply chain coordination for remote communities: A game-theoretic modeling and analysis approach. *Sustain. Cities Soc.* **2021**, *69*, 102819. [CrossRef]
109. Wang, G.; Gunasekaran, A.; Ngai, E.W.; Papadopoulos, T. Big data analytics in logistics and supply chain management: Certain investigations for research and applications. *Int. J. Prod. Econ.* **2016**, *176*, 98–110. [CrossRef]
110. Pournader, M.; Ghaderi, H.; Hassanzadegan, A.; Fahimnia, B. Artificial intelligence applications in supply chain management. *Int. J. Prod. Econ.* **2021**, *241*, 108250. [CrossRef]
111. Zhu, T.; Liu, G.C. A Novel Hybrid Methodology to Study the Risk Management of Prefabricated Building Supply Chains: An Outlook for Sustainability. *Sustainability* **2023**, *15*, 361. [CrossRef]
112. Chen, X.; Duan, J. What they talk about when they talk about urban regeneration: Understanding the concept 'urban regeneration'in PRD, China. *Cities* **2022**, *130*, 103880. [CrossRef]
113. Zakaria, S.A.S.; Gajendran, T.; Rose, T.; Brewer, G. Contextual, structural and behavioural factors influencing the adoption of industrialised building systems: A review. *Archit. Eng. Des. Manag.* **2018**, *14*, 3–26. [CrossRef]
114. Li, T.H.; Ng, S.T.T.; Skitmore, M.; Li, N. Investigating stakeholder concerns during public participation. In *Proceedings of the Institution of Civil Engineers-Municipal Engineer*; Thomas Telford Ltd.: London, UK, 2016; Volume 169, pp. 199–219.
115. Ezcan, V.; Goulding, J.S. Offsite Sustainability-Disentangling the Rhetoric through Informed Mindset Change. *Sustainability* **2022**, *14*, 4606. [CrossRef]
116. Kedir, F.; Hall, D.M. Resource efficiency in industrialized housing construction—A systematic review of current performance and future opportunities. *J. Clean. Prod.* **2021**, *286*, 125443. [CrossRef]
117. Nabi, M.A.; El-adaway, I.H. Risk-Based Approach to Predict the Cost Performance of Modularization in Construction Projects. *J. Constr. Eng. Manag.* **2021**, *147*, 04021133. [CrossRef]

118. Zhang, X.; Li, R.Y.M.; Sun, Z.; Li, X.; Samad, S.; Comite, U.; Matac, L.M. Supply Chain Integration and Its Impact on Operating Performance: Evidence from Chinese Online Companies. *Sustainability* **2022**, *14*, 14330. [CrossRef]
119. Arashpour, M.; Bai, Y.; Aranda-mena, G.; Bab-Hadiashar, A.; Hosseini, R.; Kalutara, P. Optimizing decisions in advanced manufacturing of prefabricated products: Theorizing supply chain configurations in off-site construction. *Autom. Constr.* **2017**, *84*, 146–153. [CrossRef]
120. Du, J.; Jing, H.Q.; Choo, K.K.R.; Sugumaran, V.; Castro-Lacouture, D. An Ontology and Multi-Agent Based Decision Support Framework for Prefabricated Component Supply Chain. *Inf. Syst. Front.* **2020**, *22*, 1467–1485. [CrossRef]
121. Xie, L.L.; Chen, Y.J.; Chang, R.D. Scheduling Optimization of Prefabricated Construction Projects by Genetic Algorithm. *Appl. Sci.* **2021**, *11*, 5531. [CrossRef]
122. Zhang, H.; Yu, L. Dynamic transportation planning for prefabricated component supply chain. *Eng. Constr. Archit. Manag.* **2020**, *27*, 2553–2576. [CrossRef]
123. Wang, Y.C.; Sun, R.; Ren, L.Y.; Geng, X.X.; Wang, X.M.; Lv, L. Risk Propagation Model and Simulation of an Assembled Building Supply Chain Network. *Buildings* **2023**, *13*, 981. [CrossRef]
124. Fan, C.; Deng, B.C.; Yin, Y.L. Hierarchical structure and transfer mechanism to assess the scheduling-related risk in construction of prefabricated buildings: An integrated ISM-MICMAC approach. *Eng. Constr. Archit. Manag.* **2022**, *30*, 2991–3013. [CrossRef]
125. Fang, X.; Cao, L.; Zhang, L.; Peng, B. Risk perception and resistance behavior intention of residents living near chemical industry parks: An empirical analysis in China. *Nat. Hazards* **2023**, *115*, 1655–1675. [CrossRef]
126. Antwi-Afari, P.; Ng, S.T.; Hossain, M.U. A review of the circularity gap in the construction industry through scientometric analysis. *J. Clean. Prod.* **2021**, *298*, 126870. [CrossRef]
127. Olearczyk, J.; Al-Hussein, M.; Bouferguene, A. Evolution of the crane selection and on-site utilization process for modular construction multilifts. *Autom. Constr.* **2014**, *43*, 59–72. [CrossRef]
128. Viana, D.D.; Tommelein, I.D.; Formoso, C.T. Using Modularity to Reduce Complexity of Industrialized Building Systems for Mass Customization. *Energies* **2017**, *10*, 1622. [CrossRef]
129. He, D.D.; Li, Z.F.; Wu, C.L.; Ning, X. An E-Commerce Platform for Industrialized Construction Procurement Based on BIM and Linked Data. *Sustainability* **2018**, *10*, 2613. [CrossRef]
130. Steinhardt, D.; Manley, K.; Bildsten, L.; Widen, K. The structure of emergent prefabricated housing industries: A comparative case study of Australia and Sweden. *Constr. Manag. Econ.* **2020**, *38*, 483–501. [CrossRef]
131. Salari, S.A.; Mahmoudi, H.; Aghsami, A.; Jolai, F.; Jolai, S.; Yazdani, M. Off-Site Construction Three-Echelon Supply Chain Management with Stochastic Constraints: A Modelling Approach. *Buildings* **2022**, *12*, 119. [CrossRef]
132. Ansanelli, G.; Fiorentino, G.; Tammaro, M.; Zucaro, A. A Life Cycle Assessment of a recovery process from End-of-Life Photovoltaic Panels. *Appl. Energy* **2021**, *290*, 116727. [CrossRef]
133. Li, H.X.; Al-Hussein, M.; Lei, Z.; Ajweh, Z. Risk identification and assessment of modular construction utilizing fuzzy analytic hierarchy process (AHP) and simulation. *Can. J. Civ. Eng.* **2013**, *40*, 1184–1195. [CrossRef]
134. Chen, Y.; Okudan, G.E.; Riley, D.R. Decision support for construction method selection in concrete buildings: Prefabrication adoption and optimization. *Autom. Constr.* **2010**, *19*, 665–675. [CrossRef]
135. Levander, E.; Engström, S.; Sardén, Y.; Stehn, L. Construction clients' ability to manage uncertainty and equivocality. *Constr. Manag. Econ.* **2011**, *29*, 753–764. [CrossRef]
136. Liu, J.; Gong, E.; Wang, D.; Teng, Y. Cloud model-based safety performance evaluation of prefabricated building project in China. *Wirel. Pers. Commun.* **2018**, *102*, 3021–3039. [CrossRef]
137. Han, Y.H.; Skibniewski, M.J.; Wang, L.F. A Market Equilibrium Supply Chain Model for Supporting Self-Manufacturing or Outsourcing Decisions in Prefabricated Construction. *Sustainability* **2017**, *9*, 2069. [CrossRef]
138. Yazdani, M.; Kabirifar, K.; Fathollahi-Fard, A.M.; Mojtahedi, M. Production scheduling of off-site prefabricated construction components considering sequence dependent due dates. *Environ. Sci. Pollut. Res.* **2021**, 1–17. [CrossRef]
139. Sun, X.F.; Wang, Y.Y.; Li, Y.Y.; Zhu, W.J.; Yan, D.H.; Li, J.H. Optimal pricing and carbon emission reduction decisions for a prefabricated building closed-loop supply chain under a carbon cap-and-trade regulation and government subsidies. *PLoS ONE* **2023**, *18*, e0287684. [CrossRef]
140. Zaalouk, A.; Moon, S.; Han, S. Operations planning and scheduling in off-site construction supply chain management: Scope definition and future directions. *Autom. Constr.* **2023**, *153*, 104952. [CrossRef]
141. Zhai, Y.; Chen, K.; Zhou, J.X.; Cao, J.; Lyu, Z.; Jin, X.; Shen, G.Q.; Lu, W.; Huang, G.Q. An Internet of Things-enabled BIM platform for modular integrated construction: A case study in Hong Kong. *Adv. Eng. Inform.* **2019**, *42*, 100997. [CrossRef]
142. Li, C.Z.; Zhong, R.Y.; Xue, F.; Xu, G.; Chen, K.; Huang, G.G.; Shen, G.Q. Integrating RFID and BIM technologies for mitigating risks and improving schedule performance of prefabricated house construction. *J. Clean. Prod.* **2017**, *165*, 1048–1062. [CrossRef]
143. Fang, Y.; Cho, Y.K.; Zhang, S.; Perez, E. Case study of BIM and cloud–enabled real-time RFID indoor localization for construction management applications. *J. Constr. Eng. Manag.* **2016**, *142*, 05016003. [CrossRef]
144. Alaloul, W.S.; Liew, M.S.; Zawawi, N.A.W.A.; Kennedy, I.B. Industrial Revolution 4.0 in the construction industry: Challenges and opportunities for stakeholders. *Ain Shams Eng. J.* **2020**, *11*, 225–230. [CrossRef]
145. Regona, M.; Yigitcanlar, T.; Xia, B.; Li, R.Y.M. Opportunities and Adoption Challenges of AI in the Construction Industry: A PRISMA Review. *J. Open Innov. Technol. Mark. Complex.* **2022**, *8*, 45. [CrossRef]
146. Wieland, A.; Durach, C.F. Two perspectives on supply chain resilience. *J. Bus. Logist.* **2021**, *42*, 315–322. [CrossRef]

147. Ekanayake, E.; Shen, G.Q.; Kumaraswamy, M.M.; Owusu, E.K. Identifying supply chain vulnerabilities in industrialized construction: An overview. *Int. J. Constr. Manag.* **2022**, *22*, 1464–1477. [CrossRef]
148. Abedi, M.; Fathi, M.S.; Mirasa, A.K.; Rawai, N.M. Integrated collaborative tools for precast supply chain management. *Sci. Iran.* **2016**, *23*, 429–448. [CrossRef]
149. Wang, B.; Geng, L.N.; Dang, P.; Zhang, L.H. Developing a Framework for Dynamic Organizational Resilience Analysis in Prefabricated Construction Projects: A Project Life Cycle Perspective. *J. Constr. Eng. Manag.* **2022**, *148*, 04022110. [CrossRef]
150. Yuan, Y.S.; Ye, S.D.; Lin, L.; Gen, M.S. Multi-objective multi-mode resource-constrained project scheduling with fuzzy activity durations in prefabricated building construction. *Comput. Ind. Eng.* **2021**, *158*, 107316. [CrossRef]
151. Ekanayake, E.; Shen, G.Q.; Kumaraswamy, M.; Owusu, E.K. Critical supply chain vulnerabilities affecting supply chain resilience of industrialized construction in Hong Kong. *Eng. Constr. Archit. Manag.* **2021**, *28*, 3041–3059. [CrossRef]

buildings

MDPI

Article

A Proposed Model for Variation Order Management in Construction Projects

Esam M. H. Ismaeil [1,2,*] and Abu Elnasr E. Sobaih [3,4,*]

1 Civil and Environmental Department, College of Engineering, King Faisal University, Al-Ahsa 31982, Saudi Arabia
2 Architecture and Urban Planning Department, Faculty of Engineering, Port Said University, Port Said 42526, Egypt
3 Management Department, College of Business Administration, King Faisal University, Al-Ahsa 31982, Saudi Arabia
4 Hotel Management Department, Faculty of Tourism and Hotel Management, Helwan University, Cairo 12612, Egypt
* Correspondence: emohamed@kfu.edu.sa (E.M.H.I.); asobaih@kfu.edu.sa (A.E.E.S.)

Abstract: This study aims to identify the causes of excessive change orders and their impact on public construction projects in Saudi Arabia. This should support the organizations working in the construction industry to improve variation order management (VOM) as a preventive action by dealing proactively with variation order causes. Therefore, a new methodology was proposed to minimize change orders and their impact on the successful completion of projects as well as cost during the project lifecycle. The methodology involved ten selected turnkey building projects at King Faisal University (KFU) campus, Saudi Arabia. Statistical analyses were conducted to predict the cost overrun in project size and contract value. The findings showed the most significant causes leading to variation order in public construction projects. These include the combined effect of the designer and owner technical committee, designer document, and owner stakeholder committee. Hence, a new model for VOM was developed as a best practice approach, including three stages. The first stage is the initiation process, which includes seven procedures, seven tools, and key responsibilities. The second stage deals with the course of change orders based on a certain number of procedures and weight for each parameter assigned to this phase. It supports decision processes based on a certain average ratio of weights calculation. The third stage is the decision to support decision makers in proceeding or not proceeding with the variation order. Although the present study was conducted in Saudi Arabia's public building construction project, it is envisaged that these research results are widely applicable to other developing countries. The paper presents a direction for further research to enhance the impact of cost overrun in public sector construction projects in developing countries, i.e., Saudi Arabia.

Keywords: change order; construction projects; Saudi Arabia; variation order; variation order management (VOM)

Citation: Ismaeil, E.M.H.; Sobaih, A.E.E. A Proposed Model for Variation Order Management in Construction Projects. *Buildings* **2024**, *14*, 726. https://doi.org/10.3390/buildings14030726

Academic Editors: Wenzhe Tang, Wenxin Shen and Jin Xue

Received: 14 January 2024
Revised: 3 March 2024
Accepted: 5 March 2024
Published: 8 March 2024

1. Introduction

In the construction industry, several change orders are issued during the project life cycle, mainly due to ambiguous client requirements mentioned in the contract, incomplete tender contract documents, and design errors [1,2]. However, extensive change orders result in cost overruns, schedule delays, lower productivity, and conflicts, leading to arbitration and litigation [2,3]. Construction projects address geography, site conditions, communities, physical environments, existing infrastructure, and many stakeholder requirements [1]. Construction projects are considered long-term complex projects, which are characterized by a dynamic nature [2,3]. Construction projects include multiple organizations, like the owner, the consultant for the design work, the main contractor, subcontractors, the supervision consultant, suppliers, manufacturers, and administrative and in some cases government agencies, who interact together with bidding documents, contract conditions, drawings, specifications, and bills of quantities, [1,3]. Relationships and obligations among the organizations intertwine in a way that may or may not be expected during the contracting stages, resulting in difficulties and problems leading to changes, controversial claims, and conflict among the organizations. However, this negatively affects the completion of the project by a specific time and increases the estimated cost and target quality [4,5]. Change is defined as any deviation from an agreed-upon scope and schedule [6]. Variation orders are considered common issues that most frequently occur in construction projects due to the complex nature of the construction industry, albeit they influence various aspects of construction projects [7,8].

Variation order is any contractual modification by the owner's teams related to the project. It is one of the few tools the project manager contractually has to accommodate. This includes unplanned occurrences once the project is under construction [9]. It involves an amendment and deviation from the original contract scope and often causes disputes and dissatisfaction [9,10]. Variation orders involve alteration, addition, omission, and substitution in terms of quality, quantity, and work schedule, which significantly affect cost and time overruns [11,12]. The change orders are the most frequent cause of claims, with a 55% importance index, while delay had an essential index of 52.5% [13]. It was argued that variation order management needs more formalized knowledge approaches to be well applied in the Saudi construction industry starting from the design stage [14,15]. This includes correcting and modifying the original work scope and affecting overall project performance.

1.1. Variation Order Types

There are four types of variations. The first type is directed changes from the client to the contractor to perform contract specifications modification or addition to the original work scope. The second type is constructive changes as an informal authorizing modification act to void the failure contract. The third type is beneficial variation orders to improve the quality standard and manipulate cost, schedule overrun, and optimize the client's benefits against the resource. The fourth type is compensation for negative impacts from international and local crises on enhancing the client's project value, performance, and degree of project control difficulty [8,16–19]. The purposes served by variation orders include changing contract plans and contract specifications, administrative purposes of establishing extra work, contract unit price adjustments, proposals for cost reduction incentives from value engineering method, payment after settlement of claims, payment changes, and modifications to the contract schedule, item quantities, design mistakes, and unit prices [20].

1.2. Variation Order Sources and Causes

The variations sources include the construction performance organizations, resources, environmental conditions, and contractual issues. Additionally, cost and overhead overruns, professionals' team turnover, and quality improvement lead to variations at various project stages [21]. The causes of variation order also include mismatch between contract documents and work done, unexpected problems, site conditions, inadequate design, change in specifications, preliminary working drawings, changes in design by the consultant, ambiguous design details, lack of coordination, change in methods, substituting materials and procedures, errors and omissions in design, and changes in project scope due to owner requirements [21,22]. The control of construction project change orders causes and impacts, which average 10–15% of the contract value, include the amount of change; the degradation of productivity and costs; site personnel's lack of awareness, skills, and knowledge; and substantial adjustments to the contract duration. All of these variables influence variation order, claims, and total direct and indirect costs. These variables also include the value of alteration, addition, omission, and substitution in terms of quality, quantity, and work schedule [23–26].

The negative impacts of variation orders include project cost increase, overhead expenses increase, invoice delays in logistics processes, blemishing of reputation, poor safety conditions, degradation of quality of productivity, procurement delay, official disputes, delay in completion schedule, cost overruns, owner's financial problems, impediments to the decision-making processes, design document complexity changes, insufficient working drawing details, skilled human resources shortage, contractor's financial difficulties, and source of disputes [27–29]. The most significant percentage of claims and disputes in the Gulf areas, especially Saudi Arabia, Oman, and the United Arab Emirates, are variation orders. These include new client requirements (78%), variations in quantities (74%), contractor work delay (74%), design errors or omissions (72%), and drawings and specifications inconsistencies (70%) [12,13,24,30].

1.3. Variation Order Procedures

The main goal of any change within the project lifecycle is to achieve a better state in the future than the current state. It also aims to maximize the desired benefit, improve the financial or competitive capabilities, improve the quality of procedures, achieve customers' requirements, and fulfill the contract obligations even if this increases the cost and effort [31]. Inappropriate study and evaluation of variation orders can directly lead to failure, re-work, time and money wasting, the uselessness of the project, and contract problems resulting from claims and disputes [32,33]. The most critical claims and dispute sources encountered by the project team through the project execution processes are the ongoing conflicts among tender documents [34]. The concept for executing variation orders in construction projects has international procedures, i.e., the International Federation of Consulting Engineers (FIDIC) chapter (1/13), which explains that the contractor must respond in official writing for any project engineer proposal request stating reasons for inability to comply. Otherwise, they have to provide specific requirements according to Chapter 12 unless the engineer issues his/her instructions or agrees to (7/13) from the FIDIC adjustments for changes in legislation, i.e., the proposal of the program works under Item 8/1 and affects the completion period, calculates the change value, and responds to the contractor's proposal based on Item 13/2 from the FIDIC considering modified cost increase/decrease [35–42].

The traditional process of variation orders includes three stages (Figure 1), starting with the input procedures (closed rectangular and rhombus shape), passing through the process flow procedures, and reaching the output documents. Stage 1 (with dashed rectangle) is the input stage, which includes the change request form and the change requests follow-up record form. Stage 2 is the process flow stage, which includes submitting a change request, logging in the change request, reviewing the request and determining the person who shall study it, notifying the concerned organizations of the change request, studying and technically reviewing the change request and determining its impact on the project cost and time, specifying the approval or disapproval by the concerned authority, recording the final result, and updating the tender documents. Stage 3 (with dashed rectangle) is the output stage (the wavy and cross rectangle shape), which includes change request, record follow-up, technical study of change, analytical study of costs, updated timetable, and approved change order [32,43].

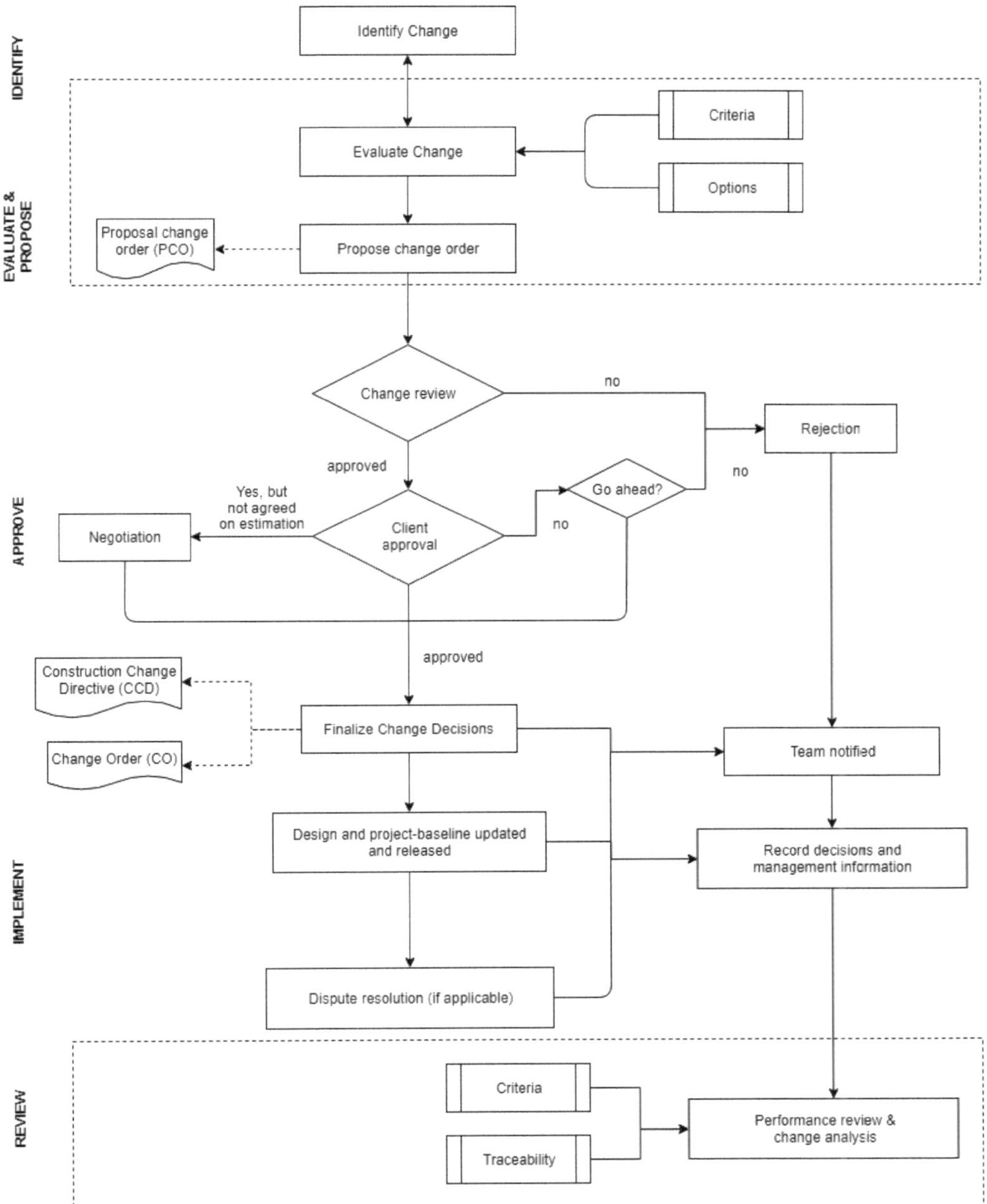

Figure 1. Common variation order procedures [32].

The causes of variation orders attributed to the construction project organizations are classified into six categories. The first category is variation orders related to the owner, i.e., additional requirements; technically disrupted or interrupted work; financial difficulties; performance expedition request; delay in actions; and value engineering study. The second includes variation orders related to the designer/consultant, i.e., document revisions; omissions or errors; insufficient details in the design document; unanticipated works; contract discrepancies; availability of the systems and materials; scope definitions; site conditions; and restrictions on the work method. Third is variation orders related to the contractor, i.e., omissions in construction procedures; work remediation; sequence of the work; late delivery for material and equipment; schedule time mistakes, financial conflicts; skilled labor lack in market; and unexpected risks. Fourth is variation orders related to project management, i.e., coordination efficiency; contract interpretation; and contract documents understanding. Fifth is variation orders related to local authorities, i.e., third-party decision; governmental laws; site access restrictions; and relocation for utility. Sixth is variation orders related to force majeure, i.e., unexpected matters from nature and human behavior [44–49].

A variation order must be a written agreement for process modification, adding or otherwise changing the work from the original contract outlined [50,51]. The change or modification ranges between a 10% increase and a 20% decrease in the contract items budget, and the limits of the contract are stipulated clearly in the contract, related to works of the same quality, and are necessary for the full completion of the project [35,52].

The procedures of the sequence of change variations when the value exceeds a specified rate of the total contract start by submitting a claim from the contractor to re-study the rates and takes place from the project start date until the primary handover, as defined in items (3–13) from the FIDIC of the general conditions for FIDIC contracts. Before issuing the primary handover certificate, the engineer can change the work through the instructions he issues or request that the contractor submit a proposal [35,53–56]. The implementation of change orders after the engineer issues instructions according to his approval of any item of work may include changes in quantities, quality, characteristics, standards, cancellation, addition, providing machines/materials/services, tests, and sequences [26,57]. The variation orders implementation must adhere to submitting an immediate notice supported by data and information [8,14,34]. This includes specifying a new price for a new item in the bill of quantities. Therefore, the new price can be derived based on the work items related to that item; otherwise, it can be estimated with a reasonable cost plus a reasonable profit margin [41,57].

The proposition of this study is as follows: change orders in projects are one of the most decisive and influential factors in delaying the implementation of construction projects and one of the most important sources of claims and disputes for compensation due to an extension of time or cost. This proposition has sub-propositions. The owner's ability to change the requirements and make amendments by deducting some contract items and adding some new ones is one of the most significant sources of claims for compensation over time and cost. The designer's lack of suitable preparation of documents, whether for design work, estimation of quantities, or conflict of documents, is one of the most important causes of variation orders and sources of claims and disputes. Identifying the sources of changes in the engineering disciplines (architectural, civil, mechanical, and electrical) of construction projects at the beginning of the project and during its implementation on a periodic basis is one of the most important approaches to controlling the sources of change orders. Presenting a specific integration procedures model for managing and controlling variation orders supports the owner's organization in avoiding overrun of time, cost, and scope, and therefore reduces the risks of claims and disputes. The study proposes a new methodology to minimize the impact of excessive change orders on the project completion cost during the project lifecycle. The study draws on an analysis of 10 recent projects conducted at a public university (KFU) in Saudi Arabia.

2. Methodology

The construction industry encounters several variation orders issued during the project lifecycle, which are often managed in a traditional way of processing by the organization management team (see Figure 1). Change orders mainly affect cost overruns and schedule delays. Hence, organizations in the construction industry need to improve variation order management processes as preventive processes to deal with the variation order causes. Therefore, the study methodology was built practically in two stages. The first stage included discovering all gaps and reasons that cause variation orders inside the ongoing case study construction project by conducting an analysis of 10 selected implemented construction projects with different functions. The second stage is the formation of a robust and practical guide model for variation order management and control to support construction industry organizations in overcoming variation order risks.

The first stage included selecting 10 construction projects (selected from 33 implemented projects on the KFU city campus, Saudi Arabia, from 2020 to 2022). The study draws on interviews with construction organization experts working on these projects. It relied on comprehensive numerical analysis using several types of software, i.e., Excel spreadsheets v11.0, Revit v 24.1.11.26, AutoCAD v 24.2, and Primavera v 23.12. The main purpose was to evaluate all change orders occurring within the execution processes of the selected 10 projects' original tender documents and as-built documents. The study precisely revealed the variation order causes and cost overrun from the original tender contract budget and conditions. The analysis classified the variation order causes in all construction disciplines. Therefore, the volume and rate of changes in each project were studied based on final invoices, which include initial tender quantities items, variation order quantities items, and the final as-built quantities items of all the disciplines implemented until the project primary handover. The analysis process with Excel spreadsheets was used to determine the types of changes for all items listed in the final invoices to identify and calculate the status of the following items:

- The items with no changes based on the tender design.
- Increased quantities of items beyond the tender design and the effect on additional changes.
- Decreased quantities of items from the tender design and the effect on deduction change.
- New items, the new items inside the tender design, and the effect on addition or deduction changes.
- Omitted items from the tender design and the effect on addition or deduction changes.

The study used Excel sheet calculation for analyzing the types of change that have been matched according to the final invoices and approved change orders. The study analyzed the changes in types, sizes, and item rates in each discipline level in the CSI bill of quantities divisions. The change rate was determined based on the financial cost of the engineering disciplines (structural/architecture/mechanical/electrical) inside each project. The second stage in methodology started based on significant procedures, which include the following:

(a) Interviews were conducted with project managers (contractors, consultants, and the project management office of the owner KFU-PMO to verify the reasons and conditions for the changes and to classify the effects of quantities, cost, and ratio that occurred in each discipline as significant findings.
(b) Classifying the phases of VOM approach in two phases includes the initiation processes phase and the variation order course processes phase based on a certain number of procedures and weight for each parameter assigned to this phase to support decision processes based on a certain average ratio of weights calculation. Applying VOM supported the decision maker of the case study campus organization in the ongoing 32 projects, which positively reduced the negative impacts of the variation order on cost and schedule overrun issues and presented a comprehensive and robust model in the construction industry that can be adopted and updated according to project conditions. Figure 2 shows the study method flowchart.

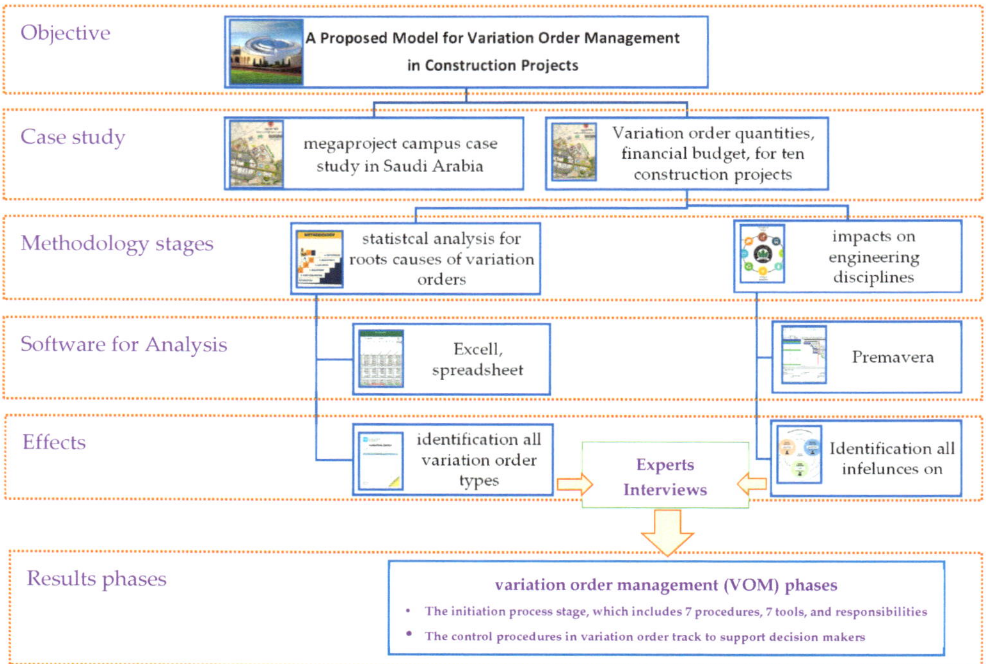

Figure 2. Study method flowchart.

The main case study of this research is the KFU city campus, which was established as an integrated service campus. The case includes 35,000 students and about 5000 employees. It also contains many colleges, service buildings, infrastructures, and housing projects for boys and girls and staff housing. The project is located in an area of 4.5 km^2. It started with the developed plan in 2004 until now, and the actual cost of the projects implemented so far has reached more than 10 billion Saudi riyals. The campus layout contains educational, infrastructure, healthcare, services, residential, and administrations projects with different types, area, and functions. Thus, it is considered as a unique mega project and not a stereotypical project. These campus construction projects had been designed by several international and local designer offices located in different countries, i.e., Europe, USA, Japan, Egypt, Jordan, China, UAE, and Saudi Arabia. Therefore, there is diversity in the contract documents with different software. On the other side, there is also diversity in the supervision and consultant offices for the projects, reaching about 15 consultant offices. In addition, there are different main contractors and subcontractors shared in the construction processes, reaching more than 70 organizations. The project execution status at the study time includes [58–60] some projects that were implemented and handed over during the period from 2015 to 2019, reaching about 33 projects; the projects under construction, totaling 21 projects; and the projects under design and tender, totaling 11 projects. Therefore, there are a total of 65 projects. The selected projects sample included 10 projects with different functions and goals. The selected projects ratio for the total implemented projects (33 projects) reached approximately 30% within the study period.

Figure 3 shows the 10 selected projects from the General Master Plan–KFU–Al-Ahsa–Saudi Arabia. The selected projects sample includes different construction project types, which include housing, service, academic, and administrative projects that have been implemented and operated to enable the study to obtain comprehensive and actual results for discovering the causes and routes of variation orders until each engineering discipline level of launching.

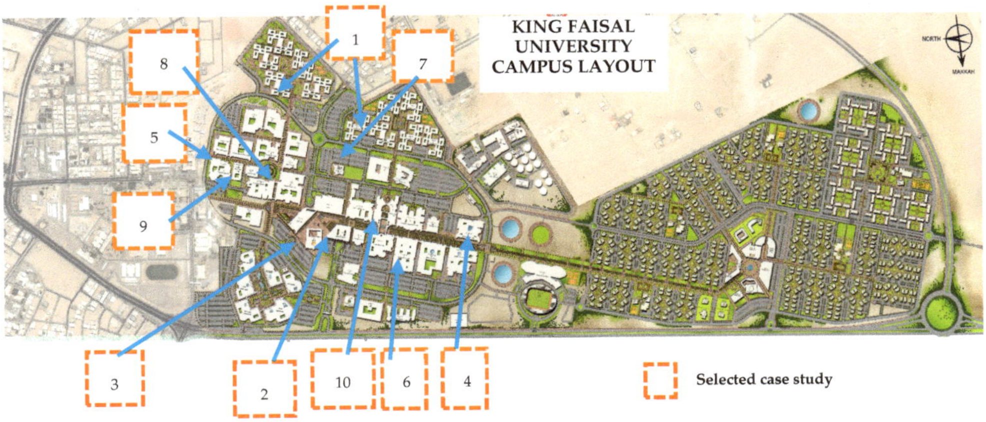

Figure 3. Selected 10 projects locations from General Master Plan–King Faisal University–Al-Ahsa–Saudi Arabia.

The financial values (according to the contract value of the bill of quantities based on the tender documents before handover processes and according to the final invoice values from the contractor after submitting the primary projects handed over) include the following:

- The financial value based on the bill of quantities = 1,196,916,993.70 Saudi riyals
- The final financial value after the handover completion = 1,282,271,123.98 Saudi riyals
- The final financial value of the change = 85,354,130.28 Saudi riyals; the final total change was 7.13%.

Table 1 shows the analysis for the variation orders for selected projects as a case study. The data in Table 1 show the case study addition quantity in a million Saudi riyals, the deduction quantity in a million Saudi riyals, the quantity of variation orders, the variation order percentage, and the quantity of invoice items.

Table 1. Analysis of variation order for 10 projects.

	Study Case Project	Variation Orders				Invoice Item Quantity
		Addition Million Saudi Riyals	Deduction Million Saudi Riyals	VO Quantity Number	Value Percent %	
1	Dormitories (Boys and Girls)	20.3	12.2	5	10	2090
2	Main Administration	2.3	6.8	8	8.4	600
3	Mosque	2.3	6.8	8	8.4	330
4	Computer sciences college (Boys)	11.2	1.4	9	8.7	610
5	Computer sciences college (Girls)	15.9	11.7	11	3.9	1156
6	Science college	11.2	11.3	11	0.3	1208
7	English language	4.9	4.1	12	0.1	1026
8	Education college	14.5	8.5	15	3.2	681
9	Activity (Boys)	17.6	7.8	12	4	326
10	Activity (Girls)	17.7	7.9	12	5.5	485

The highest addition quantity was in the dormitories (boys and girls) project with 20.3 million Saudi riyals, and the lowest quantity was in the main administration and mosque projects, with 2.3 million Saudi riyals. The highest deduction quantity was in the dormitories (boys and girls) project with 12.3 million Saudi riyals, and the lowest quantity was in the English language project with 4.1 million Saudi riyals. The highest variation order quantity was in the education college project with 15 VO, and the lowest variation order quantity was in the dormitories (boys and girls) project with 5 VO. The highest value percentage was in the dormitories (boys and girls) project, and the lowest value percentage was in the English language project. The highest value of invoice item quantity was 2090 items, and the lowest value was 326 items. This analysis explored a significant gap in dealing with variation order processes inside the campus according to different influences from all organizations involved in designing and managing these projects.

3. Results and Discussions

3.1. Analysis of Change Order Types

The data from 10 projects were analyzed to investigate the realistic variation order causes to build variation order management processes as prevention, and controlling for any issue leads to a new project variation order. The results of the five types of changes indicate that the total percentage change was 7.13%, and the change types can be clarified as follows:

Type 1: No change items implemented based on the original design and contract documents with rate 18.65% and arithmetic average 18.58%. This means that 81% of the drawing and original design for all the study sample projects has been changed.
Type 2: Quantity increase with rate + 20.72% of the total value of projects according to the bill of quantities and arithmetic average of 22.97%.
Type 3: Quantity decrease with −10.87% of the total value of projects according to the bill of quantities and arithmetic average of −10.79%.
Type 4: New items with a rate of + 8.94% of the project's total value according to the bill of quantities and arithmetic average of 9.56%.
Type 5: Deducted items with a rate of −11.66% of the total value of projects according to the bill of quantities and an arithmetic average rate of −10.47%.

Figure 4 shows the total percentage of each change type analysis in the 10-construction project case study.

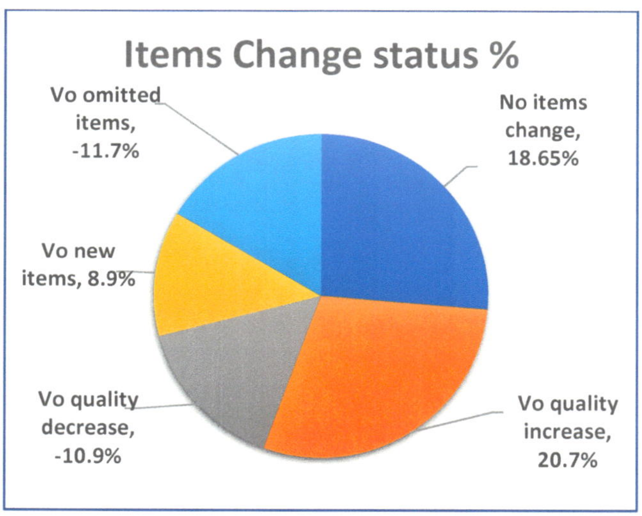

Figure 4. The total percentage of each change type analysis in the case study project.

The statistics analysis investigated each discipline deeply in each change type to investigate the details of changes that lead to the negative and/or positive variation order. The analysis led to the following results. Figure 5 shows a comparison of all total tender BOQ values, actual BOQ values, and total change percentages for each project in the study case.

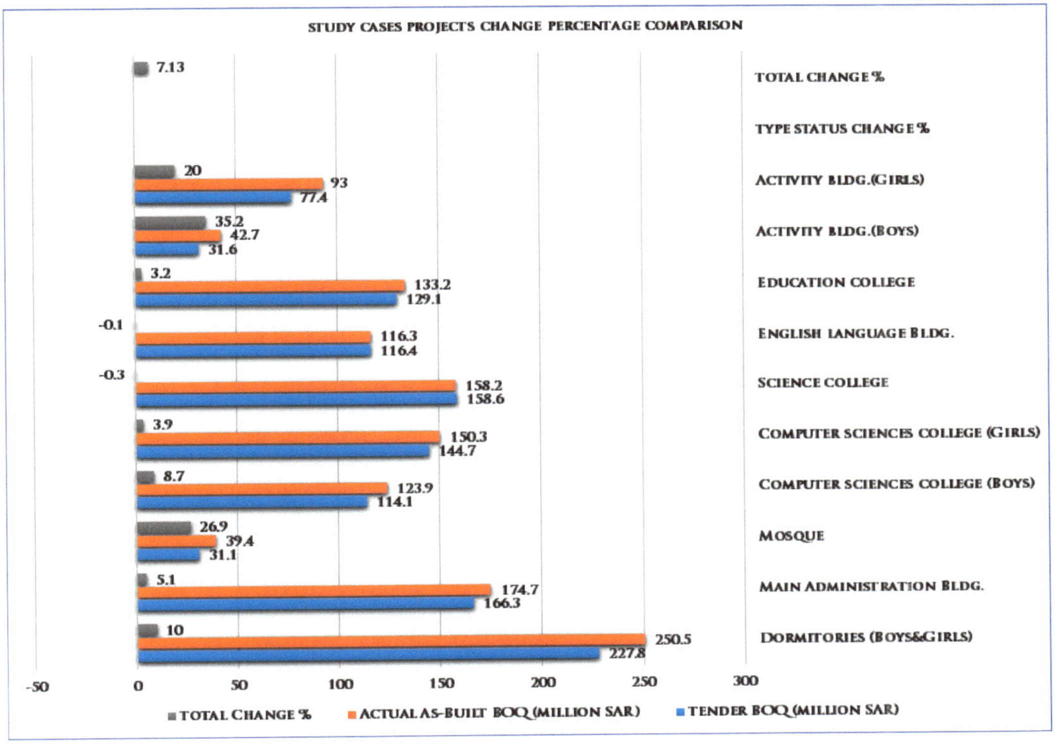

Figure 5. Tender BOQ value, the actual BOQ, and total change % for each project in the case study.

In the no change category, the highest value was in the computer sciences college (girls) project with 31.6%, and the lowest value was in the main administration project with 10.4%. With regard to the VO quantity increase percentage, the highest value was in the activity building (boys) project with 42.5%, and the lowest value was in the science college project with 9.4%. Regarding the VO quantity decrease percentage, the highest value was in the dormitories (boys and girls) project with −14.5%, and the lowest value was in the English language building project with 9.4%. Regarding the VO quantity of new items percentage, the highest value was in the main administration building project with 23.9%, and the lowest value was in the science college building project with 9.4%. Regarding the VO omitted quantity items percentage, the highest value was in the dormitories (boys and girls) project with −12.5%, and the lowest value was in the activity project (girls) with −0.3%. Figure 6 shows the analysis status for each change type for each study case project.

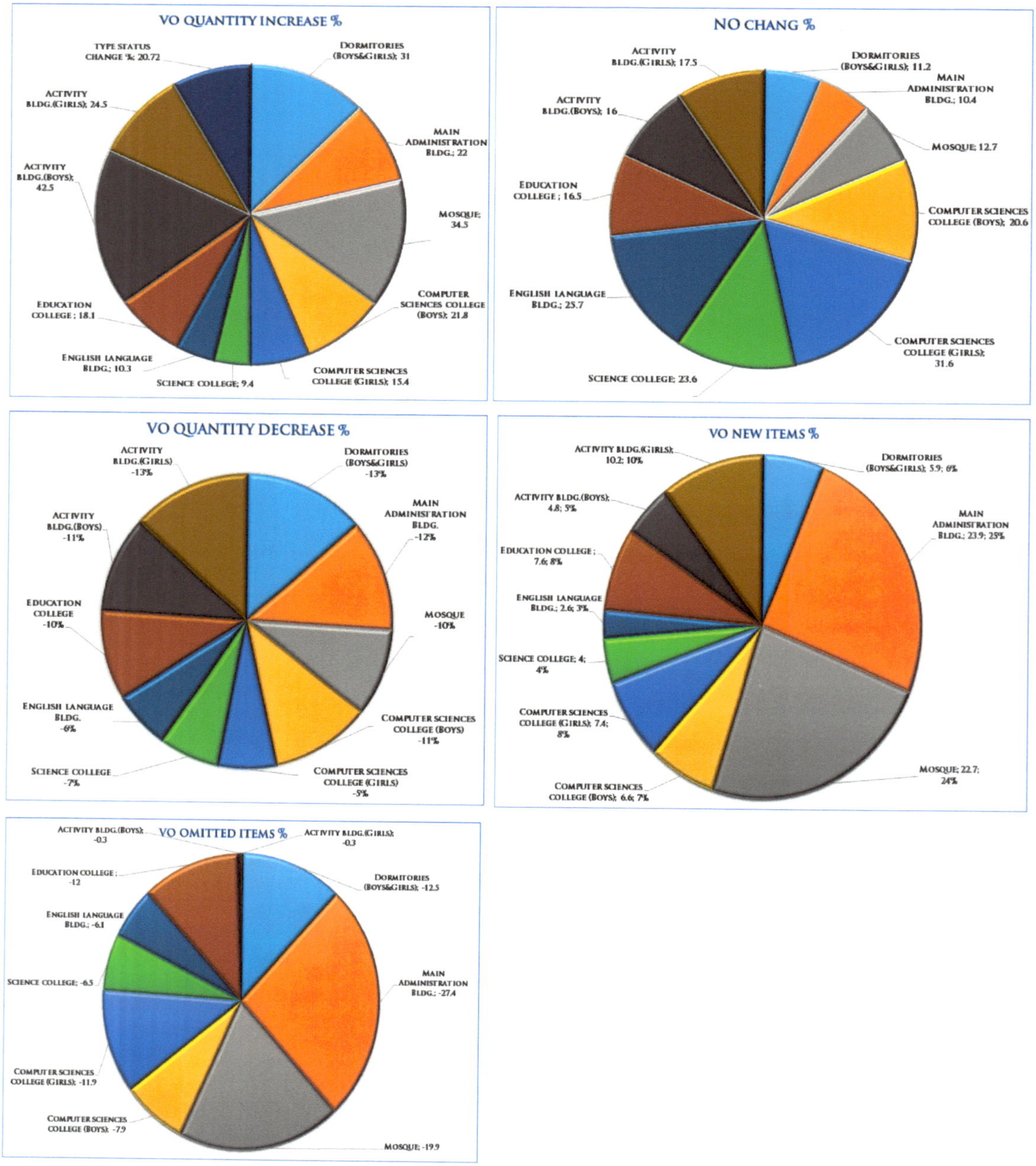

Figure 6. The analysis status for each change type for each study case project.

3.2. *The Design and Tender Document Quality*

The designer's work on the design documents influenced all case study projects in increasing and decreasing item types, which caused variation orders. The total designer effect on the design document of dormitories (boys and girls) was 45.5%, the percentage for increasing item types was 31%, and the percentage for decreasing item types was −14.5%. The total designer effect on the design document of the main administration project was 35.5%; for increasing item types, it was 22%, and for decreasing item types, it was −13.4%. The total designer effect on the design document of the mosque building project was 44.9%, the percentage for increasing item types was 34.5%, and for decreasing item types, the percentage was −10.4%. The total designer effect on the design document of the computer sciences college project (boys) was 33.8%; for increasing item types, it was 21.8%, and for decreasing item types, it was −11.9%. The total designer effect on the design document of the computer sciences college project (girls) was 22.4%; for increasing item types, it was 15.4%, and for decreasing item types, it was −7%. The total designer effect on the design document of the science college project was 16.6%; for increasing item types, it was 9.4%, and for decreasing item types, it was −7.2%. The total designer effect on the design document of the English language project was 17.2%; for increasing item types, it was 10.3%, and for decreasing item types, it was −6.9%. The total designer effect on the design document of the education college project was 28.7%; for increasing item types, it was 18.1%, and for decreasing item types, it was −10.6%. The total designer effect on the design document of the activity project (boys) was 54.2%; for increasing item types, it was 42.5%, and for decreasing item types, it was −11.7%. The total designer effect on the design document of the activity project (girls) was 38.9%; for increasing item types, it was 24.5%, and for decreasing item types, it was −14.3%. Figure 7 shows the statistical analysis for designer and design document effects on causing variation orders in the case study projects to investigate the weakness gates to be avoided.

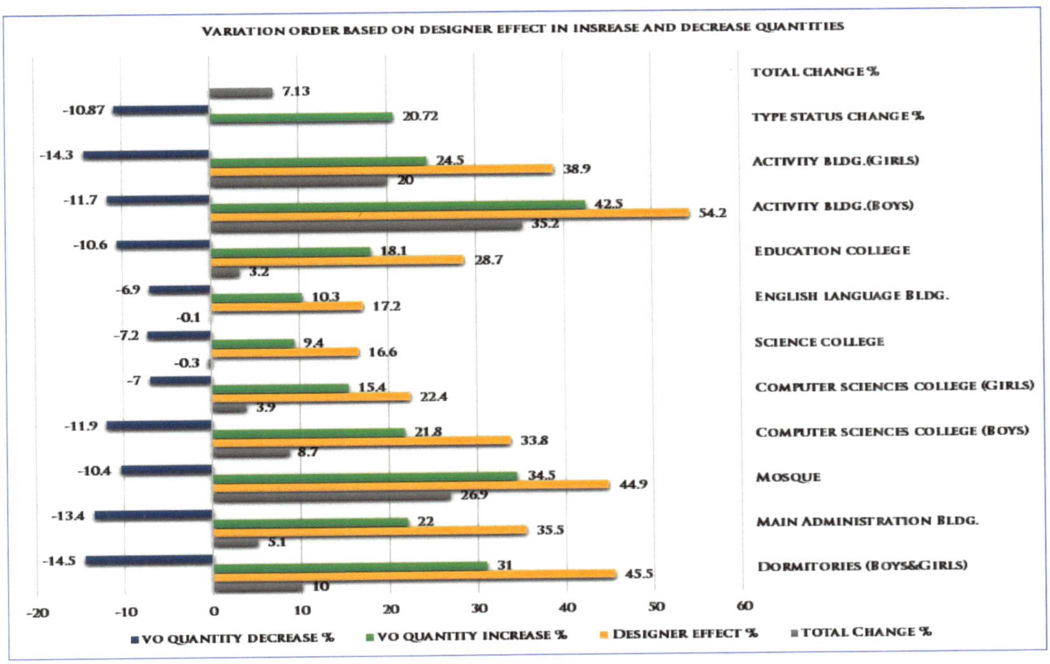

Figure 7. Statistical analysis for designer and design document effects on causing variation orders.

Therefore, the level of inefficiency and lack of preparation quality of contract documents is because of the inaccuracy of the inventory of quantities. This led to the difference, which is greater than the proportions allowed contractually and is the responsibility of the designer office and the participants in the preparation of documents for the tasks of the designer to investigate the accuracy quality of the bill of quantities and non-conflicting descriptions in the tender documents.

3.3. The Owner Team Changes

The owner's work on the design documents influenced all case study projects in the areas of new items and omitted item types, which caused variation orders. The total owner effect on the design document and execution procedures of dormitories (boys and girls) was 18.3%, the percentage for new item types was 5.9%, and the percentage for omitted item types was −12.5%. The total owner effect on the design document and execution procedures of the main administration project was 51.3%; for new item types, it was 23.9%, and for omitted item types, it was −27.4%. The total owner effect on the design document and execution procedures of the mosque building project was 42.7%; the percentage for increasing item types was 22.7%, and the percentage for decreasing item types was −19.9%. The total owner effect on the design document and execution procedures of the computer sciences college project (boys) was 14.5%; for new item types, it was 6.6%, and for omitted item types, it was −7.9%. The total owner effect on the design document and execution procedures of the computer sciences college project (girls) was 19.3%; for new item types, it was 15.4%, and for omitted item types, it was −11.9%. The total owner effect on the design document and execution procedures of the science college project was 10.5%; for increasing item types, it was 4%, and for decreasing item types, it was −6.5%. The total owner effect on the design document of the English language project was 8.7%; for new item types, it was 2.6%, and for omitted item types, it was −6.1%. The total owner effect on the design document and execution procedures of the education college project was 19.5%; for new item types, it was 7.6%, and for omitted item types, it was −12%. The total owner effect on the design document and execution procedures of the activity project (boys) was 5.1%; for new item types, it was 4.8%, and for omitted item types, it was −0.3%. The total owner effect on the design document and execution procedures of the activity project (girls) was 10.5%; for new item types, it was 10.2%, and for omitted item types, it was −0.3%. Figure 8 shows the statistical analysis for owner and design document and execution procedures effects for causing variation orders in the case study projects to investigate the weakness gates to be avoided.

The influence of the owner's changes in new and omitted items must be justified technically and financially. There is a joint effect for the designer and the technical authority team of the owner from the changes. The rate of the absolute increase in work, whether by increasing the quantity or the development according to the needs of the projects, represents (22.97% + 9.56%) = (+32.53%) for the increase in quantities, whereas the absolute rate of deduction in the work, whether by reducing the quantity or eliminating it by canceling items according to the needs of the projects, represents (10.79% + 10.47%) = (−21.26%).

3.4. The Changes in Disciplines Scope and Value

The analysis of the variation orders at the engineering disciplines level indicates significant results in civil, architectural, mechanical, and electrical engineering disciplines according to change types. The total change in the civil discipline reached about 14.14%.

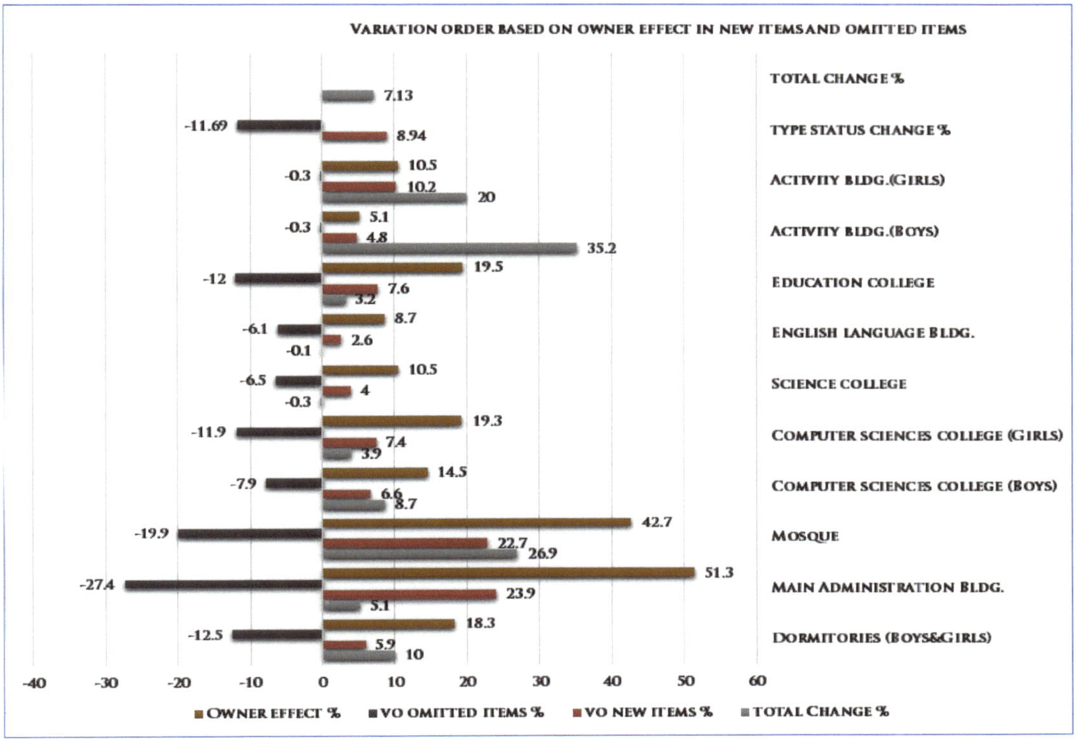

Figure 8. Statistical analysis for owner and design document and execution procedures effects for causing variation orders.

While the percentages for the civil discipline related to the no change category, quantity increase category, quantity decrease category, quantity new items category, and omitted quantity items category were, sequentially, 1.76%, 28.31%, −11.58%, 3.34%, and −5.93%. The total change in the architectural discipline reached about 9.6%, while the percentages for the architectural discipline related to the no change, quantity increase, quantity decrease, quantity new items, and omitted quantity items categories were, sequentially, 11.5%, 18.9%, −11.14%, 16%, and −14.1%. The total change in the mechanical discipline reached about −6.1%, while the percentages for the mechanical discipline related to the no change category, quantity increase category, quantity decrease category, quantity new items category, and omitted quantity items category were, sequentially, 37.58%, 16.3%, −15.2%, 36%, and −10.2%. The total change in the electrical discipline reached about 6.25%, while the percentages for the electrical discipline related to the no change category, quantity increase category, quantity decrease category, quantity new items category, and omitted quantity items category were, sequentially, 36.52%, 18.2%, −4.8%, 9.3%, and −16.5%. In Figure 9, the total change percentage for each discipline is compared with each change type in the study case's projects.

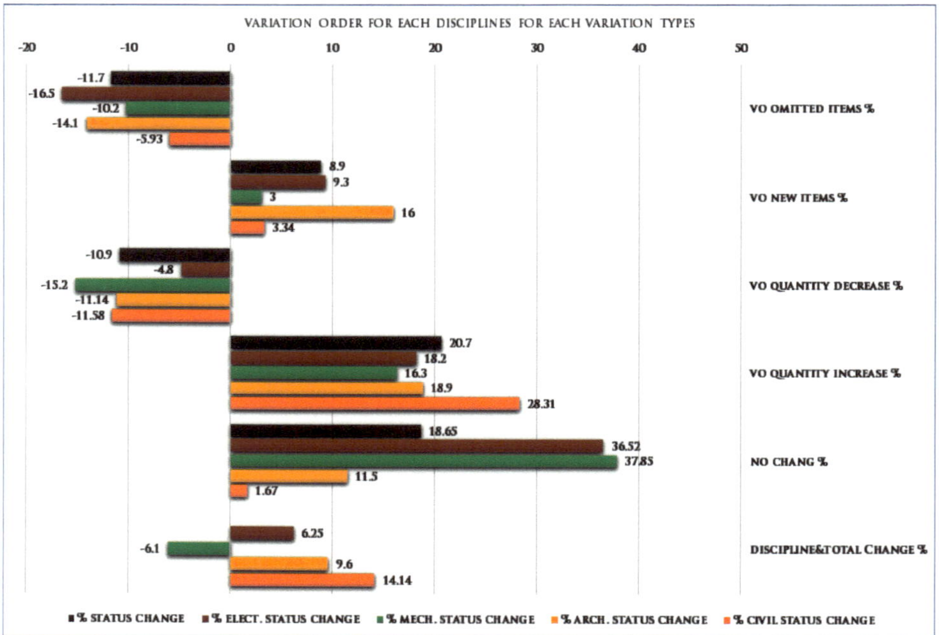

Figure 9. Total change % for each discipline compared with each change type in the study case's projects.

The change rate in the architectural discipline was according to owner requirements to be matched with new technology and modernized in the sections of metalwork, wooden works, doors and windows, finishes, special works, and furniture. The change rate in civil works was based on redesigning the foundation structure calculation and increasing quantities for excavations and backfilling. The change rate in the electrical works that encountered changes was based on redesigning the electrical loads to be matched with mechanical and electrical equipment, as well as increasing the wiring, cables, and electrical choices. The change rate in the mechanical works that encountered changes was based on redesigning the mechanical loads to be matched with mechanical equipment from the suppliers, as well as increasing the ducting, firefighting, and plumbing fixtures choices.

3.5. The Effect of Changes in Projects on Claims and Disputes

The sample of projects under study has not been subjected to any kind of suspension, force majeure, or lack of financial allocations from the owner during implementation, and there are no penalties or contractual deductions affecting the contractor. The claims for financial cost have been compensated because of the increase in quantities due to the designer or the creation of new items due to the owner according to the approved change orders, which have resulted in an increase in the cost of all projects compared to the contract value in the bill of quantities.

The time claims for the project extension have been compensated for another period. The arithmetic average of the extension period to the original project period is 42.17% as a result of the approved changes in the increase in quantities and the creation of new items. Changes in general have resulted in improving the quality of all systems and materials included in the various engineering departments, such as the architectural department and electromechanical department, in line with the contemporary developments and technology of the project implementation lifecycle due to the large period between the completion phase of the design of all tender documents, the tender and awarding phase, and the start of implementation phase, so that many systems and materials have meanwhile developed in the production market for engineering construction projects.

3.6. Variation Order Management (VOM) as a Proposed Support Methodology

Variation orders in construction projects are one of the most decisive factors affecting delays in project implementation and one of the most essential sources of claims and disputes for compensation over time or cost. The results from the analysis of 10 selected completed projects representing 30% of the ongoing construction campus projects emphasize study questions about the reasons for change orders leading to claims or disputes. The analysis shows the impacts numerically on each variation type (no change, decrease quantities, increase quantities, new items, omitted items) and on each technical discipline (architecture, civil, mechanical, electrical). This analysis takes into consideration the sequence of traditional variation control processes and proposes a new approach to make decisive control for variation orders based on overcoming the gaps found in study cases that lead to variation orders; this approach takes into consideration the results of the analysis as follows:

- Weakness of the tender document's preparation by the designer, whether for design work, quantification, or conflict between documents, is among the most important causes of changes/variation orders and sources of claims and disputes.
- An integrated and quantitative specific system of procedures helps the owner to manage and control changes and variation orders in a planned manner at the entire project level and not in a partial phase.
- Using preventive control for variation did not depend on the work progress.
- Procedures to avoid lack of technical and financial control over time, cost, quality, and scope reduce the risk of claims and disputes.

The study proposed a quantitative system to adopt VOM before or within the project start date and site preparation. The flow of processes of this model was designed in three stages as follows:

3.6.1. Procedures in the Initiation Process Stage

This stage includes a proactive process conducted and shared with all involved organizations' teams after awarding the contract to the lowest-price contractor. This process acts as the first preventative action to avoid variation orders from the start date of the project. The process is divided into procedures, consequent responsibilities, and tools for each procedure, which means preparing the procedures, responsibilities, and tools during the initiation stage of the project implementation to discover the potential changes in the execution process stage. The process is considered to be precautionary measures derived from the previous analysis of the projects selected as a documented case study of a large vital project established in Saudi Arabia. This initiation process stage includes the following:

- An inventory of the architectural, civil, electrical, and mechanical works, comparing them to the tender documents, which is the responsibility of the contractor and needs the approval of the supervising authority and the owner using the Building Information Modeling (BIM) and REVIT program.
- Presenting and approving long-term item schedules (long lead items) for all disciplines, anticipating the time of supply and installation, which are the responsibility of the contractor and require the approval of the supervising authority and the owner using the PRIMAVERA program.
- Specifying the cost of the architectural, civil, electrical, and mechanical works, which is the responsibility of the supervising authority and requires the approval of the owner using the bill of quantities and specifications.
- Specifying the influence of stakeholders on approving, which is the responsibility of the owner using meetings and specific models.

- Specifying a record of previous risks, which is the responsibility of the owner using site surveys, interviews, presentations, and site visits.
- Specifying a list of approved suppliers and sub-contractors, which is the responsibility of the owner using site surveys, interviews, presentations, and site visits.
- The approval of all calculations for structural, mechanical, and electrical systems is the responsibility of suppliers and subcontractors and requires the approval of the owner using programs such as SAP.

3.6.2. Procedures for Dealing with the Course of Change Orders Stage

This stage is divided after the need to submit the change order, based on either new items or increases in quantities. In both cases, sequential procedures are used that end with a quantitative determination that controls the approval or disapproval of the change orders.

In the case of new items, a series of procedures can be followed as follows:

- Conducting the initial determination of the reason for the request to make a variation order if it is a formal request from the stakeholders, improving the quality or technological progress or an inevitable technical necessity, and then agreeing to continue the study with technical and financial analysis or refusing to study the change order.
- Procedures for studying and analyzing the quantitative criteria, technically and financially, to determine the degree of approval or disapproval of a change order, which can be controlled through the four main focuses in the project management methodology (quality, cost, lifecycle, and project scope).
- Inspection by the specialists of technical, financial, and contractual analysis: first, the quality department to prove compliance with the technical specifications of the item with a rate of 30%; second, the scheduling department to study the impact on the project time and supply and installation time with a rate of 30%; third, other technical departments for determining the impact and technical and financial relationship with a rate of 20%; fourth, the department of accounting and finance to determine its impact on the cost of the technical department with a rate of 10%; and fifth, using value engineering to study alternatives for cost, supply time, installation time, and technical compliance with item specifications at a rate of 30%. Figure 10 illustrates the methodology of the VOM approach as a significant finding for this study.

3.6.3. Procedures for the Decision Stage

This stage is the final decision from the organization's decision maker and its committee according to the summation of the variation order qualitative impact after gathering all qualitative impacts from all related departments; therefore, the approved status for the item has a summation qualitative impact of over 75%.

The results showed that traditional analytical diagrams (Figure 1) are commonly used in most construction projects. However, there are no technical and financial procedures that can be used as a proactive step to reduce or minimize the excessive volume of change orders. In addition, there are no updated programs or software used in technical analysis and comparison in conjunction with the financial study according to market prices. Furthermore, there is no evidence of how to adjust the financial balance of previous competitors participating in the project implementation competition, which does not lead to potential judicial disputes [45]. This study focuses on change order management and controlling inside construction project processes and the difficulty encountered in the project management processes in variation order control and management. Therefore, the practical model based on a comprehensive analysis of one mega project provided a best practices approach, including three stages or phases to eliminate and control variation order influences within the lifecycle of a project. The first phase includes procedures in the initiation process stage, which includes seven procedures, seven tools, and key responsibilities. The second phase includes procedures for dealing with the course of the change orders stage. The third phase includes procedures for the decision stage to support decision makers in proceeding with the variation order or not.

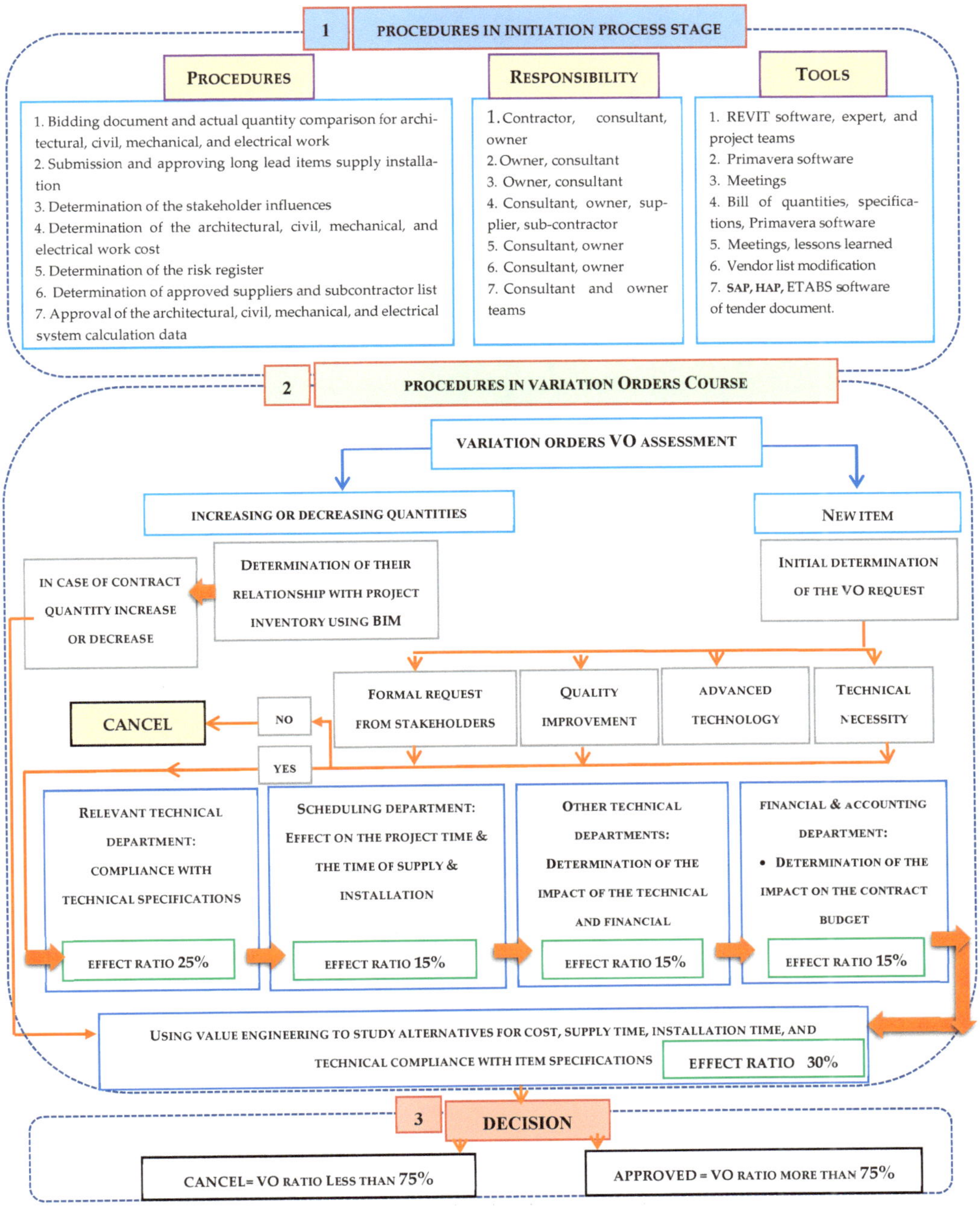

Figure 10. A proposed methodology for variation orders management.

The study results submitted a robust variation order model enabled the client technical representative team to conduct several efforts and changes related to several aspects on purpose to eliminate and control the occurrence of variation orders. These aspects include contractual aspects, including modification of the design contract document and the bidding document for consultant and contractor for the organizations working on the remaining 32 projects of the case study. These modifications include the contractual aspects, the environmental aspects, and the social aspects. The contractual aspects include the design phase and execution phase for each project throughout the main five project documents. These consist of the bill of quantities, specifications, drawings, and general/special conditions to improve the accounting of quantities, avoid document conflicts, adjust delivery time and installation time in each project's scheduled time for the long lead items, and conduct value engineering methods and processes. Therefore, the client obligated all designers to update the design software to Revit software, improve the input data in Primavera software, conduct value engineering processes, identify the risk items in long lead items supply and installation, identify authorized stakeholders, improve tracking procedures in the project schedule from purchase order until installation for each item, and identify and approve the deliverables processes and commitment with project scope procedures. Thus, the client committee, after applying VOM, discovered significant numerical results in variation order quantity and its influence on the deduction of the remaining project budget, with a total average of 68%. Whereas, in the study case sample project (10 projects), the additional items causing the additional project budget was 11.79% (from Table 1), the deduction items causing the reduction in project budget was 7.85% (from Table 1), and the average of the total VO quantity was 10.3. On the other side, in the study case remaining project (32 projects), the client committee, after applying VOM, discovered that the additional items caused the additional project budget to be 3.2% (from Table 1), and the deduction items caused the reduction in the project budget to be 2.5% (from Table 1). The average of the total VO quantity was 3.3. The environmental aspects improvement appeared in the cohesive specifications between the manufacturing source in the vendor list qualifications and the installing method in the all-discipline items. The social aspects appeared in developing robust communication channels for all team members for the seven organizations working on the campus, which include client, designer, consultant, main contractor, subcontractor, supplier, and manufacturer.

The study provided the educational organizations in construction project management with a robust model as lessons learned to mitigate the risks associated with construction projects. This study provided the international and local organizations working in the construction project management industry with a comprehensive model that can be updated according to project conditions. This study provided the nonprofit international and local organizations working on issuing project management guidelines with an update on the processes of risk and integration management and chapters working in the construction projects management industry with a comprehensive model that can be adjusted according to project conditions. This study opens the gate for other studies to update the variation order controlling models in commercial and healthcare projects, merging global economic and environmental cost impacts on variation order, and compromising the global and local crises with codes and contractual conditions can be another area of research.

These study results can contribute effectively to the Saudi Vision 2030, which has three main pillars: a vibrant society, a thriving economy, and an ambitious nation. The study results can contribute to the thriving economic pillar of the construction industry. The study supports diversifying the economy in public investment with minimum risks in the construction project budget, enhancing assets growth of the public investment fund, and localizing edge technology and knowledge through the public investment fund with a robust system for minimizing the occurrence of variation orders in construction projects. The study contributes effectively to enhancing government effectiveness by presenting a comprehensive method to control the construction project budget and the public balance construction projects budget, which improves the performance of the government apparatus in the construction industry. In addition, the study results contribute to enhancing the effectiveness of financial planning and the efficiency of government spending in the construction industry by enhancing the performance of government entities working in the construction industry by developing a robust matrix for controlling the causes of cost overruns in construction projects.

The study encountered several challenges, such as environmental culture. These included discussions and negotiations about technical and financial solutions for variation orders in the construction industry, stockholder interpretation in the financial and technical procedures of the construction projects, compensation difficulties from the related ministries, technical weakness of the main contractor organization, and lack of updating the construction project information from the consultant office. Despite the study being limited and designed for specific conditions and types, VOM processes and procedures contain flexibility and the possibility to be updated and adjusted for other construction projects according to construction project conditions, location, time, client nature, financial source, and organization type.

4. Conclusions

Variation orders in construction project management are one of the most complex challenge processes for project success because of their direct influence on scope, time, cost, and quality processes. The processes for prediction, prevention, and controlling the variation orders in construction project management within the project lifecycle, starting from the design phase until the handover phase, are a great concern for all construction project organizations. Therefore, the study built a variation order management model applicable to construction project processes based on financial and technical analysis for the 10 selected implemented construction project types and functions that constitute 30% of the implemented construction project cost in the megaproject case study. The selected projects in the case study contained 103 variation orders with a total cost of 86 million Saudi riyals. Therefore, the study, based on interviews with experts, provided a comprehensive and numerical analysis for the selected projects of the case study using several software and documents, i.e., Excel spreadsheets, Revit, AutoCAD, and Primavera. Final invoices include initial tender quantities of items, variation order quantities of items for all disciplines, and the final as-built quantities of items to discover the variation order causes implemented in all engineering disciplines until the project primary handover.

The results enabled the study to establish, develop, and classify a new model for the VOM approach as a best practice approach. This approach includes three stages. The first phase is the initiation process, which includes seven procedures, seven tools, and key responsibilities. The second stage deals with the course of change orders based on a certain number of procedures and weights for each parameter assigned to this stage to support decision processes based on a certain average ratio of weights calculation to support decision makers in proceeding with a variation order or not. The third stage is the decision stage to support decision makers in proceeding or not proceeding with a variation order. The study tested the results by applying VOM in the remaining 32 projects, which supported the decision maker of the case study campus organization to eliminate and control the variation order. The study contributes to the Saudi Vision 2030 in thriving the Kingdom's economy, albeit it opens the gates for several studies in variation order management in the construction industry.

Author Contributions: Conceptualization, E.M.H.I. and A.E.E.S.; methodology, E.M.H.I.; software, E.M.H.I.; validation, E.M.H.I. and A.E.E.S.; formal analysis, E.M.H.I.; investigation, E.M.H.I.; resources, E.M.H.I. and A.E.E.S.; data curation, E.M.H.I.; writing—original draft preparation, E.M.H.I. and A.E.E.S.; writing—review and editing, E.M.H.I. and A.E.E.S.; visualization, E.M.H.I. and A.E.E.S.; supervision E.M.H.I. and A.E.E.S.; project administration, E.M.H.I.; funding acquisition, E.M.H.I. and A.E.E.S. All authors have read and agreed to the published version of the manuscript.

Funding: This work was supported by the Deanship of Scientific Research, Vice Presidency for Graduate Studies and Scientific Research, King Faisal University, Saudi Arabia [Project No. GRANT5447].

Data Availability Statement: Data are available upon request from the first author.

Conflicts of Interest: The authors declare no conflicts of interest.

References

1. Construction Extension to the Pmbok®Guide, PMI Institute. 2021. Available online: http://faspa.ir/wp-content/uploads/2017/03/faspa.ir-Constrauction.pdf (accessed on 10 January 2024).
2. *AIA Document A201–2017 General Conditions of the Contract for Construction*; American Institute of Architects: Washington, DC, USA, 2017.
3. Mahamid, I. Effect of change orders on rework in highway projects in Palestine. *J. Financ. Manag. Prop. Constr.* **2017**, *22*, 62–76. [CrossRef]
4. Hussein, A.F.F.; Al-Mamary, Y.H. S, Conflicts: Their types, and their negative and positive effects on organizations. *Int. J. Sci. Technol. Res.* **2019**, *8*, 10–13.
5. Hughes, W.; Champion, R. *Construction Contracts: Law and Management*; Routledge: London, UK, 2021; Volume 87.
6. Balbaa, A.A.K.; El-Nawawy, O.A.M.; El-Dash, K.M.; El-Mageed Badawy, M.B.A. Risk assessment for causes of variation orders for residential projects. *J. Eng. Appl. Sci.* **2019**, *14*, 701–708.
7. Mohammad, N.; Ani, A.I.C.; Rakmat, R.A.O. Causes and effects of variation orders in the construction of terrace housing projects: A case study in the state of Selangor, Malaysia. *Int. J. Supply Chain Manag.* **2017**, *6*, 226–232.
8. Naji, K.K.; Gunduz, M.; Naser, A.F. The effect of change-order management factors on construction project success: A structural equation modeling approach. *J. Constr. Eng. Manag.* **2022**, *148*, 04022085. [CrossRef]
9. Msallam, M.; Abojaradeh, M.; Jrew, B.; Zaki, I. Controlling of variation orders in highway projects in Jordan. *J. Eng. Archit.* **2015**, *3*, 95–104. [CrossRef]
10. Khanzadi, M.; Nasirzadeh, F.; Dashti, M.S. Fuzzy cognitive map approach to analyze causes of change orders in construction projects. *J. Constr. Eng. Manag.* **2018**, *144*, 04017111. [CrossRef]
11. Mohammad, K.H.; Ali, N.S.; Najm, B.M. Assessment of the cost and time impact of variation orders on construction projects in Sulaimani governorate. *J. Eng.* **2021**, *27*, 106–125. [CrossRef]
12. Alsuliman, J.A. Effective Stakeholder Engagement in Variation Order Management at the Design Stage of Public Sector Construction Projects in Saudi Arabia. Ph.D. Thesis, Heriot-Watt University, Edinburgh, UK, 2014.
13. El-Sayegh, S.; Ahmad, I.; Aljanabi, M.; Herzallah, R.; Metry, S.; El-Ashwal, O. Construction disputes in the UAE: Causes and resolution methods. *Buildings* **2020**, *10*, 171. [CrossRef]
14. Hayati, K.; Latief, Y. Risk analysis and prevention system to minimize claim and dispute on construction projects. *IOP Conf. Ser. Earth Environ. Sci.* **2019**, *365*, 012030. [CrossRef]
15. Illankoon, I.M.C.S.; Tam, V.W.; Le, K.N.; Ranadewa, K.A.T.O. Causes of disputes, factors affecting dispute resolution and effective alternative dispute resolution for Sri Lankan construction industry. *Int. J. Constr. Manag.* **2019**, *22*, 1–11. [CrossRef]
16. Nachatar, J.S.; Hussin, A.A.; Omran, A. Variations in government contract in Malaysia. *Manag. J.* **2010**, *12*, 40–53.
17. Oyewobi, L.O.; Jimoh, R.; Ganiyu, B.O.; Shittu, A.A. Analysis of causes and impact of variation order on educational building projects. *J. Facil. Manag.* **2016**, *14*, 139–164. [CrossRef]
18. Mahmoud, S.Y.; Elshaikh, E.A. The potential effects of variation orders on building projects in Khartoum State-Sudan. *Int. J. Constr. Eng. Manag.* **2019**, *8*, 70–79.
19. Gobana, A.B.; Thakur, A.S. Critical review on causes and effects of variation order on construction project. *Crit. Rev.* **2017**, *4*, 1602–1606.
20. Enshassi, A.; Arain, F.; Al-Raee, S. Causes of variation orders in construction projects in the Gaza Strip. *J. Civ. Eng. Manag.* **2010**, *16*, 540–551. [CrossRef]
21. Sadewo, A.P. Identifying Causes, Effects, Benefits, and Remedies of Change Orders on Building Projects in Yogyakarta. Ph.D. Thesis, Uajy, Sleman, Indonesia, 2018.
22. Demirkesen, S.; Ozorhon, B. Impact of integration management on construction project management performance. *Int. J. Proj. Manag.* **2017**, *35*, 1639–1654. [CrossRef]
23. Diekmann, J.E.; Nelson, M.C. Construction claims: Frequency and severity. *J. Constr. Eng. Manag.* **1985**, *111*, 74–81. [CrossRef]
24. Assaf, S.A.; Al-Hejji, S. Causes of delay in large construction projects. *Int. J. Proj. Manag.* **2006**, *24*, 349–357. [CrossRef]
25. Desai, N.; Pitroda, J.; Bhavsar, J. A review of change order and assessing causes affecting change order in construction. *Int. Acad. Res. Multidiscip.* **2015**, *2*, 152–162.
26. Bakhary, N.A.; Adnan, H.; Ibrahim, A. A study of construction claim management problems in Malaysia. *Procedia Econ. Financ.* **2015**, *23*, 63–70. [CrossRef]
27. Arain, F.M. Strategic management of variation orders for institutional buildings: Leveraging on information technology. *Proj. Manag. J.* **2005**, *36*, 66–77. [CrossRef]
28. Memon, A.H.; Rahman, I.A.; Hasan, M.F.A. Significant causes and effects of variation orders in construction projects. *Res. J. Appl. Sci. Eng. Technol.* **2014**, *7*, 4494–4502. [CrossRef]
29. Elziny, A.A.; Mohamadien, M.A.; Ibrahim, H.M.; Fattah, M.A. An expert system to manage dispute resolutions in construction projects in Egypt. *Ain Shams Eng. J.* **2016**, *7*, 57–71. [CrossRef]
30. Alnuaimi, A.; Taha, R.; Mohsin, M.; Alharthi, A. Causes, Effects, Benefits, and Remedies of Change Order on Public Construction Projects in Oman. *J. Constr. Eng. Manag.* **2010**, *136*, 615–622. [CrossRef]
31. Ndihokubwayo, R. An Analysis of the Impact of Variation Orders on Project Performance. Ph.D. Thesis, Cape Peninsula University of Technology, Cape Town, South Afirca, 2008.

32. Tran, N.N.; Do, S.T.; Nguyen, T.A.; Le, L.H. *Variation Order Management in Vietnam Construction Projects*; Springer: Singapore, 2020; pp. 1007–1014.
33. Soliman, E. Recommendations to mitigate delay causes in Kuwait construction projects. *Am. J. Civil Eng. Archit.* **2017**, *5*, 253–262. [CrossRef]
34. Ammar, Y. *Management of Change in Construction Projects—Study of Its Causes and Analysis of Its Effects on Claims and Disputes*; Faculty of Engineering in Matarya—Helwan University: Cairo, Egypt, 2016.
35. International Federation of Consulting Engineers. *FIDIC Conditions of Contract for Construction for Building and Engineering Works Designed by the Employer*; International Federation of Consulting Engineers: Geneva, Switzerland, 1999.
36. Khalifa, W.M.; Mahamid, I. Causes of change orders in construction projects. *Eng. Technol. Appl. Sci. Res.* **2019**, *9*, 4956–4961. [CrossRef]
37. Rashid, I.; Elmikawi, M.; Saleh, A. The impact of change orders on construction projects sports facilities case study. *J. Am. Sci.* **2012**, *8*, 628–631.
38. Zawawi, N.A.W.A.; Azman, N.F.I.N.; Kamar, S.; Shamil, M. Sustainable construction practice: A review of change orders (CO) in construction projects. In Proceedings of the International Conference on Environment, Pulau Pinang, Malaysia, 23–25 March 2010.
39. Arun, C.O.; Rao, B.N. Knowledge based decision support tool for duration and cost overrun analysis of highway construction projects. *J. Inst. Eng.* **2007**, *88*, 27–33.
40. Wu, C.; Hsieh, T.; Cheng, W. Statistical Analysis of Causes for Design Change in Highway Construction in Taiwan. *Int. J. Proj. Manag.* **2005**, *23*, 554–563. [CrossRef]
41. Alaryan, A.; Beltagi, E.; Elshahat, A.; Dawood, M. Causes and effects of change orders on construction projects in Kuwait. *Int. J. Eng. Res. Appl.* **2014**, *4*, 1–8.
42. Soomro, F.A.; Memon, M.J.; Chandio, A.F.; Sohu, S.; Soomro, R. Causes of time overrun in construction of building projects in Pakistan. *Eng. Technol. Appl. Sci. Res.* **2019**, *9*, 3762–3764. [CrossRef]
43. Dastyar, B.; Esfahani, A.F.; Askarifard, M.; Abbasi, A.M. Identification, Prioritization and Management of Construction Project Claims. *J. Eng. Proj. Prod. Manag.* **2018**, *8*, 90–96. [CrossRef]
44. Hanna, A.; Camlic, R.; Peterson, P.; Lee, M. Cumulative effect of project changes for electrical and mechanical construction. *J. Constr. Eng. Manag.* **2004**, *130*, 762–771. [CrossRef]
45. Perkins, R.A. Sources of changes in design/build contracts for a governmental owner, management of engineering and technology. *Portland Int. Cent. Publ.* **2009**, *5–9*, 2148–2153.
46. Hwang, B.G.; Zhu, L.; Ming, J.T.T. Factors affecting productivity in green building construction projects: The case of Singapore. *J. Manag. Eng.* **2017**, *33*, 04016052. [CrossRef]
47. Johnson, J. Practitioner's Forum: Construction Quality Assurance under Change Conditions. *J. Archit. Eng.* **2000**, *6*, 103–104. [CrossRef]
48. McEniry, G.; Ibbs, W. The cumulative effect of change orders on labour productivity: The Leonard study 'reloaded'. *Revay Rep* **2007**, *26*, 1–8.
49. Sun, M.; Sexton, M.; Aouad, G.; Fleming, A.; Senaratne, S.; Anumba, C.; Chung, P.; El-Hamalawi, A.; Motawa, I.; Yeoh, M.L. Managing changes in construction projects. *EPSRC Ind. Rep.* **2004**, 185–190.
50. Love, P.; Irani, Z.; Edwards, D. A Rework Reduction Model for Construction Projects. *IEEE Trans. Eng. Manag.* **2004**, *51*, 426–440. [CrossRef]
51. Osama, M.; Assem, I.; El-Rayes, K. Change order impacts on labor productivity. *J. Constr. Eng. Manage.* **2005**, *13*, 354–359.
52. Ibbs, W.; Nguyen, L.D.; Lee, S. Quantified impacts of project change. *J. Prof. Issues Eng. Educ. Pract.* **2007**, *133*, 45–52. [CrossRef]
53. Assbeihat, J.M.; Sweis, G.J. Factors affecting change orders in public construction projects. *Int. J. Appl.* **2015**, *5*, 56–63.
54. Hao, Q.; Shen, W.; Neelamkavil, J.; Thomas, R. Change management in construction projects. In Proceedings of the International Conference on Information Technology in Construction, Beijing, China, 16–18 October 2008.
55. Yadeta, A.E. The impact of variation orders on public building projects. *Int. J. Constr. Eng. Manag.* **2016**, *5*, 86–91.
56. Alnaas, K.A.A.; Khalil, A.H.H.; Nassar, G.E. Guideline for preparing comprehensive extension of time (EoT) claim. *HBRC J.* **2014**, *10*, 308–316. [CrossRef]
57. Thani, H.A.; Saleh, M.E.; Amer, N.H. Identification and analysis of main reasons of variation orders and their impacts during construction project life cycle. In Proceedings of the International Conference on Civil and Architecture Engineering, Cairo, Egypt, 3–5 April 2018; pp. 1–19.
58. *Quarter Report, Quality Control Department*; New Administration Project; King Faisal University: Al-Hofuf, Saudi Arabia, 2017.
59. *Accountant Department Report*; New Administration Project; King Faisal University: Al-Hofuf, Saudi Arabia, 2017.
60. *Final Reports, Change Orders, Invoices, (APTEC–ALKIFAH–ALARAB—ALKHODARY–SOUTH STAR) Contractor Head Quarter*; New Administration Project; King Faisal University: Al-Hofuf, Saudi Arabia, 2017.

buildings

MDPI

Article

Navigating Complexity: Enhancing Infrastructure Megaproject Performance Through Effective Alliance Management Capability

Xiaoyan Chen [1,*], Daoan Fan [2], Yan Liu [3] and Xinyue Zhang [4]

[1] School of Urban Economics and Management, Beijing University of Civil Engineering and Architecture, Beijing 102616, China
[2] School of Engineering, Shanghai Sanda University, Shanghai 201209, China; fandaoan@sandau.edu.cn
[3] School of Management and Engineering, Nanjing University, Nanjing 210008, China; ly@nju.edu.cn
[4] School of Project Management/John Grill Institute for Project Leadership, The University of Sydney, Sydney, NSW 2037, Australia; xinyue.zhang@sydney.edu.au
* Correspondence: chenxiaoyan92@bucea.edu.cn

Abstract: While extensive research has delved into the impact of management strategies on project outcomes, the specific influence of alliance management on infrastructure megaprojects remains less explored. This study delves into the relationship between alliance management capability and performance in the context of infrastructure megaprojects. Central to our investigation is the hypothesis that collaborative conflict management mediates this relationship, with leader–member exchange playing a moderating role. Based on 205 surveys collected from 13 megaprojects in China, regression analysis and bootstrapping methods were used to test the research hypotheses. The findings reveal a positive correlation between alliance management capability and infrastructure megaproject performance, mediated by collaborative conflict management. Significantly, leader–member exchange acts as a moderator in the alliance management capability–collaborative conflict management nexus. These insights underscore the pivotal role of alliance management capability in elevating project performance, meriting increased attention in future empirical research. By shedding light on the mediating and moderating mechanisms at play, this paper unravels the complexities of how alliance management capability impacts project performance, offering practical guidance for industry practitioners.

Keywords: alliance management capability; collaborative conflict management; leader–member exchange; megaproject performance; regression analysis

Academic Editor: Pramen P. Shrestha

Received: 12 November 2024
Revised: 11 December 2024
Accepted: 16 December 2024
Published: 1 January 2025

Citation: Chen, X.; Fan, D.; Liu, Y.; Zhang, X. Navigating Complexity: Enhancing Infrastructure Megaproject Performance Through Effective Alliance Management Capability. *Buildings* **2025**, *15*, 118. https://doi.org/10.3390/buildings15010118

1. Introduction

Infrastructure megaprojects are large-scale complex public projects that provide basic public services for economic growth, social development, and people's lives [1]; typical infrastructure megaprojects include but are not limited to high-speed rail lines, airports, seaports, energy projects, airports, dams, and communication technology (ICT) systems. The complexity of infrastructure megaprojects, stemming from their size, diversity of required expertise, risk profile, and the need for efficient resource utilization, makes alliances not only beneficial but often essential for successful project execution. Considering that the traditional practice of separation from the design, construction, and operation and maintenance processes of a project prevents effective teamwork and collaboration [2], alliance management capability has been identified as one of the most critical competencies for infrastructure megaproject participants, especially the project leaders, to possess in the complex and ever-changing environment [3]. Alliance management capability cannot be underestimated in productive organizations where teamwork is essential to project and organizational success [4,5]. Existing studies show that enhancing alliance management capability is especially beneficial in an organization that requires professional competence, effective teamwork, and working experience [6,7]. The reason may be that the stronger the alliance management capability, the more efficiently organizations can access the network resources embedded in the alliance relationship, which is not only conducive to resisting the risk of uncertainty in the collaborative innovation process but also serves as a critical driver to activate technological innovation among organizations. Despite alliance management capability being regarded as an important factor in obtaining professional competence [8,9], questions about the processes through which alliance management capability contributes to performance, especially in the context of infrastructure megaprojects, remain unanswered.

In the infrastructure megaproject, not only does cooperation exist, but also different types of conflict may occur, such as the intra-organization conflict between organization employees and leaders and inter-organization conflict among different participants [10,11]. Different conflict management strategies or styles can be adopted to manage it. Collaborative conflict management is one of the conflict management styles, which refers to alliance members coordinating or cooperating to solve the conflict to obtain win–win results and satisfy the interests of alliance members [12]. When alliance members adopt collaborative attitudes in dealing with conflict, they mainly focus on solving common problems by sharing their resources, which would affect the overall performance. For example, existing studies proposed collaborative methods to enhance stakeholder engagement and identified drivers and obstacles of collaborative practices on project performance [13]. However, in the infrastructure megaproject context, how collaborative conflict management and alliance management capability are related to project performance remains to be explored.

The relationship between leaders and organizational members has been regarded as an essential factor that can help manage conflict and facilitate organizational performance [14]. Leader–member exchange (LMX) explains interactions between leaders and employees in an organization or alliance [15]. The positive attitudes and behaviors between leaders and employees might change the members' and leaders' awareness, attitudes, and even actions. A high-quality LMX is connected with diverse beneficial work-related outcomes, for instance, job satisfaction, organizational commitment, work environment, organization performance, etc. [16]. As such, LMX may work as a moderator that illustrates the complex relationships among alliance management capability, collaborative conflict management, and project performance. This is a non-negligible gap in the existing research as alliance management capability is necessary but insufficient in infrastructure megaprojects performance; thus, efficient relationship management is also required to further improve the performance.

Even though alliance management capability and project performance have received researchers' attention, it still calls for the exploration of how alliance management capability relates to project performance in the infrastructure megaproject context. By shedding light on the mechanism that takes place between alliance management capability and infrastructure megaproject performance, the present study aims to explore (1) whether collaborative conflict management mediates the relationship between alliance management capability and infrastructure megaproject performance and (2) whether and how LMX moderates the relationship between alliance management capability and infrastructure megaproject performance.

The research contributes to infrastructure megaproject alliance management research by revealing the mechanisms and processes of how alliance management capability influences project performance. This study extends the impact of alliance management capability on performance specifically to the temporary context of infrastructure megaprojects. It also innovates by applying LMX theory to uncover its moderating effect on conflict resolution and alliance dynamics, thereby deepening our grasp of how leadership navigates the intricate web of inter- and intra-organizational relationships crucial for megaproject success. The research findings also equip project leaders with actionable insights for adeptly managing the complex interplay of relationships within and between organizations in megaprojects.

2. Theory and Hypotheses

2.1. Alliance Management Capability

Alliances are essential for leveraging resources and gaining competitive advantages, accessing new markets, sharing risks, enhancing learning, and fostering innovation in projects [17]. However, there is no universal, unanimous definition of alliance management capability [18]. Usually, alliance management capability is regarded as a comprehensive relational ability of organizations within an alliance to dynamically establish, govern, and coordinate multiple alliance partnerships, to harness and manage alliance cooperation issues by stimulating synergistic effects among organizations within the alliance, to integrate, coordinate and optimize resources to achieve alliance goals, and to obtain alliance benefits and realize win–win situations [19,20]. Accumulating evidence indicates that alliance management capability is grounded on the resource-based view (RBV) perspective because the advantage of one organization over other organizations comes from resources and capabilities [9,21]. In this respect, alliance management capability is a dynamic capability for organizations to form alliances and obtain advantageous resources that could help them gain more advantages.

The different dimensions of alliance management capability are part of a seamless and mutually supportive process, which can explain the relationships between alliance management capabilities and other factors. Existing studies explore what makes up alliance management capability. For example, paper [22] showed that alliance management capability is the ability of organizations to effectively deal with the various governance issues that arise in the alliance and categorize it into three dimensions: communication capability, coordination capability, and bonding capability. Communication capability is the ability to share information and exchange ideas in the process of cooperation. Coordination capability entails the amalgamation of resources from alliance members, viewed from a worldwide standpoint, and the synchronization of their strategies and actions. Bonding capability represents an organization's skill in cultivating and nurturing robust connections with its partners in the alliance. Similarly, Schilke and Goerzen [23] classified alliance management capability into four types of routines, including coordination capability, learning capability, sensing capability, and transformation capability. More specifically, coordination capacity means coordinating resources within a separate alliance (i.e., inter-organizational coordination) and a whole alliance portfolio (i.e., alliance portfolio coordination). Learning capacity reveals the routines that maintain inter-organizational knowledge transfer and knowledge creation. Sensing capacity demonstrates an organization's capability to identify opportunities to join a strategic alliance. Transformation capacity is the routine needed to adapt and restructure the alliance during the alliance process. Other management capabilities, such as alliance scanning capability, governance capability, partner selection capability, pro-activeness capability, and alliance portfolio coordination, are standard dimensions in existing alliance management capability studies [24,25].

Schilke and Goerzen's framework is one of the first to conceptualize and operationalize alliance management capability as a second-order construct, which is crucial for understanding how organizations effectively manage their strategic alliances. Schilke and Goerzen's framework's broad applicability across different types of alliances and industries makes it a valuable resource for researchers and practitioners working in various sectors, such as the manufacturing industry, construction industry, service industry, agricultural industry, transportation industry, and others. In this research, Schilke and Goerzen's framework on alliance management capability is adopted and adapted due to its ability to capture the dynamic and interactive nature of alliances within the megaproject environment.

High coordination capability ensures that all partners are working towards common objectives with minimal conflict or redundancy [26]. It facilitates efficient resource utilization, timely decision-making, and smooth execution of project tasks, leading to better project outcomes and reduced risks of delays or cost overruns. Learning capability enables the alliance to adapt to changing environments, integrate diverse expertise, and innovate [27]. It leads to continuous improvement in processes and strategies, enhancing the overall quality and competitiveness of the project. Especially, learning capability is crucial for long-term sustainability and success. Effective sensing capability allows the project alliance to anticipate owner needs and potential risks [28]. It enables proactive adjustments in project strategies and operations, ensuring that the project remains relevant and viable in a dynamic business landscape. Transformation capability is key to adapting and evolving in response to new challenges and opportunities in projects [29]. It allows the alliance to modify its structure, processes, or products to better meet project goals and market demands. High transformation capability can lead to breakthrough innovations and significant improvements in project outcomes.

In a word, alliance management capability is essential for organizations to achieve competitive advantage in the context of globalization. Organizations can be invincible in fierce competition and close cooperation by profoundly understanding the essence of alliance management capability and how it influences performance.

2.2. Alliance Management Capability and Project Performance

Infrastructure megaprojects are typically large-scale endeavors that involve extensive construction, significant financial investments, and long timeframes. Their sheer size and scope can surpass the capabilities of a single organization, making alliances necessary to pool resources, expertise, and technologies. Anecdotal evidence suggests that alliance management is essential to a successful project because alliance enables effective integration between diverse participants (e.g., owner, designer, contractor). Integration management is included as a knowledge area in the Project Management Body of Knowledge (PMBoK) and involves the process of combining, harmonizing, and coordinating project management. Paper [30] showed that integration or its attributes play an important role in achieving higher levels of performance and success in project alliances. Existing studies also confirm that project alliances can share both risk and reward among alliance members [31] and promote inter-organizational collaboration among different participants [32]. Collaboration in infrastructure projects requires integrating the goal, information, performance, and actions and keeping tight interaction/interoperation [33]. By collaboration in project alliance, a coordinated effort can help provide a clear understanding of project workflow and task allocation, which is critical to maintaining project performance.

Based on the analysis above, it could be inferred that alliance management capability enhances project capability [34], a crucial capability for achieving project goals. Specifically, alliance management capability led by augmenting the overall project performance tends to remove barriers to maximize project alliance members' contributions and success and generate competitive capacity that is conducive to project performance. Thus, the following hypothesis is put forward.

Hypothesis 1. *There is a positive relationship between alliance management capability and infrastructure megaproject performance.*

2.3. Mediating Role of Collaborative Conflict Management

The collaborative conflict management style developed by paper [12] incorporates the impact of cultural diversity on conflict management styles, recognizing the heterogeneity of different organizations and the influence of individualism–collectivism orientations on conflict management. This is suitable for megaprojects where heterogeneous stakeholders with different interests are involved in megaprojects and conflicts occur often.

Conflicts are common in project alliances and can bring both opportunities and challenges. When conflicts among infrastructure project alliances occur, an organization will judge whether and to what extent its goals are aligned with other alliance members, which will influence its attitudes and behaviors towards the conflict, and then choose different conflict management styles to cope with the conflict [35]. Specifically, when alliance members perceive that their goals are aligned with each other, they believe that the achievement of the alliance's goals also contributes to their benefits and thus adopt a positive and collaborative attitude (e.g., proactive communication and dynamically adjusting their cooperation strategies) to deal with and manage conflict [36]. The collaborative approach means that alliance members who value others' abilities and options, resulting in collaborative goal achievement and valuable communication to resolve conflict, can maintain a long-term partnership with alliance members. In addition, a collaborative attitude contributes to compatible goals and high-quality solutions to disputes and gives other members more confidence in effective collaboration [37]. Thus, the influence of collaborative conflict management on project performance within an adversarial environment cannot be ignored.

Alliance management capabilities are necessary for accomplishing tasks in project-laden settings where cooperation and competition coexist. With organizations increasingly becoming involved in alliances, management capacity plays an active role in alliance management. Collaborative initiatives among alliance members can give organizations the right to prioritize alliance partners, helping them find the partners most aligned with achieving their goals, thus forming healthy project alliances and further reducing potential conflicts. What is more, when conflicts occur, an organization with strong management capabilities will choose to communicate with its partners promptly about their vision and methods. It can gain an understanding of its partners, acquire new knowledge and information, and promote a positive method to resolve the conflict. The positive attitude of alliance members promotes the collaborative approach to conflict management [38].

Despite the potential benefits of collaborative conflict management, few studies have examined factors that may decrease the likelihood of conflict management related to the relationship between alliance management capability and project performance. Based on the above arguments, we hypothesize that:

Hypothesis 2. *Collaborative conflict management mediates the relationship between alliance management capability and infrastructure megaproject performance.*

2.4. Moderating Role of Leader–Member Exchange

LMX theory is a concept in organizational behavior and leadership research that shows the dyadic relationship between leaders and members (employees) rather than viewing leadership as a role that influences all subordinates uniformly [39]. LMX enhances internal team dynamics and external interactions, both of which are essential for navigating the complexities and challenges of large-scale project endeavors [40]. Regrettably, no studies were found to examine the moderating role of LMX between alliance management capability and conflict management in an infrastructure megaproject context. As stated, LMX indicates support between supervisor and subordinate within an organization. High-quality LMX, characterized by a high level of mutual trust, opportunity, support, and respect [41,42], promotes the completion of project work efficiently.

Infrastructure megaprojects involve diverse teams with members from various disciplines. The success of megaprojects heavily relies on the motivation and engagement of every team member. High-quality LMX could promote the relationship between alliance management capability and collaborative conflict management in different ways. It could (1) provide more opportunities for communication between employees and leaders; (2) engender more trust, interaction, support, and formal and informal rewards, which leads to greater attention to more substantial work support and supervisor responsiveness; (3) contribute to role expansion and responsibility-taking, and strengthens intrinsic motivation to improve the work environment and high ethical standards; (4) facilitate the leader–member relationship to a partnership level; (5) enable leaders to empower employees and provide them with a reasonable degree of accountability based on their abilities, needs, and input [43,44]. High-quality LMX can create a positive and supportive environment where trust, communication, empowerment, and commitment thrive. These elements are crucial for enhancing alliance management capabilities and fostering a collaborative approach to conflict management. Existing studies also indicated that LMX plays a significant role in team workers' attitudes and behaviors [45–47] and employees' work environments and influences team workers' feelings of mutual obligation and reciprocity, which may positively impact the relationship between alliance management capability and collaborative conflict management. Based on the above arguments, it is hypothesized that:

Hypothesis 3. *LMX moderates the relationship between alliance management capability and collaborative conflict management, so high-quality LMX strengthens this relationship.*

Figure 1 shows the conceptual model that indicates relationships among alliance management capability, LMX, collaborative conflict management, and infrastructure megaproject performance.

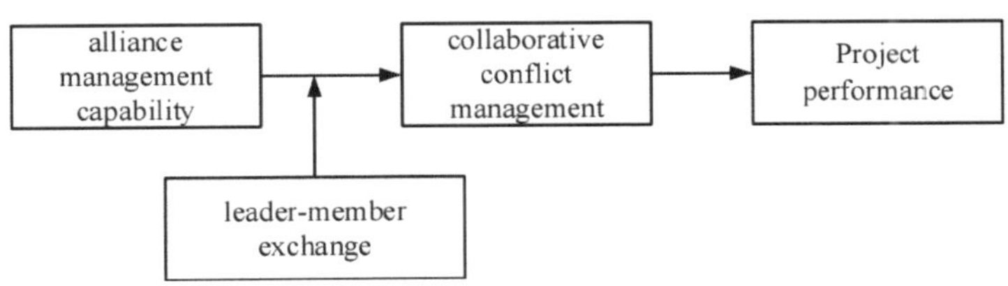

Figure 1. Conceptual model.

3. Methods

3.1. Research Sample

Researchers adopted a questionnaire survey to collect data. The questionnaire was initially written in English according to the measures developed in existing studies and then translated into Chinese to conveniently distribute it to Chinese infrastructure megaproject practitioners. A two-way translation method minimizes cross-cultural information asymmetry as much as possible [48]. The questionnaire survey is composed of three sections. The first section collects basic information about the infrastructure megaproject and the firm to which the respondents belong. The second section shows the measurement for alliance management capability, collaborative conflict management, project performance, and LMX using the 5-point Likert scale. The third section aims to obtain basic information about the respondents, such as age, gender, educational background, working experience, etc.

Before the formal survey, a pilot test was conducted to check whether the questionnaire survey was easily understood and whether the validity and reliability could be assured. We distributed the questionnaire to 45 megaproject practitioners, including project managers, consulting firm personnel, contractor personnel, suppliers, professors who had connections with practitioners in megaprojects, etc. All of them finished the questionnaire. Based on the feedback, researchers revised the description of the questionnaire survey.

In the formal survey, researchers initially contacted the project manager in different megaprojects, explained the purpose of the research, and inquired about the possibility of involving the personnel from their project in the survey. With permission, researchers distributed the survey to workers and managers on the infrastructure megaproject site to ensure respondents and researchers communicated promptly about the research aims and the meaning of the questions, which aimed to enhance the accuracy of the responses. To gain more representative data, researchers adopted a mixed sampling technique, including purposive and snowball sampling, to ensure sample diversity and maximize the number of qualified respondents. To ensure that all respondents have experience with alliance work, researchers clearly stated the background and significance of the study at the beginning of the questionnaire, and once again explained the purpose of our research to the respondents, indicating that only those with relevant experience are eligible to complete the survey. Moreover, the respondents were guaranteed confidentiality to ensure the reliability of their answers.

After all, researchers collected 247 completed data from 13 infrastructure megaproject teams in China, such as Shanghai West Coast Media Port, Shenzhen Qianhai Infrastructure Project, Jinan High-Tech Park Infrastructure Project et al. These megaprojects are distributed in different regions in China and have been listed as key development megaprojects by local governments. These projects have had a significant impact on the economic and social development of the local areas. Researchers and their research team members have participated in these megaprojects and are familiar with project managers in these megaprojects.

After receiving the questionnaire, researchers then checked them and deleted invalid answers, for example, missed answering the questions. Finally, 205 surveys were used to examine our research questions. Table 1 lists the demographic distribution of the respondents. The abundant background of the respondents guarantees the validity and reliability of the survey.

Table 1. Respondent's demographics.

		Number	Percentage
Gender	Male	140	68.3%
	Female	65	31.7%
Educational background	Senior high school or below	59	28.78%
	Bachelor	81	39.51%
	Master	61	29.76%
	PhD	4	1.95%
Working experience in megaprojects (years)	<5	6	2.92%
	5–10	47	22.93%
	10–20	101	49.27%
	20–30	46	22.44%
	>30	5	2.44%
Roles in megaprojects	Contractor	78	38.05%
	Designer	19	9.27%
	Supplier	24	11.71%
	Supervisor	36	17.56%
	Consultant	27	13.17%
	Client	21	10.24%

3.2. Measures

Alliance management capability. The nine-item scale was used to measure alliance management capability. The detailed measurements are shown in Table 2. Response options ranged from 1, "strongly disagree" to 5, "strongly agree". Researchers conducted a confirmatory factor analysis (CFA) at the project level to ensure the suitability of the measurement. The fit indexes for three first-order factors plus one second-order factor fell within a reasonable range, with $\chi^2 = 32.92$. The GFI value is 0.96, exceeding the acceptable threshold of 0.9, while the AGFI score is 0.93, also above the minimum standard of 0.9. Additionally, the RMSEA value is 0.04, which is below the benchmark of 0.05.

Table 2. Measurement scales and items.

Measure	Sub-Measures	Questions
Alliance management capability	Alliance scanning	We actively monitor our environment to identify partnering opportunities.
		We routinely gather information about prospective partners from various forums.
		We are alert to market developments that create potential alliance opportunities.
	Alliance coordination	We actively coordinate with other organizations in an alliance.
		We systematically coordinate our strategies across different alliances.
		We have processes to systematically transfer knowledge across alliance partners.
	Alliance learning	We conduct periodic reviews of our alliances to understand what we are doing right and where we are going wrong.
		We periodically collect and analyze field experiences from our alliances.
		We modify our alliance-related procedures as we learn from experience.
Collaborative conflict management		We tried to bring all our concerns out in the open so that the issues could be resolved in the best possible way.
		We tried to investigate an issue with alliance members to find a solution acceptable to us.
		We exchanged accurate information with alliance members to solve a problem together.
		We tried to work with alliance members to find solutions to a problem that satisfied our expectations.
		We collaborated with teammates to come up with decisions acceptable to us.
Leader–member exchange		My leader understands my job problems and needs/I understand my leader's job problems and needs.
		My leader is satisfied with what I do.
		My leader recognizes my potential.
		Regardless of how much formal authority my manager has built into his/her position, he/she would use that power to help me solve problems in my work.
		Regardless of the amount of formal authority my leader has, he/she would "bail you out," at his/her expense.
		I have enough confidence in my leader that I would defend and justify his/her decision if he/she were not present to do so.
		I keep a wonderful working relationship with my manager.
Infrastructure megaproject performance		Both owner and contractor were satisfied with the project outcomes.
		The project is completed on time.
		The quality of the project is in line with the designed objectives.
		There were no cost overruns on the project.

Collaborative conflict management. Researchers adopted the five-item scale to measure collaborative conflict management. Response options ranged from 1, "strongly disagree" to 5, "strongly agree".

Leader–member exchange. Researchers used the seven-item scale developed by Graen and Uhl-Bien [49] to measure leader–member exchange. This multi-level, multi-domain perspective acknowledges that leadership is not a one-size-fits-all phenomenon but is instead shaped by the unique interactions and relationships within different organizational contexts. Response options ranged from 1, "strongly disagree" to 5, "strongly agree".

Infrastructure megaproject performance. Researchers used the four-item scale developed by Demirkesen and Ozorhon [50] to measure project performance. Response options ranged from 1, "strongly disagree" to 5, "strongly agree".

Control variables. In the analysis, researchers included firm size and firm age as control variables to better estimate the effect range of the hypothesized variables.

3.3. Reliability and Validity Testing

Researchers first assessed the sample adequacy based on the Kaiser–Meyer–Olkin (KMO) value and Sphericity of Bartlett's test. The analysis shows that the KMO result is 0.853 and reached the satisfactory standard according to the minimum threshold of 0.7. The result for Bartlett's test was 0.00, demonstrating that the data significance at the 95.0% confidence level can be assured.

The Cronbach's alpha was used to check reliability. The alpha reliability was 0.90 for alliance scanning, 0.90 for alliance coordination, 0.89 for alliance learning, and 0.91 for alliance management capability. The alpha reliability was 0.84 for collaborative conflict management, 0.87 for leader–member exchange, and 0.85 for project performance. All results are larger than the criteria of the recommended value of 0.7, indicating that reliability was achieved.

Researchers conducted confirmatory factor analysis (CFA) to evaluate whether these variables captured distinct constructs using Amos 20.0 [51]. Researchers used the nine items of alliance management capability, five items of collaborative conflict management, seven items of leader–member exchange, and four items of project performance to conduct CFA based on the maximum likelihood method. Researchers tested $\chi 2$ differences among a four-factor measurement model, a three-factor, and a single-factor model to see which model better fit the data. The four-factor measurement model fits the data well, as shown in Table 3. The result showed that the four-factor measurement model was better than any alternative factor measurement model. The results also provide support for the discriminant validity of our measures.

Table 3. Results of Confirmatory Factor Analysis.

Measurement Model	χ^2	df	GFI	AGFI	RMSEA
Four-factor model	453.70	318	0.92	0.91	0.04
Three-factor model-1	506.53	324	0.85	0.82	0.05
Three-factor model-2	564.04	321	0.83	0.80	0.06
Three-factor model-3	721.82	321	0.77	0.72	0.08
One-factor model	868.95	324	0.70	0.65	0.09

4. Results

4.1. Descriptive Statistical Analysis

Table 4 demonstrates descriptive statistics and correlations for all variables, including means, standard deviations (SD), and correlations. Table 4 shows that means for alliance management capability, collaborative conflict management, LMX, and infrastructure megaproject performance range from 3.90 to 4.20 and SD range around 0.4. The results indicate the difference and variance between the measures investigated. In addition, all correlations are in the expected direction and provide evidence to test our hypotheses further. The alliance management capability correlates significantly with collaborative conflict management (r = 0.646, $p < 0.01$), LMX (r = 0.201, $p < 0.01$), and infrastructure megaproject performance (r = 0.762, $p < 0.01$). Collaborative conflict management correlates significantly with LMX (r = 0.184, $p < 0.01$) and infrastructure megaproject performance (r = 0.752, $p < 0.01$).

Table 4. Descriptive statistics results.

Variables	Mean	SD	1	2	3	4	5	6
Firm age	1.86	0.66	1					
Firm size	1.43	0.49	−0.154 *	1				
AMC	4.12	0.39	0.068	−0.058	1			
CCM	4.12	0.44	0.034	−0.021	0.646 **	1		
LMX	4.07	0.38	−0.088	−0.076	0.201 **	0.184 **	1	
Infrastructure megaproject performance	3.90	0.41	0.045	−0.032	0.762 **	0.752 **	0.214 **	1

Notes: N = 205; ** $p < 0.01$, * $p < 0.05$, AMC refers to "alliance management capability", CCM refers to "collaborative conflict management", and LMX refers to "leader–member exchange".

4.2. Hypotheses Testing

Following the paper [52], this study adopted hierarchical regression analysis to test our moderated mediation model. To reduce multi-collinearity, researchers centered LMX, alliance management capability, collaborative conflict management, and infrastructure megaproject performance before calculating interaction results.

As the results are shown in Table 5 (M4 and M5), after controlling firm age and firm size, alliance management capability is positively and significantly related to infrastructure megaproject performance (β = 0.76, $p < 0.01$). The additional proportion of variance in infrastructure megaproject performance explained by alliance management capability is also significant (ΔR2 = 0.58, $p < 0.01$), thus Hypothesis 1 is supported.

Table 5. Results of regression analysis.

	CCM (T2)			Infrastructure Megaproject Performance (T3)			
	M1	M2	M3	M4	M5	M6	M7
Control variables							
firm size	0.03	0.01	0.04	0.04	−0.01 *	0.18	−0.01
firm age	−0.17	0.02	0.17	−0.03	0.01	−0.01	0.01
Independent variables							
AMC (T1)		0.65 **	0.63 **		0.76 **		0.48 **
Moderator							
LMX (T1)			0.25				
Interactions							
AMC × LMX			0.33 *				
Mediation							
CCM (T2)						0.75 **	0.45 **
R2	0.00	0.42	0.42	0.03	0.58	0.57	0.70
F	0.14	48.07 **	29.13 **	2.27 **	93.03 **	87.43 **	114.90 **
ΔR2	−0.01	0.41	0.02	0.03	0.58	0.56	0.13
ΔF	0.14	143.70 **	0.60 *	3.27 **	277.80 **	261.05 **	86.18 **

Notes: N = 205; ** $p < 0.01$, * $p < 0.05$, AMC refers to "alliance management capability", CCM refers to "collaborative conflict management", and LMX refers to "leader–member exchange".

Table 5 also provides evidence of the mediating effect of collaborative conflict management. The results indicate that (1) alliance management capability is positively related to collaborative conflict management (β = 0.65, $p < 0.01$, M2); (2) alliance management capability is positively related to infrastructure megaproject performance (β = 0.76, $p < 0.01$, M5); (3) collaborative conflict management is positively related to infrastructure megaproject performance (β = 0.75, $p < 0.01$, M6), and (4) after entering collaborative conflict management, the relationship between alliance management capability and infrastructure megaproject performance is still significant and this influence becomes weaker (β = 0.48, $p < 0.01$, M7). Researchers further tested the significance of the indirect effects using the PRODCLIN program. The PRODCLIN program results indicate that the indirect effect of alliance management capability on project performance via collaborative conflict management is significant. Specifically, the 95% confidence interval of the indirect effect is [0.03, 0.08], which does not include zero. Therefore, Hypothesis 2 is partially supported.

To test the moderating effect of LMX, researchers adopt multiple regression analysis. As Table 5 illustrates, the interaction between the alliance management capability and LMX positively relates to collaborative conflict management ($\beta = 0.33$, $p < 0.05$, M3). Thus, hypothesis 3 is supported.

A moderated path analysis has been conducted to examine the mediating and moderating effects together [53], which adopted bootstrapped 1000 samples to compute bias-corrected confidence intervals. An indirect effect is significant if the 95% confidence interval excludes zero [53]. Specifically, as Table 6 indicates, the indirect effect of collaborative conflict management on the relationship between the alliance management capability and project performance is significantly higher at a low ($\beta = 0.47$, $p < 0.01$) rather than a high ($\beta = 0.53$, $p < 0.05$) level of LMX. In addition, the results presented in Table 6 indicate that the moderator moderates the direct effect of alliance management capability on collaborative conflict management ($\Delta\beta = 0.10$, $p < 0.01$), providing further support for our theoretical argument that high-quality LMX tends to strengthen this relationship. Overall, Hypotheses 3 receive further support.

Table 6. Results of the moderated path analysis.

Moderating Variable	AMC (X) → CCM (M) → PP (Y)				
	Stage		Effect		
	First	Second	Direct Effects	Indirect Effects	Total Effects
	PMX	PYM	PYX	PYM PMX	PYX + PYM PMX
Low-LMX	0.66 **	0.40 **	0.47 **	0.27 **	0.74 **
High-LMX	0.76 **	0.44 **	0.53 *	0.33 **	0.86 **
Differences	0.10 **	0.04	0.06	0.06 **	0.12

Notes: N = 205, ** $p < 0.01$, * $p < 0.05$. PMX is the path from AMC to CCM; PYM is the path from CCM to PP; PYX is the path from AMC to PP. Low-LMX refers to one standard deviation below the mean of LMX; high-LMX refers to one standard deviation above the mean of LMX. Tests of indirect and total effect differences are based on bias-corrected confidence intervals derived from bootstrap estimates.

5. Discussion and Conclusions

5.1. Discussion

In today's global environment, strategic alliances represent an important source of growth and competitive advantage; strong alliance management capability allows one to improve competitive position and rapidly enter a new market. Existing research investigates the influence of alliance management capability on performance in different contexts, such as the manufacturing and logistics industry; however, how alliance management capability influences project performance in the context of infrastructure megaprojects remains to be explored. Motivated by this, this study examined the relationship between alliance management capability and project performance and the mechanism to generate better project performance.

Previous studies have overlooked the fact that the value of alliance management capability may vary across different contexts and alliance capabilities that are very effective in one context may be less successful in another [54]. Also, the paper [55] indicated that the context-dependent nature of alliance management capability needs to be considered in research. Following this, this research extends the direct influence of alliance management capability on performance to a temporary (i.e., infrastructure megaproject) context. Based on data collected in the Chinese infrastructure megaproject, the results demonstrate that alliance management capability significantly positively impacts project performance. Indeed, the infrastructure megaproject is achieved in a temporary organization composed of many permanent organizations [56]. The unique characteristics indicate that alliance in the infrastructure megaproject context is often one-time, and cooperation among alliance members is challenging as they have different requirements and interests. Therefore, when organizations form alliances, effective management of these alliances becomes crucial for achieving project success. In addition, alliance management in the context of infrastructure megaprojects is implemented at the project level rather than at the firm level [57]. By effectively managing alliances at the infrastructure megaproject level, organizations can leverage their partners' expertise, resources, and networks, leading to enhanced project performance [58]. This collaboration allows for sharing of knowledge, best practices, and innovation, resulting in improved project outcomes. Also, strong alliance management capabilities enable organizations to proactively identify and address challenges, mitigate risks, and adapt to changing project requirements, ultimately leading to increased project success. Thus, high-quality alliance management capability is critical for the success of infrastructure megaprojects.

In addition, this study also found that collaborative conflict management partially mediates the relationship between alliance management capability and project performance. This indicates that alliance management capability could directly and indirectly, through collaborative conflict management, influence project performance. In other words, a strong alliance management capability alone may not directly lead to improved project performance. Instead, how conflicts are managed within the alliance through collaborative approaches plays a crucial role in determining the project's ultimate success [13]. In this sense, by collaboratively managing conflicts, project organizations can enhance communication, build trust, and foster stronger relationships within the alliance. This, in turn, can lead to improved cooperation, greater alignment of goals and strategies, and the effective resolution of challenges and disagreements. Finally, these factors contribute to better project performance. The research findings extend existing research as most studies focus on the mediating effect of knowledge acquisition and sharing [59], alliance control and strength of ties [60], inter-organizational learning capabilities [55], and so on, ignoring that conflict management is also an important factor impacting the alliance performance. Based on the findings, it can be inferred that the relationship between alliance management capability and project performance is influenced by how conflicts are managed within the alliance, and effective conflict management can help prevent or minimize disruptions, ensure smooth collaboration, and maintain focus on project goals.

Moreover, the relationships between alliance management capability, collaborative conflict management, and project performance are likely to be complex and influenced by various contextual factors, for example, LMX. Existing LMX research mainly focuses on intra-organizational relationships [61], ignoring the inter-organizational relationship. This study explores the moderated mediating effect of LMX in the inter-organizational relationship among infrastructure megaproject organizations. Based on analysis, it is clear that LMX works as a moderator (Hypothesis 3). LMX is not just a measure of the quality of relationships but also a reflection of the trust, respect, and mutual understanding that exist between a leader and their team [62]. When LMX is intense and characterized by high levels of trust, respect, open communication, interaction, and support, project team members are more likely to engage in collaborative conflict management. They feel empowered to share their concerns, opinions, and ideas openly, which facilitates the resolution of conflicts in a constructive manner. On the other hand, when LMX is weak or characterized by low levels of trust and open communication, team members may hesitate to express their concerns or engage in collaborative conflict management. This can hinder the organization's ability to effectively manage conflicts within alliances, potentially leading to unresolved issues, deteriorating relationships, or even alliance failure.

5.2. Theoretical and Managerial Contributions

By shedding light on the mechanism influence of alliance management capability on project performance in the infrastructure megaprojects context, this research makes contributions to infrastructure megaproject management research.

First, the research findings extend the direct influence of alliance management capability on performance to a temporary (i.e., infrastructure megaproject) context. The megaprojects, being temporary, are unique combinations of permanent and temporary organizational layers, which require dynamic and adaptive management routines that enable effective coordination and collaboration among various stakeholders. The research results are particularly relevant in the context of megaprojects, where megaproject success is contingent upon the ability to manage a diverse portfolio of alliances and partnerships effectively.

Second, our study has broadened the research on how alliance management capability influences infrastructure megaproject performance by exploring the mediating mechanism and further explaining the boundary condition for this indirect relationship. The findings indicate why collaborative conflict management is a direct predictor of project performance and a mediator of relationships between alliance management capability and infrastructure megaproject performance. These findings add new insights into infrastructure megaproject research, in general, and alliance management capability in particular.

Third, this study's exploration of the moderating role of LMX in the context of collaborative conflict management within inter-organizational relationships, specifically in infrastructure megaprojects, significantly broadens the theoretical scope of LMX research. It addresses the intricate dynamics of leadership and team interactions in large-scale projects, offering a detailed view of the impact that the quality of leader–team relationships can have on conflict resolution and alliance management. The findings underscore the importance of trust, respect, and mutual understanding in facilitating or hindering effective management within project alliances. This understanding is vital for crafting leadership strategies that can navigate the complexities of multi-organizational project environments, where project success is contingent upon the adept handling of diverse interests and conflicts.

The present study also has some managerial implications for promoting project performance. First, in infrastructure megaprojects, enhancing alliance management capability in project alliances is important for the project implementation and success. It is essential to adopt a structured approach that encompasses strategic partner selection, clear alliance strategies, and a culture of collaboration and trust. By leveraging alliance management capabilities, organizations can optimize resource utilization and drive technological innovation, leading to improved project outcomes. To further bolster project success, creating a dedicated alliance function can provide knowledge management, internal coordination, and alliance assessment.

Second, alliance managers should motivate members toward collaborative behaviors to deal with conflicts and provide necessary emotional and material support to help them achieve infrastructure megaproject goals. By integrating conflict resolution strategies that emphasize trust-building and open communication, managers can foster an environment conducive to collaborative behaviors. This approach not only addresses immediate conflicts but also strengthens the alliance's ability to adapt to changes and achieve project goals, as underscored by the importance of coordination in alliance management. Furthermore, providing emotional and material support, as suggested, is crucial for maintaining high-performance cycles and preventing the escalation of conflicts, which is vital for the success of strategic alliances.

Moreover, LMX's moderating role guides leaders to cultivate high-quality relationships with team members, which enhances collaborative conflict management and alliance capabilities. By doing so, it improves project performance through increased trust, open communication, and effective resource allocation, ultimately leading to better team cohesion and success in achieving project goals.

5.3. Limitations and Future Research

This study, like all studies, also has several limitations. Firstly, this study is cross-sectional, which means it cannot rule out the causal effect of LMX on project performance. Hence, future research can use longitudinal studies or experiments to establish further the causality suggested by present research. Secondly, the elements of alliance management capability involve multiple dimensions and may change at different times; therefore, developing a multiple and dynamic construct of alliance management capability is another frontier for future research. Third, the samples used in this study are limited to Chinese megaprojects. The relationships between alliance management capability, project performance, collaborative conflict management, and LMX may be distinct for different countries. Cross-culture comparisons of their relationships could be an interesting research topic.

Author Contributions: Conceptualization, X.C. and D.F.; methodology, Y.L.; validation, X.Z.; writing—original draft preparation, X.C. and D.F.; writing—review and editing, Y.L. and X.Z.; funding acquisition, X.C. All authors have read and agreed to the published version of the manuscript.

Funding: This research was funded by National Natural Science Foundation of China (Grant No. 72401020 and 72201125).

Institutional Review Board Statement: Not applicable.

Informed Consent Statement: Verbal informed consent was obtained from the participants. Verbal consent was obtained rather than written because of privacy.

Data Availability Statement: Data are available from the corresponding author upon reasonable request.

Conflicts of Interest: The authors declare no conflicts of interest.

References

1. Juarez Cornelio, J.R.; Sainati, T.; Locatelli, G. Digging in the megaproject's graveyard: Why do megaprojects die, and how to check their health? *Int. J. Proj. Manag.* **2023**, *41*, 102501. [CrossRef]
2. AlHarmoodi, S.M.; Dulaimi, M. The alignment of the open innovation process and the project lifecycle. *Built Environ. Proj. Asset Manag.* **2022**, *12*, 556–572. [CrossRef]
3. Yang, W.; Meyer, K.E. Alliance proactiveness and firm performance in an emerging economy. *Ind. Mark. Manag.* **2019**, *82*, 226–237. [CrossRef]
4. Heimeriks, K.H.; Klijn, E.; Reuer, J.J. Building capabilities for alliance portfolios. *Long Range Plan.* **2009**, *42*, 96–114. [CrossRef]
5. Leischnig, A.; Geigenmüller, A. Examining alliance management capabilities in university-industry collaboration. *J. Technol. Transf.* **2020**, *45*, 9–30. [CrossRef]
6. Shi, C.; Chen, Y.; Hua, Y.; Tang, Y. Understanding subcontracting organizational arrangements for construction projects in China: Integrating capabilities and uncertainty. *Eng. Constr. Archit. Manag.* **2023**, *30*, 2381–2399. [CrossRef]
7. Omar, A.-T.; Desmond, L.; Zaheer, K. Examining alliance management capabilities in cross-sector collaborative partnerships. *J. Bus. Res.* **2019**, *101*, 268–284.
8. Dubey, R.; Bryde, D.J.; Graham, G.; Foropon, C.; Kumari, S.; Gupta, O. The role of alliance management, big data analytics and information visibility on new-product development capability. *Ann. Oper. Res.* **2024**, *333*, 743–767. [CrossRef]
9. Dhaundiyal, M.; Coughlan, J. Extending alliance management capability in individual alliances in the post-formation stage. *Ind. Mark. Manag.* **2022**, *102*, 12–23. [CrossRef]
10. Ruijter, H.; van Marrewijk, A.; Veenswijk, M.; Merkus, S. 'Filling the mattress': Trust development in the governance of infrastructure megaprojects. *Int. J. Proj. Manag.* **2021**, *39*, 351–364. [CrossRef]
11. Ning, Y.; Zwikael, O. Effective Combinations of Control Strategies in Inter-Organizational Projects. *IEEE Trans. Eng. Manag.* **2024**, *71*, 3062–3075. [CrossRef]
12. Paul, S.; Samarah, I.M.; Seetharaman, P.; Mykytyn, P.P., Jr. An empirical investigation of collaborative conflict management style in group support system-based global virtual teams. *J. Manag. Inf. Syst.* **2004**, *21*, 185–222. [CrossRef]

13. Cheng, F.-F.; Wu, C.-S.; Chang, J.Y.T. Interproject Conflict Management Through Cooperation in an Enterprise System Implementation Program. *Proj. Manag. J.* **2020**, *51*, 582–598. [CrossRef]
14. Xu, H.; Wayne, S.-J.; Wang, L.; Pan, J. LMX differentiation as a double-edged sword: A social hierarchy perspective for understanding the beneficial and detrimental effects of LMX differentiation on team performance. *Pers. Psychol.* **2024**, *77*, 713–745. [CrossRef]
15. Buch, R. Leader–member exchange as a moderator of the relationship between employee–organization exchange and affective commitment. *Int. J. Human Resour. Manag.* **2015**, *26*, 59–79. [CrossRef]
16. Martin, R.; Guillaume, Y.; Thomas, G.; Lee, A.; Epitropaki, O. Leader–member exchange (LMX) and performance: A meta-analytic review. *Pers. Psychol.* **2016**, *69*, 67–121. [CrossRef]
17. Pargar, F.; Kujala, J.; Aaltonen, K.; Ruutu, S. Value creation dynamics in a project alliance. *Int. J. Proj. Manag.* **2019**, *37*, 716–730. [CrossRef]
18. Forkmann, S.; Henneberg, S.C.; Mitrega, M. Capabilities in business relationships and networks: Research recommendations and directions. *Ind. Mark. Manag.* **2018**, *74*, 4–26. [CrossRef]
19. Gonçalves, R.G. The role of the alliance management capability. *Serv. Ind. J.* **2011**, *31*, 1961–1978. [CrossRef]
20. Al-Tabbaa, O.; Zahoor, N. Alliance management capability and SMEs' international expansion: The role of innovation pathways. *J. Bus. Res.* **2024**, *171*, 114384. [CrossRef]
21. Kauppila, O.P. Alliance management capability and firm performance: Using resource-based theory to look inside the process black box. *Long Range Plan.* **2015**, *48*, 151–167. [CrossRef]
22. Schreiner, M.; Kale, P.; Corsten, D. What really is alliance management capability and how does it impact alliance outcomes and success? *Strateg. Manag. J.* **2009**, *30*, 1395–1419. [CrossRef]
23. Schilke, O.; Goerzen, A. Alliance management capability: An investigation of the construct and its measurement. *J. Manag.* **2010**, *36*, 1192–1219. [CrossRef]
24. Kandemir, D.; Yaprak, A.; Cavusgil, S.T. Alliance orientation: Conceptualization, measurement, and impact on market performance. *J. Acad. Mark. Sci.* **2006**, *34*, 324–340. [CrossRef]
25. Leischnig, A.; Geigenmueller, A.; Lohmann, S. On the role of alliance management capability, organizational compatibility, and interaction quality in interorganizational technology transfer. *J. Bus. Res.* **2014**, *67*, 1049–1057. [CrossRef]
26. Tee, R.; Davies, A.; Whyte, J. Modular designs and integrating practices: Managing collaboration through coordination and cooperation. *Res. Policy* **2019**, *48*, 51–61. [CrossRef]
27. Tiwana, A. Do bridging ties complement strong ties? An empirical examination of alliance ambidexterity. *Strateg. Manag. J.* **2008**, *29*, 251–272. [CrossRef]
28. Floricel, S.; Miller, R. Strategizing for anticipated risks and turbulence in large-scale engineering projects. *Int. J. Proj. Manag.* **2001**, *19*, 445–455. [CrossRef]
29. Inigo, E.A.; Albareda, L. Sustainability oriented innovation dynamics: Levels of dynamic capabilities and their path-dependent and self-reinforcing logics. *Technol. Forecast. Soc. Chang.* **2019**, *139*, 334–351. [CrossRef]
30. Che Ibrahim, C.K.I.; Costello, S.B.; Wilkinson, S. Making sense of team integration practice through the "lived experience" of alliance project teams. *Eng. Constr. Archit. Manag.* **2018**, *25*, 598–622. [CrossRef]
31. Zhu, F.; Jiang, M.; Yu, M. The role of the lead firm in exploratory projects: How capabilities enable exploratory innovation of project alliances. *Int. J. Manag. Proj. Bus.* **2020**, *13*, 312–339. [CrossRef]
32. Love, P.E.D.; Mistry, D.; Davis, P.R. Price competitive alliance projects: Identification of success factors for public clients. *J. Constr. Eng. Manag.* **2010**, *136*, 947–956. [CrossRef]
33. Serrano, V.; Fischer, T. Collaborative innovation in ubiquitous systems. *J. Intell. Manuf.* **2007**, *18*, 599–615. [CrossRef]
34. Davies, A.; Brady, T. Explicating the dynamics of project capabilities. *Int. J. Proj. Manag.* **2016**, *34*, 314–327. [CrossRef]
35. Lam, P.-K.; Chin, K.-S.; Pun, K.-F. Managing conflict in collaborative new product development: A supplier perspective. *Int. J. Qual. Reliab. Manag.* **2007**, *24*, 891–907.
36. Yin, J.; Qu, M.; Li, M.; Liao, G. Team Leader's Conflict Management Style and Team Innovation Performance in Remote R&D Teams—With Team Climate Perspective. *Sustainability* **2022**, *14*, 10949. [CrossRef]
37. Wong, A.; Wei, L.; Wang, X.; Tjosvold, D. Collectivist values for constructive conflict management in international joint venture effectiveness. *Int. J. Confl. Manag.* **2018**, *29*, 126–143. [CrossRef]
38. Liu, J.; Cui, Z.; Feng, Y.; Perera, S.; Han, J. Impact of culture differences on performance of international construction joint ventures: The moderating role of conflict management. *Eng. Constr. Archit. Manag.* **2020**, *27*, 2353–2377. [CrossRef]
39. Dansereau, F. A dyadic approach to leadership: Creating and nurturing this approach under fire. *Leadersh. Q.* **1995**, *6*, 479–490. [CrossRef]
40. Zheng, J.; Gu, Y.; Xie, H.; Wu, G. Linking innovation, empowerment to facilitate project performance: A mediated moderation model. *J. Eng. Technol. Manag.* **2023**, *68*, 101750. [CrossRef]

41. Dienesch, R.M.; Liden, R.C. Leader-member exchange model of leadership: A critique and further development. *Acad. Manag. Rev.* **1986**, *11*, 618–634. [CrossRef]
42. Maslyn, J.M.; Uhlbien, M. Leader-member exchange and its dimensions: Effects of self-effort and other's effort on relationship quality. *J. Appl. Psychol.* **2001**, *86*, 697–708. [CrossRef] [PubMed]
43. Saeed, B.B.; Afsar, B.; Cheema, S.; Javed, F. Leader-member exchange and innovative work behavior. *Eur. J. Innov. Manag.* **2019**, *22*, 105–124. [CrossRef]
44. Cong, W.; Goh, Y.M.; Zhang, S.; Liang, H.; Zhou, Z.; Guo, N. Linking Leader–Member Exchange and Informal Safety Communication of Construction Workers in China: A Cognitive Psychology Perspective. *J. Constr. Eng. Manag.* **2023**, *149*, 04023092. [CrossRef]
45. Henderson, D.J.; Wayne, S.J.; Shore, L.M.; Bommer, W.H.; Tetrick, L.E. Leader--member exchange, differentiation, and psychological contract fulfillment: A multilevel examination. *J. Appl. Psychol.* **2008**, *93*, 1208–1219. [CrossRef]
46. Tse, H.H.M.; Ashkanasy, N.M.; Dasborough, M.T. Relative leader–member exchange, negative affectivity and social identification: A moderated-mediation examination. *Leadersh. Q.* **2012**, *23*, 354–366. [CrossRef]
47. Vidyarthi, P.R.; Liden, R.C.; Anand, S.; Erdogan, B.; Ghosh, S. Where do i stand? examining the effects of leader-member exchange social comparison on employee work behaviors. *J. Appl. Psychol.* **2010**, *95*, 849–861. [CrossRef]
48. He, Q.; Chen, X.; Wang, G.; Zhu, J.; Yang, D.; Liu, X.; Li, Y. Managing social responsibility for sustainability in megaprojects: An innovation transitions perspective on success. *J. Clean. Prod.* **2019**, *241*, 118395. [CrossRef]
49. Graen, G.B.; Uhl-Bien, M. Relationship-based approach to leadership: Development of leader-member exchange (lmx) theory of leadership over 25 years: Applying a multi-level multi-domain perspective. *Leadersh. Q.* **1995**, *6*, 219–247. [CrossRef]
50. Demirkesen, S.; Ozorhon, B. Impact of integration management on construction project management performance. *Int. J. Proj. Manag.* **2017**, *35*, 1639–1654. [CrossRef]
51. Anderson, J.C.; Gerbing, D.W. Structural equation modeling in practice: A review and recommended two-step approach. *Psychol. Bull.* **1988**, *103*, 411–423. [CrossRef]
52. Muller, D.; Judd, C.M.; Yzerbyt, V.Y. When moderation is mediated and mediation is moderated. *J. Personal. Soc. Psychol.* **2005**, *89*, 852–863. [CrossRef] [PubMed]
53. Edwards, J.R.; Lambert, L.S. Methods for integrating moderation and mediation: A general analytical framework using moderated path analysis. *Psychol. Methods* **2007**, *12*, 1. [CrossRef] [PubMed]
54. Wang, Y.; Rajagopalan, N. Alliance Capabilities:Review and Research Agenda. *J. Manag.* **2015**, *41*, 236–260.
55. Cabello-Medina, C.; Carmona-Lavado, A.; Cuevas-Rodriguez, G. A contingency view of alliance management capabilities for innovation in the biotech industry. *BRQ Bus. Res. Q.* **2020**, *23*, 2340944420901050. [CrossRef]
56. Winch, G.M. Three domains of project organising. *Int. J. Proj. Manag.* **2014**, *32*, 721–731. [CrossRef]
57. Hietajärvi, A.-M.; Aaltonen, K.; Haapasalo, H. What is project alliance capability? *Int. J. Manag. Proj. Bus.* **2017**, *10*, 404–422. [CrossRef]
58. Ninan, J.; Mahalingam, A.; Clegg, S. External stakeholder management strategies and resources in megaprojects: An organizational power perspective. *Proj. Manag. J.* **2019**, *50*, 625–640. [CrossRef]
59. Niesten, E.; Jolink, A. The impact of alliance management capabilities on alliance attributes and performance: A literature review. *Int. J. Manag. Rev.* **2015**, *17*, 69–100. [CrossRef]
60. Mamavi, O.; Meier, O.; Zerbib, R. Alliance management capability: The roles of alliance control and strength of ties. *Manag. Decis.* **2015**, *53*, 2250–2267. [CrossRef]
61. Mumtaz, S.; Rowley, C. The relationship between leader–member exchange and employee outcomes: Review of past themes and future potential. *Manag. Rev. Q.* **2020**, *70*, 165–189. [CrossRef]
62. Le Blanc, P.M.; González-Romá, V. A team level investigation of the relationship between Leader–Member Exchange (LMX) differentiation, and commitment and performance. *Leadersh. Q.* **2012**, *23*, 534–544. [CrossRef]

buildings

MDPI

Article

Network-Based Modeling of Lean Implementation Strategies and Planning in Prefabricated Construction

Pei Dang [1], Linna Geng [2], Zhanwen Niu [3], Shan Jiang [4,*] and Chao Sun [1]

1 School of Economics and Management, Tianjin Chengjian University, No. 26, Jinjing Road, Xiqing District, Tianjin 300384, China
2 School of Engineering, Design & Built Environment, Western Sydney University, Sydney 2150, Australia; l.geng@westernsydney.edu.au
3 College of Management and Economics, Tianjin University, No. 92 Weijin Road, Nankai District, Tianjin 300072, China
4 School of Economics, Wuhan University of Technology, No. 122, Luoshi Road, Wuhan 430070, China
* Correspondence: jiang_shan@whut.edu.cn

Abstract: Prefabricated construction (PC) is increasingly promoted in the construction sector for its potential benefits, including reduced resource assumption and improved quality. Accordingly, Lean methods are popularly applied to PC projects for optimizing operational processes and enhancing their performance in line with strategic objectives. A key factor in effectively implementing Lean to improve strategic control is developing specific strategies and planning that consider their complex interactions. Thus, this paper aims to propose a quantitative network-based model by integrating Interpretive Structural Modeling (ISM) and Matrix Impact Cross-Reference Multiplication Applied to a Classification (MICMAC) under complex network theory to develop a Lean implementation framework for effective strategy formulation. Specifically, 17 Lean implementation strategies for PC in the context of the Chinese prefabrication industry were identified via an extensive literature review and expert interviews. Then, ISM-MICMAC quantitatively identifies the direct and indirect relationships among strategies, while subsequent analysis of Topological Structure Weight (TSW) and Structural Degree Weight (SDW), as complex network parameters, is used to evaluate the importance of each strategy. The findings show that the strategic planning for Lean implementation in PC consists of four levels, i.e., foundation, organizational, technical, and control. Selecting appropriate Lean tools and technologies is crucial for PC implementation, which must be built on a top-level management team and foster a Lean culture. Moreover, it involves building a standardized system of processes and activities, enhancing both internal and external collaboration, and continuously improving processes in response to changes. On one hand, this in-depth network-based analysis offers practical insights for PC stakeholders, particularly in China, on Lean implementation in line with PC performance and strategic control and objectives. On the other hand, the network-based model can be future-implemented globally. Additionally, this study expands the current body of knowledge on Lean in PC by exploring the interrelationships of Lean implementation strategies.

Keywords: prefabricated construction; Lean implementation; strategy planning; ISM-MICMAC; complex network

Citation: Dang, P.; Geng, L.; Niu, Z.; Jiang, S.; Sun, C. Network-Based Modeling of Lean Implementation Strategies and Planning in Prefabricated Construction. *Buildings* **2024**, *14*, 3182. https://doi.org/10.3390/buildings14103182

Academic Editors: Wenzhe Tang, Wenxin Shen and Jin Xue

Received: 5 September 2024
Revised: 25 September 2024
Accepted: 2 October 2024
Published: 6 October 2024

1. Introduction

Prefabricated construction (PC), characterized by standardized design, factory production, on-site assembly, and life-cycle data management [1] is a sustainable approach with a great increasing popularity. For example, the Ministry of Housing and Urban–Rural Development of the People's Republic of China (MHURD) has aimed to develop PC to comprise over 30% of all new buildings by 2025 [2]. It differs from traditional in situ cast construction, where the operation management of internal and external processes and the overall supply chain play a crucial role in determining project performance [3]. In this context, Lean has been particularly effective, demonstrating substantial improvements in industrial chain management, strategic management, construction capabilities, and overall project performance [4–7]. It refers to the tools and practices for precisely defining value, reducing unnecessary interference to increase efficiency, doing more and more with less and less to gear products more towards user needs, and eliminating waste while creating value [8]. The effective implementation of Lean needs an applicable planning comprised of implementation strategies that reflect the socio-cultural and operational contexts, which will serve as a guideline to identify relevant Lean practices and specify step-by-step procedures to implement Lean in line with PC performances and strategic missions and objectives [9,10].

However, ignoring the interrelationships among Lean implementation strategies primarily leads stakeholders in PC to fail the effective implementation of lean [11]. Particularly speaking, Lean implementation can be seen as a systematic engineering involving multiple strategies and factors [12,13]. These strategies are not isolated but are interacted in complex interactions, significantly impacting the effectiveness of their implementation. It demands that stakeholders in PC develop systematic strategies following a certain plan rather than merely applying Lean tools and methods [11,14]. Therefore, it is imperative to construct a plan for Lean implementation by thoroughly exploring the interrelationships between Lean implementation strategies.

Current research has proposed a series of methods to explore the interrelationships among strategies, including Factor Analysis (FA), Analytic Network Process (ANP), Structural Equation Modeling (SEM), and Interpretive Structure Modeling (ISM) [15]. Among them, ISM is considered particularly effective for depicting the mutual relationships and hierarchical structures between various factors [16]. It is often integrated with Matrix Impact Cross-Reference Multiplication Applied to a Classification (MICMAC) to more clearly and quantitatively describe the relationships and their effects on project implementation [17]. To effectively formulate the planning of Lean implementation, it is vital to determine the importance of these strategies within the complex system as well, which not only recognizes their relative importance but also considers their interacted structures. Complex network theory is therefore well-suited for exploring characteristics such as Topological Structure Weight (TSW) and Structural Degree Weight (SDW) in complex systems composed of nodes with intricate interactions [18]. Consequently, a systematic and quantitative network-based ISM-MICMAC model is developed to explore structural and mutually influential interrelationships among Lean implementation strategies, further formulating a plan for Lean implementation for PC.

The objectives of this study are: (1) identifying Lean implementation strategies for advancing prefabricated construction (PC); (2) quantifying the direct and indirect relationships between these strategies; and (3) exploring their priorities. This will facilitate the effective implementation of Lean practices within PC contexts through an innovative network-based analysis of interrelationships among Lean strategies. Moreover, this study not only applies existing theories of Lean, PC, ISM-MICMAC, and complex network theory but also contributes theoretical enrichment to the field of project management.

2. Literature Review

2.1. Lean Implementation Strategies

Current research has focused on how to effectively implement Lean in the construction sector from various perspectives, regarding influencing factors, obstacles, challenges, and strategies [8,19–21]. For example, Yunus et al. [13] identified 31 key factors for implementing Lean in Malaysian Industrialized Building Systems, particularly on management support, process management, and education and training. Similarly, Hussein and Zayed [11] employed meta-analysis to determine the top seven influential factors in modular construction projects, with respect to management, technology, culture, knowledge, finance, government, skills, logistics, and communication. For obstacles or challenges in implementing Lean, Mano et al. [22] identified 83 obstacles and determined eight key obstacles involving culture, leadership, and structure in Lean construction. These types of studies have laid the groundwork for developing appropriate Lean strategies. For instance, Ahmed et al. [20] proposed Lean implementation strategies to enhance the level of Lean implementation by identifying the main 41 challenges. In earlier 2015, Bashir et al. [23] proposed 13 Lean implementation strategies by identifying implementation obstacles in the UK construction industry.

These previous studies enable stakeholders to understand the key elements of successful Lean implementation in PC projects. However, they have paid less attention to examining the full perspective of implementation frameworks for Lean strategies, with a few studies shedding light on this aspect. For example, Gao and Low [12] proposed a four-tier Lean implementation framework based on 14 principles of the "Toyota-way" model, involving philosophy, process, people, partners, and problem-solving aspects. Despite the pilot conceptual frameworks, there is a lacking specific Lean implementation framework, focusing on strategy planning for PC, especially in the context of China. Moreover, the interrelationships between Lean implementation strategies are ignoring [11]. This limits potential process management and improvement of PC projects through Lean. In fact, the successful implementation of Lean is systematic engineering. On one hand, the implementation strategies are interconnected. Neglecting these interrelationships means stakeholders cannot formulate effective Lean planning and strategies [3]. On the other hand, Lean not only focuses on the selection of tools and technologies but also needs to address the current demands of construction projects and the industry.

2.2. Models for Interrelationships Analysis

In recent years, researchers from the construction management field have focused on exploring the interrelationships among factors or strategies and their impact on project implementation. Several quantitative models have been proposed [15,24,25], including but not limited to FA, ANP, SEM, and ISM. Among of them, FA, ANP, and SEM primarily focus on classifying factors but fail to further break them into a logically progressive hierarchical structure based on their interrelationships [26,27]. Notably, ISM is seen as an effective method for clearly depicting the interrelationships and hierarchical structure among factors, which is widely used in systems engineering and particularly suitable for system analysis involving numerous variables, complex factor relationships, and unclear hierarchical structures [24,28,29]. It helps simplify complex systems and assists in identifying the structure within the system [30]. This is because it not only establishes direct and indirect relationships among factors but also identifies the extent of their impact on the target, thereby facilitating the formulation of effective implementation plans [31].

Notably, when exploring structural relationships among factors using ISM, it is often integrated with MICMAC model [17,32]. This integrated ISM-MICMAC model provides a precise depiction to validate the factors and their relationships, as well as the roles of different factors in ISM, thus promoting subsequent clearer planning measures [33–35]. For example, Gan et al. [36] used ISM to explore the interrelationships among barriers to the transformation of the Chinese construction industry towards PC and further utilized MICMAC to classify these barriers to identify the key barriers. This integrated model provides insight into how these barriers influence one another, with the potential for future research to quantify these interrelationships on a larger scale. Then, Sarhan et al. [10] employed ISM-MICMAC to develop a Lean implementation framework for Saudi Arabia's construction industry. The ISM technique in this study was to specify the hierarchical relationships among the 12 critical success factors that contribute to the successful implementation of Lean construction. From these studies, the ISM-MICMAC model has been validated as an effective tool for understanding the relationships among numerous elements within a system by developing a structured model of these relationships [37,38]. This helps to impose order on and direction to the relationships among elements in a system, such that their influence can be analyzed.

Despite its merits, this ISM-MICMAC model is criticized for overlooking the relative importance among factors [39]. Put another way, the interrelationships among Lean implementation strategies are complex and multifaceted [40]. In such a complex network structure, the degree of connection, the position of strategies, and interactions have a significant impact on the importance of strategies. For instance, Wang et al. [41] have considered the impact of out-degree, in-degree, and network hierarchy of nodes on the importance of nodes. In the ISM-MICMAC model of Lean strategies, there are both driving and dependent strategies. Therefore, their importance is not only related to adjacent strategies but also connected to strategies with indirect relationships. However, existing ISM-MICMAC studies paid less attention to this aspect. Complex network methods are thus introduced owing to their advantages in both focusing on the relative importance of factors and exploring the impact of the structural characteristics of complex networks on factor significance [42]. The mutual relationships between nodes and their importance can be determined through topological structure parameters within complex networks [43].Regarding this, TSW is a favorable parameter used in complex network analysis to assess the relative importance of nodes [44]. It evaluates the contribution of a node based on its position in the overall network topology. In studies involving network theory, TSW is often used alongside other metrics, like SDW, to provide deeper insights into the hierarchical and relational structure within the network. Liu and Xu [45] have combined the ISM with the complex network method to identify critical factors in manufacturing systems. However, they did not conduct MICMAC analysis for the path analysis.

Therefore, this paper integrates a network-based ISM-MICMAC model to systematically explore the interrelationships of the strategies and their relative importance to developing the Lean implementation framework. This aims to help stakeholders in PC develop more effective Lean strategies and paths, thereby enhancing strategic control and project performance.

3. Methodology

The four-step research design is presented as Figure 1.

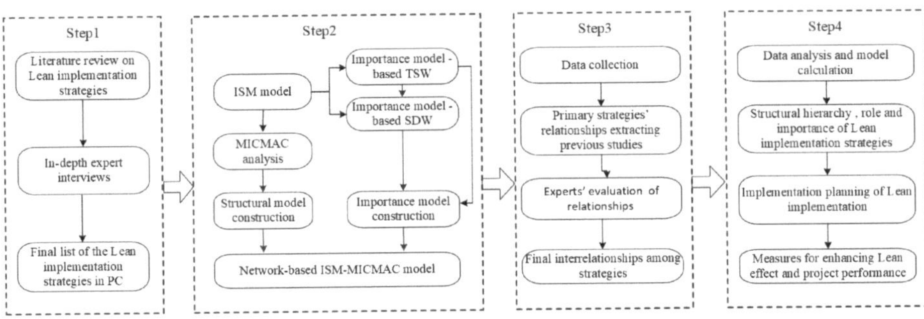

Figure 1. Research design for Lean planning.

3.1. *Identifying the Lean Implementation Strategies*

Firstly, a list of Lean implementation strategies in PC was identified via a detailed literature review and further examined through in-depth expert interviews to obtain the final list. Two rounds of literature review were conducted to identify potential Lean implementation strategies. The literature databases include global databases, i.e., "Google Scholar", "Web of Science", "Scopus", and Chinese databases such as "CNKI to include full relevant sources. The 1st round of searching criteria was set as Title/Abstract/Keyword = ("lean" OR "just-in-time") AND ("prefabricated construction" OR "prefabrication" OR "precast" OR "off-site construction" OR "industrial building system") AND ("strategies") with no time limitations. As peer-reviewed papers follow a rigorous review process as compared to the conference papers, document type was set to Article, and language was set to English. Only about 20 journal papers are retrieved, indicating that research on Lean implementation strategies in PC is relatively limited. Then, the 2nd round of literature review mainly involves expanding keyword searches, as the successful implementation of Lean is also related to various factors, including barriers, risks, drivers, or challenges. The searching criteria was reset as Title/Abstract/Keyword = ("lean" OR "just-in-time") AND ("prefabricated construction" OR "prefabrication" OR "precast" OR "off-site construction" OR "industrial building system") AND ("strategies OR "factors" OR "barriers" OR "risks" OR "drivers" OR "challenges"). This search brought forth nearly 100 papers at first, after excluding duplicates. Next, the papers went through visual filtering: the Abstract was scanned to remove irrelevant studies. Finally, there were 25 publications remaining for a full-text review, which detailed the specific strategies, factors, barriers, risks, drivers, and challenges of implementing Lean. Based on their frequency of occurrence and importance, 33 Lean implementation strategies in PC are identified, which are mentioned more than 3 times. These strategies refer to four aspects of process: technology, organization, and culture [11,46].

Then, in-depth experts' interviews with key stakeholders in PC were conducted to examine the comprehensiveness and effectiveness of the 33 strategies from the literature. The key stakeholders in PC are consisted of developer, designer, producer, general contractor, subcontractor, supplier, and PC consultant [2,47]. The interviewees were selected and invited according to the stakeholder-based sampling principle: (1) with sufficient knowledge regarding Lean implementation and more than 5 years of related experience; (2) having undertaken important tasks of implementing Lean in PC projects; (3) holding senior positions in the project teams [48]. These principles can ensure that the selected interviewees are qualified to answer the questions pertaining to the Lean implementation strategies to ensure validity and accuracy. Interviewees were recommended by the MOHURD and contacted whether they could be participants via email or telephone [47,49]. A total of 30 interviewees who met the selected principles were selected, which consisted of 5 developers, 3 designers, 5 producers, 5 general contractors, 4 subcontractors, 3 suppliers, and 5 PC consultants. Table 1 shows the interviewee profiles.

Table 1. Interviewee profiles.

Stakeholder Group	Number	Main Position	Education Level	Years of Experience
Developer	1	General manager	Ph.D.	≥5
	4	Business manager	Master	
Designer	3	Business manager	Master	≥5
Producer	1	General manager	Ph.D.	≥5
	4	Factory manager	Master	
General contractor	2	General manager	Master	≥5
	3	Project manager		
Subcontractor	1	General manager	Bachelor	≥5
	3	Project manager		
Supplier	1	General manager	Master	≥5
	2	Business manager	Bachelor	
PC consultant	2 3	Professor Project manager	Ph.D. Master	≥5

Prior to the interviews, research background and purpose were sent to each interviewee via email. After obtaining their consent, interviewees were invited by telephone or face-to-face to discuss whether the Lean implementation strategies identified from the literature existed in actual PC projects and could potentially affect the performance. Each interview lasted 1–2 h to confirm the reliability of the identified strategies. The interviewees also proposed additional strategies and described them in detail based on their own experiences. The opinions of different interviewees had the same weight. When disagreements among interviewees existed, these interviewees were contacted for further discussion. After three rounds of discussion, the interviewees reached an agreement for all questions. Ultimately, a list of 17 Lean implementation strategies could be generated after in-depth interviews, as shown in Table 2.

Table 2. Final Lean implementation strategies.

No.	Lean Implementation Strategies	Sources
S1	Establish a top-level plan and leadership team	[50]
S2	Develop an efficient decision-making organizational structure	[51,52]
S3	Implement a comprehensive performance evaluation and incentive mechanism	[11,19]
S4	Focus on customer needs and value	[53]
S5	Establish a comprehensive resource management system	[54,55]
S6	Enhance internal communication and collaboration within the enterprise	[56]
S7	Strengthen external communication and collaboration with the enterprise	[56]
S8	Develop a comprehensive risk management system	[57]
S9	Accurately formulate project implementation and scheduling plans	[58]
S10	Select appropriate Lean and information tools and technologies	[59]
S11	Strengthen process management and continuously optimize project plans and workflows	[60,61]
S12	Establish a standardized structural system and operational activities	[46]
S13	Develop a comprehensive system of standards, rules, and regulations	[62]
S14	Foster a lean and intelligent culture and employee awareness	[63]
S15	Emphasize employee knowledge acquisition and skills training	[64]
S16	Utilize external consulting firms and academic institutions for assistance	[50]
S17	Establish a continuous improvement mechanism and culture	[11,65]

3.2. The Network-Based ISM-MICMAC Model

A network-based ISM-MICMAC model integrating ISM, MICMAC, and complex networks is constructed. In the model, ISM-MICMAC aims to identify the strategies' structural relationships and roles, while TSW and SDW aim to evaluate strategies' importance through considering structural and mutual impact.

3.2.1. Structural Model Part: ISM-MICMAC for Lean Implementation Strategies

1. The ISM model of Lean implementation strategies in PC

ISM is a model that formulates a complex system into a visualized hierarchical structure and helps to understand the direct and indirect relationships among the strategies [66]. The steps for the ISM model are discussed below [3,10,45]:

Step 1: Lean implementation strategies are identified and listed through the extensive review of literature and experts' opinions.

Step 2: A contextual relationship among the identified strategies is developed to examine as to which pairs of Lean implementation strategies should be checked.

Step 3: A structural self-interaction matrix (SSIM) is developed that indicates pairwise relationships among strategies of the system.

In this step, the direct relationships among strategies can be evaluated from four aspects: W, X, Y, and Z, which are indicated as follows:

- W indicates that strategy *i* has a direct impact on strategy *j*, but the reverse is not true.
- X indicates that strategy *j* has a direct impact on strategy *i*, but the reverse is not true.
- Y indicates that there is a direct interaction between strategy *i* and strategy *j*.
- Z indicates that there is no direct interaction between strategy *i* and strategy *j*.

Step 4: A reachability matrix (RM) is constructed from the SSIM by replacing each cell entry of the SSIM by 1 and 0, and the matrix is checked for transitivity. The transitivity of the contextual relation is a basic assumption made in the ISM. It states if a strategy U is related to strategy V and strategy V is related to strategy W, then U is necessarily related to W. Thus, a final RM is developed.

Step 5: The final RM developed in Step 4 is categorized into different levels.

Step 6: A directed graph or digraph is drawn based on the contextual relationships given above in the reachability matrix, and then the transitive links are removed from the digraph.

Step 7: By substituting variable nodes with relationship statements, an ISM model is generated from the resultant digraph.

Step 8: The ISM model of Lean implementation strategies generated in Step 7 is reviewed to find out that any conceptual inconsistency and necessary modifications are considered through experts' opinions.

2. The MICMAC analysis for Lean implementation strategies in PC

MICMAC analysis complements the ISM by exploring constraints that usually are embedded within the ISM network [37]. In this analysis, the objective of MICMAC analysis is to identify the key strategies that drive the ISM model based on their driving power and dependence power [33]. The driving power and the dependence power of each strategy in MICMAC are obtained by summing the entries of possibilities of interactions in its row and column of the final reachability matrix of ISM. Seventeen Lean implementation strategies can be further classified into four categories based on the value of the driving power and the dependence: autonomous strategies (AUSs), dependent strategies (DESs), linkage strategies (LISs), and driving strategies (DRSs), which reflect their various roles and impacts for implementing Lean [33].

AUSs have minimal driving influence and dependency on other elements in the ISM network, resulting in a relatively small impact on achieving the network's objectives, making them of secondary consideration and focus [33]. DESs exhibit high dependency and relatively weak driving power, typically positioned at the upper levels of the ISM, which have a direct impact on the objectives and require close monitoring to assess the effectiveness of the lower-level strategies [38]. LISs possess high intensity in both driving power and dependency, typically located in the middle of the ISM, acting as a critical link between upper and lower levels. It plays a key role in achieving objectives and therefore requires focused attention and management [35]. DRSs have strong driving power and weak dependency, typically located at the lower levels of the ISM, serving as the foundation for the implementation of other strategies, resulting in them being the most critical strategies in the ISM and requiring priority attention and assurance [67].

3.2.2. Importance Evaluation Model Part: TSW and SDW of Lean Implementation Strategies in PC

To enhance the precision in assessing the importance of Lean implementation strategies, this paper integrates the influence of both the topological features of complex networks and the hierarchical structure of the ISM model on strategy importance [45]. By calculating the TSW and SDW, this paper computes the comprehensive weights of each Lean strategy to evaluate its relative importance. The steps for the TSW and SDW models are discussed below, according to Congliang et al. [68], Huang et al. [69], Yu et al. [70], Chen et al. [43], and Liu and Xu [45].

1. Importance evaluation model of Lean implementation strategies based on TSW

Step 1: Degree calculation of each node in the ISM network of Lean implementation strategies

In the ISM network of Lean implementation strategies, the degree of the node S_i set as D_{Si} refers to the number of nodes that have a direct impact relationship with it. Due to the directional influence relationship between various strategies, they may have an impact on other ones and may also be influenced. Among them, the number of edges pointing from S_i to other strategies is the output of S_i, set as ${D_{Si}}^{out}$. The number of edges pointing from other strategies to S_i is the in degree of S_i, set as ${D_{Si}}^{into}$. The calculation formula for the degree D_{Si} of S_i is shown in Equation (1) [43].

$$D_{Si} = \sum_{j=1}^{n} (a_{ij} + a_{ji}) - b = \sum_{j=1}^{17} (a_{ij} + a_{ji}) - b \tag{1}$$

where n is the number of Lean implementation strategies; S_i and S_j are two strategy nodes in the Lean strategy network; a_{ij} is the value between S_i and S_j in the AM illustrated in Appendix B; and b is the number of other strategies corresponding to strategy S_i, where both the row and column strategies are set to "1".

Step 2: Average degree calculation in the ISM network of Lean implementation strategies

The average degree (AD) of the ISM network is the average degree of all strategy nodes in the network, calculated using the formula shown in Equation (2) [43].

$$AD = \sum_{i=1}^{n} D_{si}/n = \sum_{i=1}^{17} D_{si}/17 \tag{2}$$

where n is the number of Lean implementation strategies and D_{Si} is the degree of S_i.

Step 3: Calculation of strategy node distance in the ISM network of Lean implementation strategies

In the ISM network of Lean implementation strategies, the distance between two strategy nodes S_i and S_j is defined as the minimum number of edges involved from S_i to S_j along the direction of the transmission relationship, denoted as d_{SiSj}. The calculation formula is presented as Equation (3) [43]. If there is no direct or indirect relationship between two strategy nodes S_i and S_j, their distance is infinity (∞).

$$d_{S_iS_j} = \min_{i,j}(l_{S_iS_j}) \tag{3}$$

where l_{SiSj} represents the length of all paths between strategy node S_i and node S_j, that is, the number of edges.

Step 4: Efficiency calculation of strategy nodes in the ISM network of Lean implementation strategies

In the ISM network of Lean implementation strategies, the efficiency of a strategy node S_i, denoted as E_{Si}, is a metric reflecting the speed at which the strategy influences other strategies. It is inversely proportional to the node distance d_{SiSj}, with the calculation formula provided in Equation (4) [43].

$$E_{S_i} = \frac{1}{n-1}\sum_{j=1, j\neq i}^{n}\frac{1}{d_{S_iS_j}} = \frac{1}{16}\sum_{j=1, j\neq i}^{17}\frac{1}{d_{S_iS_j}} \tag{4}$$

where n is the number of Lean implementation strategies and d_{SiSj} is the distance between strategy nodes S_i and S_j.

Step 5: Construction of a node importance contribution matrix in the ISM network of Lean implementation strategies

Through Steps 1 to 4, the basic topological structure parameters of the ISM network of Lean implementation strategies, including node degree, network degree, node distance, and node efficiency, have been calculated. Further consideration is given to the interrelationships between strategies, particularly the impact on the importance of adjacent strategies. This mutual influence of importance between strategies can be measured through the Importance Contribution Value (*ICV*) of strategy nodes. *ICV* can be calculated according to Equation (5), representing the importance contribution of the Lean strategy S_i to its adjacent strategies, denoted as CV_{Si} [43,69,70].

$$ICV_{S_i} = D_{S_i}/AD^2 \tag{5}$$

In Equation (5), D_{Si} denotes the degree of strategy node S_i, and AD is the average degree of the ISM network of Lean implementation strategies.

Based on the *ICV* of strategy nodes, an importance contribution matrix (*ICM*) for the entire ISM network of Lean implementation strategies can be constructed. Strategy nodes only contribute importance to adjacent nodes that have a direct impact on them, and the constructed *ICM* is shown in Equation (6) [43,69,70].

$$ICM = \begin{bmatrix} 1 & a_{12}ICV_{S_2} & \cdots & a_{1n}ICV_{S_n} \\ a_{21}ICV_{S_1} & 1 & \cdots & a_{2n}ICV_{S_n} \\ \vdots & \vdots & \ddots & \vdots \\ a_{n1}ICV_{S_1} & a_{n2}ICV_{S_2} & \cdots & 1 \end{bmatrix} \tag{6}$$

In Equation (6), a_{ij} is an element in the adjacency matrix of the ISM network of Lean implementation strategies, with a value of "1" or "0". ICV_{Si} is the node importance contribution matrix of S_i.

Step 6: Construction of a node importance evaluation matrix in the ISM network of Lean implementation strategies

Based on *ICM* in Step 5, to further consider the impact of efficiency between strategy nodes, the *ICV* and efficiency values of strategy nodes are integrated to define the Importance Evaluation Value (*IEV*) of strategy nodes, as shown in Formula (7) [43,69,70].

$$IEV_{S_i} = E_{S_i} * ICV_{S_i} \tag{7}$$

In Formula (7), E_{Si} represents the efficiency of strategy S_i; ICV_{Si} represents the contribution value of the importance of strategy S_i.

Based on this, it is feasible to construct an importance evaluation matrix (*IEM*) for the ISM network of Lean implementation strategies. The computational formula of IEM is illustrated in Equation (8) [43,69,70].

$$IEM = \begin{bmatrix} 1 & a_{12}E_{S_2}ICV_{S_2} & \cdots & a_{1n}E_{S_n}ICV_{S_n} \\ a_{21}E_{S_1}ICV_{S_1} & 1 & \cdots & a_{2n}E_{S_n}ICV_{S_n} \\ \vdots & \vdots & \ddots & \vdots \\ a_{n1}E_{S_n}ICV_{S_1} & a_{n2}E_{S_2}ICV_{S_2} & \cdots & 1 \end{bmatrix} \tag{8}$$

Where a_{ij} is an element in the adjacency matrix of the ISM network of Lean implementation strategies, with its value of "1" or "0"; E_{Si} is the efficiency of strategy S_i; and ICV_{Si} is the contribution value of the importance of S_i.

Step 7: Calculating the importance weight of each strategy in the ISM network of Lean implementation strategies

Based on *IEM*, by comprehensively considering the efficiency E_{Si} of the strategy node S_i and the importance evaluation value IEV_{Sj} of the strategy S_j affected by it, its importance weight w_{si} in the network can be calculated, shown as in Equation (9) [43,69,70].

$$w_{S_i} = E_{S_i} \sum_{j=1, j\neq i}^{n} IEV_{S_j} \tag{9}$$

2. Importance evaluation model of Lean implementation strategies based on SDW

Based on calculating the importance weight values of Lean implementation strategies based on TSW, it is necessary to further comprehensively consider the impact of the structural hierarchy of strategy nodes in the ISM network on their own importance. Thereby, this paper further uses a strategy node importance evaluation method that considers the out degree, in degree, and ISM model hierarchy of strategy nodes to determine their SDW. The calculation process based on SDW mainly consists of three steps, namely calculating the level weights of the ISM model, determining the impact coefficients of the out degree and in degree of the strategy nodes, and calculating the structured weights of the strategy nodes.

Step 1: Level weight calculation of the ISM network of Lean implementation strategies

Based on the ISM model and MICMAC analysis, Lean implementation strategies at different levels have different roles and importance in the successful implementation of Lean in PC. To this end, the network level weight (*LW*) is used to represent the importance level of each level in the ISM model, and the specific calculation formula is shown in Equation (10) [45].

$$LW_{L_i} = 1/i \Big/ \sum_{i=1}^{N} (1/i) \tag{10}$$

In Formula (10), i is the structural hierarchy value of the ISM model, i = 1, 2, 3. . ., N.

Step 2: Determining the impact coefficients of the out degree and in degree of the strategy nodes in the ISM network

In analyzing the importance of strategy nodes based on the ISM model, it is not only necessary to consider the network level located in the strategy node but also to consider the level of other strategy nodes that have a direct impact on strategies. As mentioned above, strategies affect or are influenced by other strategies, corresponding to the out-degree nodes and in-degree nodes in the ISM model, whose importance influence coefficients can be expressed as O and I, respectively, with $O < I$ and $O + I = 1$.

Step 3: Structured weight calculation of implementation strategy nodes in the ISM network

By calculating the level weights of the ISM network and impact coefficients of the out degree and in degree of the strategy nodes, the structural weight (*SW*) of each strategy node can be further calculated. The calculation formula is shown in Equation (11) [45]

$$SW_{S_i} = LW_{S_i}(I\sum_k LW_{Sk \to Si} N_{Sk \to Si} + O\sum_j LW_{S_i \to S_j} N_{S_i \to S_j}) \tag{11}$$

In Equation (11), LW_{Si} represents the level weight of the ISM model where the strategy node S_i is located; S_j and S_k represent the strategy nodes S_i points to and is directed to, respectively; and $LW_{Si \to Sj}$, $N_{Si \to Sj}$, $LW_{Sk \to Si}$, and $N_{Sk \to Si}$, respectively, represent the level weights and their own numbers of strategy nodes S_j and S_k in the ISM model.

3. Comprehensive importance evaluation of Lean implementation strategies in PC

Based on the importance weights based-TSW w_{si} and based-SDW SW_{Si}, it is necessary to integrate the w_{si} and SW_{Si} to scientifically evaluate the importance of Lean implementation strategies. The comprehensive weight (*CW*) by multiplying the w_{si} with SW_{Si} is shown in Equation (12) [45].

$$CW_{S_i} = w_{S_i} * SW_{S_i} \tag{12}$$

3.3. Data Collection

The interrelationships among strategies in the ISM model are determined through experts' evaluation, which is also the data source of MICMAC analysis and importance calculation. In other words, the interrelationships are initially identified through literature review. Then, 30 experts from Table 1 are invited to evaluate the importance of each mutual relationship between strategies using a five-point Likert scale and applying the principle of "the minority yielding to the majority" [3]. Specifically, if the number of experts who evaluate the relationship as "4" and "5" exceeds 15, then the relationship between the strategies is considered to exist. Thus, all mutual relationships among strategies can be constructed to establish SSIM and further calculate the processes of the model.

3.4. Data Analysis and Model Caculation

Finally, the structural hierarchy, the roles and importance of Lean implementation strategies in PC, and corresponding planning were determined, and effective measures for enhancing project performance for stakeholders were identified by emphasizing the crucial roles of important Lean implementation strategies.

4. Results and Discussion

4.1. Structural Analysis of Lean Implementation Strategies-Based ISM

1. Constructing the ISM model of Lean implementation strategies in PC

Step 1: Establishing the structural relationships among Lean implementation strategies and SSIM

The interrelationships among strategies are determined through a literature review and 30 experts' evaluations, which are described in Section 3.2.1. Then, the final SSIM is constructed, shown as in Appendix A.

Step 2: Establishing AM of Lean implementation strategies in PC

The structural relationships of W, X, Y, and Z in SSIM are converted into a binary matrix represented by "0" or "1" to construct the AM of Lean implementation strategies, as shown in Appendix B.

Step 4: Constructing RM of Lean implementation strategies in PC

The *RM* is obtained through Boolean operations based on *AM*. Specifically, the *AM* is added to the identity matrix I to obtain the matrix $(AM + I)$. Boolean operations are then performed on it until the matrix no longer changes, resulting in *RM*. The formula is shown as follows, where k represents the number of iterations of the matrix.

$$(AM + I)^1 \neq (AM + I)^2 \neq (AM + I)^3 \neq \ldots \neq (AM + I)^{k-1} = (AM + I)^k = RM \quad (13)$$

It was revealed that when $k = 7$, $(AM + I)^6 = (AM + I)^7$, RM can be obtained as $RM = (AM + I)^6 = (AM + I)^7$, which is shown in Appendix C.

Step 5: Level division of Lean implementation strategies

The 17 Lean implementation strategies can be further divided into different levels based on the RM, according to the steps of hierarchical division in (24; 16). The results of level division are as follows: $L_1 = [S_8]$, $L_2 = [S_4, S_5, S_9, S_{11}]$, $L_3 = [S_6, S_7]$, $L_4 = [S_{17}]$, $L_5 = [S_{10}, S_{12}, S_{16}]$, $L_6 = [S_2, S_3, S_{13}]$, $L_7 = [S_{14}, S_{15}]$, $L_8 = [S_1]$.

Step 6: Constructing the ISM model of the Lean implementation strategies in PC

A preliminary diagram illustrating the interrelationships of 17 strategies based on AM is initially created, as shown in Figure 2. Additionally, this figure demonstrates that the complex relationships between strategies form a complex network. This provides a starting point for exploring the mutual influence and importance of strategies using complex network theory.

The ISM model is accordingly constructed, clearly reflecting the interaction and hierarchy of Lean implementation strategies, which is shown in Figure 3.

As in Figure 3, the ISM model for Lean implementation strategies is a multi-level progressive network system with an eight-level hierarchical structure. The Lean implementation strategies positioned at different structural levels are closely connected and mutually influential, highlighting the differentiated roles of these strategies in the successful Lean implementation. The strategies located at the lower levels of the ISM model directly or indirectly impact the upper-level strategies while simultaneously supporting and promoting the successful LC implementation.

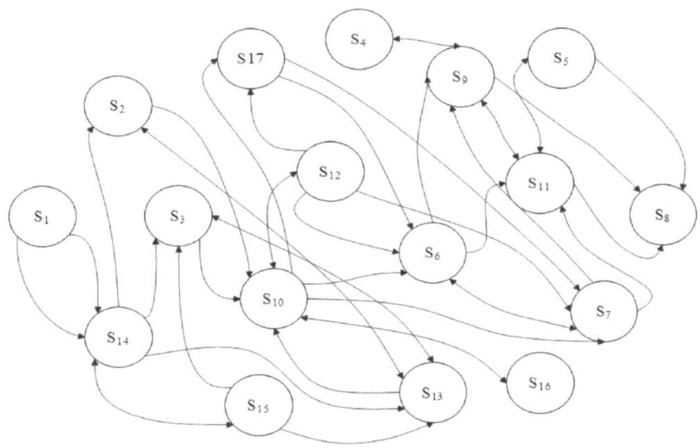

Figure 2. Structural relationship network among 17 Lean implementation strategies.

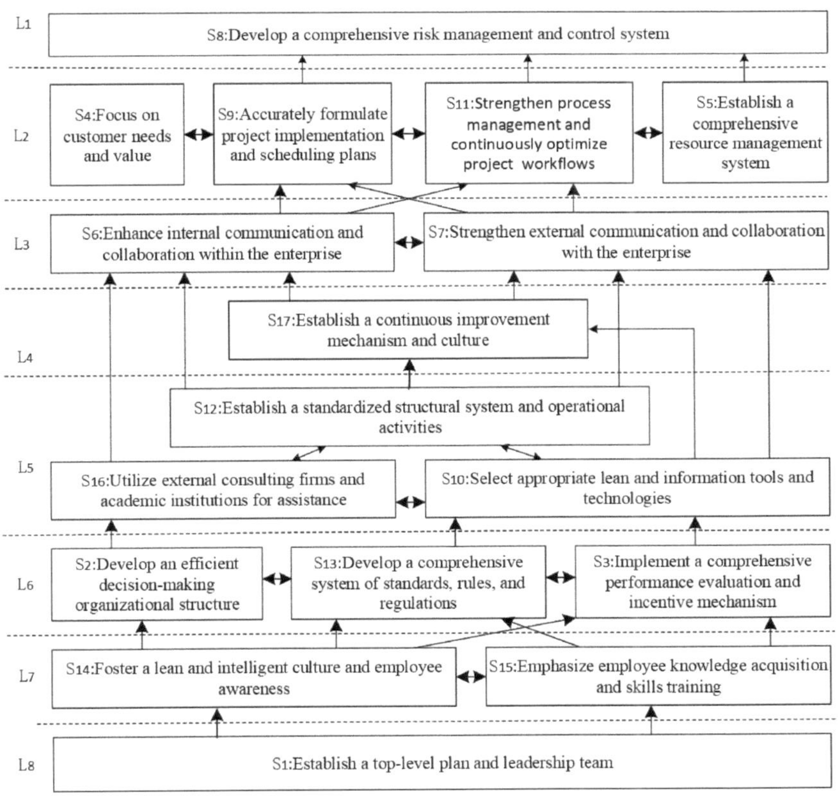

Figure 3. The ISM model of Lean implementation strategies in PC.

2. MICMAC analysis of the ISM model of Lean implementation strategies in PC

As aforementioned in Section 3.2.1, MICMAC is used to analyze the role of strategies in the ISM network. The 17 Lean implementation strategies can be classified into four categories: AUSs, DESs, LISs, and DRSs, based on the DRP and DEP values [33]. The DRP and DEP values can be calculated, as shown in Appendix D. The results can be plotted in quadrants based on the DRP as the horizontal axis and DEP as the vertical axis, as shown in Figure 4.

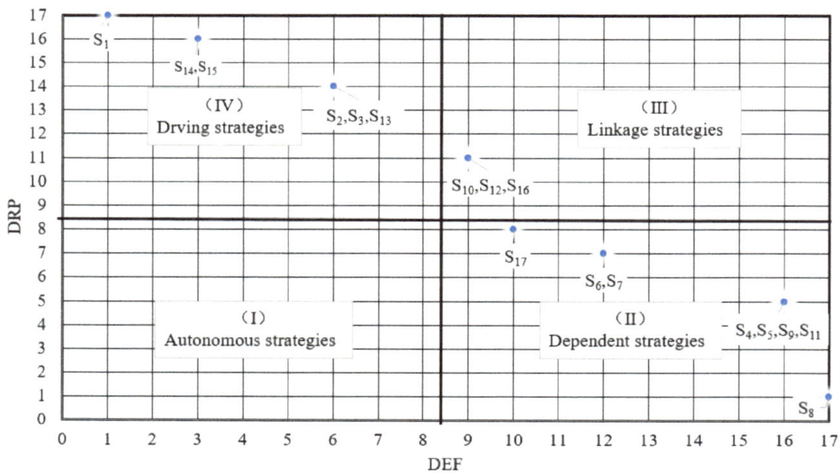

Figure 4. Category distribution of the 17 Lean implementation strategies based on MICMAC.

In Figure 4, there are no Lean implementation strategies distributed in the bottom-left quadrant (Quadrant I) of AUSs, which indicates that they are not independent of each other, rather having complex interrelationships and mutual influences. S_{17}, S_6, S_7, S_4, S_5, S_9, S_{11}, and S_8 are in the bottom-right quadrant (Quadrant II) as DESs, which are generally positioned at the upper levels of the ISM model and have a strong dependency on the lower-level strategies. S_{10}, S_{12}, and S_{16} are in the top-right quadrant (Quadrant III) as LISs, which are generally positioned in the middle levels of the ISM model, characterized by importance on other strategies and the successful Lean implementation. S_1, S_{14}, S_{15}, S_2, S_3, and S_{13} are in the bottom-right quadrant (Quadrant IV) as DRSs, which are generally positioned at the lower levels of the ISM model and have a significant influence on other strategies.

4.2. Importance Analysis

1. Importance evaluation of Lean implementation strategies based on TSW

The weights of Lean implementation strategies are calculated based on network topology to evaluate the importance of each strategy. The evaluation process primarily involves analyzing network characteristics such as strategy node degree D_{Si}, average degree AD, node distance d_{SiSj}, and node efficiency E_{Si}. The ICM and IEM are constructed to determine the weights of the strategy nodes, according to Equations (1)–(9).

Step 1: Degree of strategy nodes in the ISM network

The degree of the 17 Lean strategy nodes within the ISM network is calculated by Equation (1). The calculation results are detailed in Table 3.

Table 3. Degree of the 17 Lean implementation strategies.

Strategy	Node	Degree of Node	Strategy	Node	Degree of Node
S_1	D_{S1}	2	S_{10}	D_{S10}	8
S_2	D_{S2}	3	S_{11}	D_{S11}	5
S_3	D_{S3}	3	S_{12}	D_{S12}	4
S_4	D_{S4}	1	S_{13}	D_{S13}	5
S_5	D_{S5}	2	S_{14}	D_{S14}	5
S_6	D_{S6}	6	S_{15}	D_{S15}	4
S_7	D_{S7}	6	S_{16}	D_{S16}	1
S_8	D_{S8}	3	S_{17}	D_{S17}	4
S_9	D_{S9}	5		—	

Step 2: Average degree of the ISM network of Lean implementation strategies

The AD of the Lean implementation strategies is calculated by Equation (2). The AD is $67/17 = 3.9$, which indicates each lean construction implementation strategy has an averagely direct influence on approximately four other strategies.

Step 3: Node distance of strategies in the ISM network

The node distance (D) between two strategies S_i and S_j (d_{SiSj}) is defined as the minimum number of edges involved from S_i to S_j along the direction of the transmission relationship. The D between 17 Lean implementation strategies is determined by Equation (3), shown in Appendix E.

Step 4: Efficiency of strategy nodes in the ISM network

The efficiency (E) of strategy nodes is a metric reflecting the speed at which the strategy influences other strategies. The E of strategy nodes S_1–S_{17}, denoted as E_{S1}–E_{S17}, are calculated by Equation (4) and are provided in detail in Table 4.

Table 4. E of the 17 Lean implementation strategies.

Strategy	Node	E of Node	Strategy	Node	E of Node
S_1	E_{S1}	0.374	S_{10}	E_{S10}	0.438
S_2	E_{S2}	0.401	S_{11}	E_{S11}	0.219
S_3	E_{S3}	0.401	S_{12}	E_{S12}	0.375
S_4	E_{S4}	0.104	S_{13}	E_{S13}	0.432
S_5	E_{S5}	0.177	S_{14}	E_{S14}	0.454
S_6	E_{S6}	0.281	S_{15}	E_{S15}	0.423
S_7	E_{S7}	0.281	S_{16}	E_{S16}	0.276
S_8	E_{S8}	0.000	S_{17}	E_{S17}	0.250
S_9	E_{S9}	0.219		—	

Step 5: Node importance contribution matrix for the ISM network

The influencing interrelationships between strategies can be measured through the ICV of strategy nodes. The ICV of 17 Lean implementation strategies can be calculated by Equation (5). Then, the ICM for the entire ISM network can be further constructed according to Equation (6), which is shown in detail in Appendix F.

Step 6: Node importance evaluation matrix for the ISM network

The ICV and E of 17 Lean implementation strategy nodes are integrated to define IEV, which can be calculated by Equation (7). Then, the IEM of the 17 strategies can be further constructed according to Equation (8), which is shown in detail in Appendix G.

Step 7: The Topological Structure Weight of strategy nodes in the ISM network

The Topological Structure Weight (w) of the 17 Lean implementation strategies can be finally determined by Equation (9), by considering the E_{Si} and IEV_{Sj} of S_j affected by it. The results are shown in Table 5.

Table 5. The w of 17 Lean implementation strategies in PC.

Strategy	Code of Weight	Value of Weight	Strategy	Code of Weight	Value of Weight
S_1	w_{S1}	0.097	S_{10}	w_{S10}	0.177
S_2	w_{S2}	0.149	S_{11}	w_{S11}	0.021
S_3	w_{S3}	0.149	S_{12}	w_{S12}	0.194
S_4	w_{S4}	0.007	S_{13}	w_{S13}	0.168
S_5	w_{S5}	0.013	S_{14}	w_{S14}	0.187
S_6	w_{S6}	0.072	S_{15}	w_{S15}	0.157
S_7	w_{S7}	0.072	S_{16}	w_{S16}	0.064
S_8	w_{S8}	0.000	S_{17}	w_{S17}	0.056
S_9	w_{S9}	0.017		—	

2. Importance evaluation of Lean implementation strategies based on SDW

The structural weight (SW)-based ISM network can be calculated by Equations (10) and (11). The SWs of 17 Lean implementation strategies are shown in Table 6.

Table 6. The SW of 17 Lean implementation strategies in PC.

S_i	L_i	LW_{Si}	S_k	S_j	$I\sum_k LW_{Sk\to Si}N_{Sk\to Si}$	$O\sum_j LW_{S_i\to S_j}N_{S_i\to S_j}$	SW_{Si}
S_1	L_8	0.046	0	S_{14}, S_{15}	0.0000	0.0530	0.0024
S_2	L_6	0.061	S_{14}, S_{13}	S_{13}, S_{10}	0.1710	0.0675	0.0145
S_3	L_6	0.061	S_{14}, S_{13}, S_{15}	S_{13}, S_{10}	0.3758	0.0675	0.0270
S_4	L_2	0.184	S_9	S_9	0.1380	0.0460	0.0339
S_5	L_2	0.184	S_{11}	S_{11}, S_8	0.1380	0.2760	0.0762
S_6	L_3	0.123	$S_{10}, S_{12}, S_{17}, S_7$	S_7, S_{11}, S_9	1.0890	0.3683	0.1792
S_7	L_3	0.123	$S_6, S_{17}, S_{10}, S_{12}$	S_6, S_{11}, S_9	1.0890	0.3683	0.1792
S_8	L_1	0.368	S_9, S_5, S_{11}	0	1.2420	0.000	0.4570
S_9	L_2	0.184	S_4, S_6, S_7, S_{11}	S_4, S_{11}, S_8	1.8420	0.5520	0.4405
S_{10}	L_5	0.074	$S_{16}, S_2, S_{13}, S_3, S_{12}$	$S_{16}, S_{12}, S_6, S_{17}, S_7$	1.2413	0.6075	0.1368
S_{11}	L_2	0.184	S_9, S_6, S_7, S_5	S_9, S_5, S_8	1.8420	0.5520	0.4405
S_{12}	L_5	0.074	S_{10}	S_{10}, S_6, S_7, S_{17}	0.0555	0.4120	0.0346
S_{13}	L_6	0.061	S_2, S_{14}, S_{15}, S_3	S_2, S_3, S_{10}	0.6840	0.1470	0.0507
S_{14}	L_7	0.053	S_1, S_{15}	S_{15}, S_2, S_{13}, S_3	0.1485	0.2360	0.0204
S_{15}	L_7	0.053	S_1, S_{14}	S_{14}, S_{13}, S_3	0.1485	0.1313	0.0148
S_{16}	L_5	0.074	S_{10}	S_{10}	0.0555	0.0185	0.0055
S_{17}	L_4	0.092	S_{10}, S_{12}	S_6, S_7	0.2220	0.1230	0.0317

3. Comprehensive importance weights of Lean implementation strategies in PC

Based on the importance weights of w and SW, it is integrating them to obtain CW to more scientifically and comprehensively evaluate the importance of Lean implementation strategies. CWs of 17 Lean implementation strategies can be calculated by Equation (12) and are shown in Table 7.

4.3. Structural and Importance Analysis of Lean Implementation Strategies in PC

The comprehensive structural and importance analyses of 17 Lean implementation strategies in PC are described in Table 8.

As shown in Table 8, Lean implementation strategies positioned at higher levels of ISM correspond to DESs identified through MICMAC analysis. These strategies have the most direct impact on successful Lean implementation. They are supported by lower-level strategies and are crucial for monitoring and evaluating the effectiveness of other strategies. For example, the Lean strategy S_8: Develop a comprehensive risk management system exhibits the strongest dependency and weakest driving force, which aligns with its position at the top level (L_1) of the ISM model, relying on the implementation of lower-level strategies. Lean implementation strategies positioned at intermediate levels of the ISM correspond to LISs identified through MICMAC analysis. These strategies play a pivotal role in connecting the lower and upper levels, meaning they must inherit the outcomes of lower-level strategies while also providing support for the implementation of upper-level strategies. Therefore, these strategies are the most critical and require the greatest attention.

Table 7. The CW of 17 Lean implementation strategies in PC.

S_i	w_{Si}	SW_{Si}	*CW_{Si}	Rank	Ordering Rank	S_i
S_1	0.097	0.0024	0.000236486	16	1	S_{10}
S_2	0.149	0.0145	0.002167727	11	2	S_6
S_3	0.149	0.0270	0.004028699	8	3	S_7
S_4	0.007	0.0339	0.000236992	15	4	S_{11}
S_5	0.013	0.0762	0.000990288	13	5	S_{13}
S_6	0.072	0.1792	0.012905406	2	6	S_9
S_7	0.072	0.1792	0.012905406	3	7	S_{12}
S_8	0.000	0.4570	0.00000000	17	8	S_3
S_9	0.017	0.4405	0.007488432	6	9	S_{14}
S_{10}	0.177	0.1368	0.024214928	1	10	S_{15}
S_{11}	0.021	0.4405	0.009250416	4	11	S_2
S_{12}	0.194	0.0346	0.006711430	7	12	S_{17}
S_{13}	0.168	0.0507	0.008516088	5	13	S_5
S_{14}	0.187	0.0204	0.003810780	9	14	S_{16}
S_{15}	0.157	0.0148	0.002327800	10	15	S_4
S_{16}	0.064	0.0055	0.000350464	14	16	S_1
S_{17}	0.056	0.0317	0.001777440	12	17	S_8

Note: "*" indicates that more decimal places have been retained to better distinguish the strategy weights and rankings, due to the similarity in the weight values of some Lean construction strategies.

Table 8. Comprehensive structural and importance analyses of 17 Lean implementation strategies.

S_i	Importance	ISM	ISM-Based Role Analysis	MICMAC	MICMAC-Based Role Analysis
S_{10}	1	L_5	Positioned in the middle level, it belongs to the technical aspect, supporting upper-level strategies through technical implementation based on lower-level strategies	Linkage Strategy	Has strong driving and dependency relationships with other strategies, making it critical and requiring close management
S_6	2	L_3	Located at the third-highest level, it has a direct impact on LC implementation but also serves as a key link that transmits the effects of lower-level strategies and supports upper-level	Dependent Strategy	Has high dependency on other strategies but also possesses some driving force, requiring monitoring and some attention
S_7	3	L_3	Located at the third-highest level, it has a direct impact on the successful implementation of LC but also serves as a key link that transmits the effects of lower-level strategies and supports upper-level strategies	Dependent Strategy	Has high dependency on other strategies but also possesses some driving force, requiring monitoring and some attention
S_{11}	4	L_2	Located at the second-highest level, it has a more direct impact on the successful implementation of LC but also serves as a key link that monitors the effects of lower-level strategies and supports the implementation of upper-level strategies	Dependent Strategy	Has the second-highest dependency on other strategies and a relatively weaker driving force, requiring monitoring and some attention to assess the effectiveness of lower-level strategies
S_{13}	5	L_6	Located at the third-lowest level, it provides relatively fundamental support for LC implementation	Driving Strategy	Has the third-strongest driving force and weaker dependency on other strategies, forming the foundation for successful LC and requiring early attention

Table 8. *Cont.*

S_i	Importance	ISM	ISM-Based Role Analysis	MICMAC	MICMAC-Based Role Analysis
S_9	6	L_2	Located at the second-highest level, it has a more direct impact on the successful implementation of LC but also serves as a key link that monitors the effects of lower-level strategies and supports upper-level strategies	Dependent Strategy	Has the second-highest dependency on other strategies and a relatively weaker driving force, requiring monitoring and some attention to assess the effectiveness of lower-level strategies
S_{12}	7	L_5	Positioned in the middle level, it belongs to the technical aspect of LC implementation, supporting upper-level strategies through technical implementation based on lower-level	Linkage Strategy	Has strong driving and dependency relationships with other strategies, making it critical and requiring close management
S_3	8	L_6	Located at the third-lowest level, it provides relatively fundamental support for LC implementation	Driving Strategy	Has the third-strongest driving force and weaker dependency on other strategies, forming the foundation for successful LC and requiring early attention
S_{14}	9	L_7	Located at the second-lowest level, it provides important foundational support for LC implementation and is necessary for early-stage implementation	Driving Strategy	Has the second-strongest driving force and weaker dependency on other strategies, forming the foundation for successful Lean and requiring early attention
S_{15}	10	L_7	Located at the second-lowest level, it provides important foundational support for LC implementation and is necessary for early-stage implementation	Driving Strategy	Has the second-strongest driving force and weaker dependency on other strategies, forming the foundation for successful Lean and requiring early attention
S_2	11	L_6	Located at the third-lowest level, it provides relatively fundamental support for LC implementation	Driving Strategy	Has the third-strongest driving force and weaker dependency on other strategies, forming the foundation for successful LC and requiring early attention
S_{17}	12	L_4	Positioned in the middle level, it belongs to the technical aspect of LC implementation, supporting upper-level strategies through technical implementation based on lower-level	Dependent Strategy	Has high dependency on other strategies but also possesses some driving force, requiring monitoring and some attention
S_5	13	L_2	Located at the second-highest level, it has a more direct impact on the successful implementation of LC but also serves as a key link that monitors the effects of lower-level strategies and supports upper-level strategies	Dependent Strategy	Has the second-highest dependency on other strategies and a relatively weaker driving force, requiring monitoring and some attention to assess the effectiveness of lower-level strategies
S_{16}	14	L_5	Positioned in the middle level, it belongs to the technical aspect of LC implementation, supporting upper-level strategies through technical implementation based on lower-level strategies	Linkage Strategy	Has strong driving and dependency relationships with other strategies, making it critical and requiring close management for the successful implementation of LC

Table 8. *Cont.*

S_i	Importance	ISM	ISM-Based Role Analysis	MICMAC	MICMAC-Based Role Analysis
S_4	15	L_2	Located at the second-highest level, it has a more direct impact on the successful implementation of LC but also serves as a key link that monitors the effects of lower-level strategies and supports upper-level strategies	Dependent Strategy	Has the second-highest dependency on other strategies and a relatively weaker driving force, requiring monitoring and some attention to assess the effectiveness of lower-level strategies
S_1	16	L_8	Located at the lowest level, it is the most fundamental Lean strategy and should be prioritized	Driving Strategy	Has the strongest driving force and weakest dependency on other strategies, forming the most basic part of LC implementation and requiring top priority
S_8	17	L_1	Located at the highest level, it has the most direct impact on the successful implementation of LC and requires the most focused monitoring	Dependent Strategy	Directly impacts LC implementation, has the highest dependency on other strategies, and is the weakest driving force, requiring focused monitoring to assess the effectiveness of other strategies

Additionally, it is important to note that when evaluating the importance of Lean implementation strategies using complex networks, the implementation sequence of strategies is overlooked. For instance, strategy S_{16}: Utilize external consulting firms and academic institutions for assistance ranks 14th, and strategy S_6: Enhance internal communication and collaboration within the stakeholders ranks 2nd. Moreover, S_6 needs to be implemented first in terms of importance. However, S_{16} serves as the foundation for implementing S_6. According to the sequence derived from ISM and MICMAC analysis, S_{16} should be prioritized for implementation.

Thereby, this demonstrates that the network-based ISM-MICMAC analysis of Lean implementation strategies is complementary. It not only clearly demonstrates the hierarchical order of priorities for the successful implementation of lean but also establishes the key strategies within each level. This provides a solid foundation for accurately defining and formulating the implementation planning for Lean.

4.4. The Planning of Lean Implementation in PC

Accordingly, a structured and clearly prioritized plan for the successful Lean implementation in PC can be developed through exploring the strategies' interrelationships, roles, and importance. As illustrated in Figure 5, Lean implementation strategies highlighted with a light blue background hold higher importance and require more attention at their respective levels. These highlighted key strategies are selected according to the rank of comprehensive weights from first to eighth. The implementation planning of Lean in PC can be divided into four levels: the foundation level (Level 1), the organizational level (Level 2), the technical level (Level 3), and the control level (Level 4).

Level 1 of the planning is mainly composed of strategies at lower levels of the ISM model, including S_1 in level L_8, and S_{14} and S_{15} in level L_7. These strategies are all "driving strategies", which support the implementation of other strategies. Level 2 of the planning consists of strategies at slightly lower levels of the ISM model, including S_2, S_3, and S_{13} in level L_6. Although these strategies are categorized as "driving strategies" in the MICMAC analysis, they are not classified as the foundation level, considering they have dependence on S_1 in L_8, and S_{14} and S_{15} in L_7. They are considered the organizational level due to the important support of organizational structure in Lean implementation. Level 3 of the planning is composed of strategies near the middle of the ISM, including S_{16}, S_{10}, and S_{12} in L_5, S_{17} in L_4, and S_6 and S_7 in L_3, which are key technical strategies for the Lean successful implementation. Among them, S_{16}, S_{10}, and S_{12} are "linkage strategies"; although S_{17}, S_6, and S_7 are "dependency strategies", they have relatively high driving forces, which are classified as the technical level. Level 4 of planning is composed of strategies at higher levels of ISM, including S_8 in L_1 and S_4, S_9, S_{11}, and S_5 in L_2. These strategies are all "dependency strategies" that directly impact the Lean successful implementation. Since the implementation of these strategies depends on the support of other strategies, their effectiveness serves as an indicator of the overall success of these related strategies and necessitates focused control.

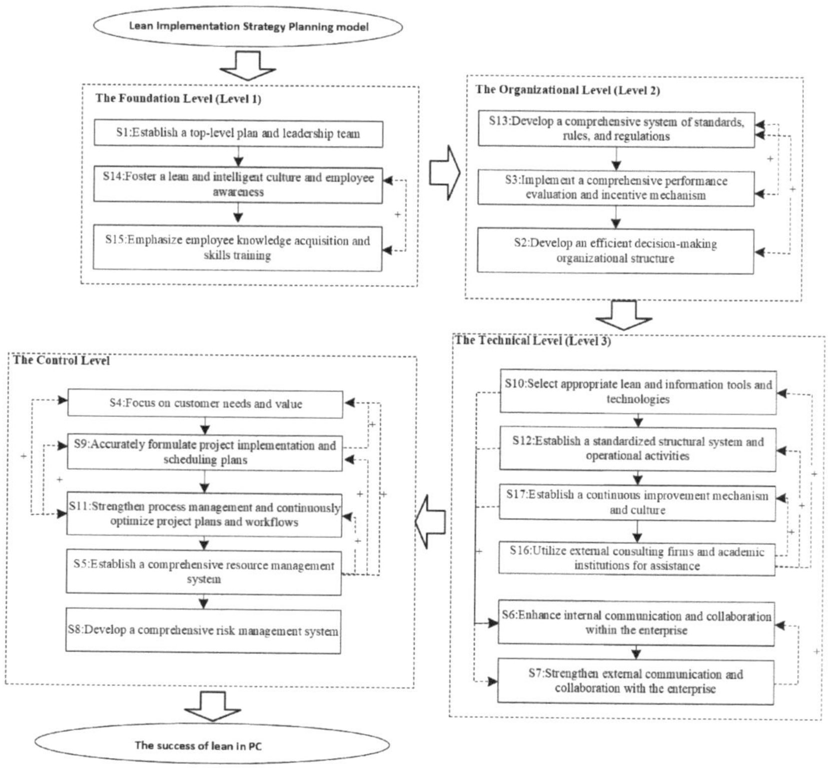

Figure 5. The planning of Lean implementation strategies.

4.5. Discussion

A network-based ISM-MICMAC model was utilized to explore the planning of Lean implementation strategies in the PC context. This model comprises two main parts: the structural model part and the importance analysis part. A total of 17 Lean implementation strategies for Chinese PC were analyzed, leading to four levels of planning of Lean implementation strategies.

The foundation-level strategies consist of Lean implementation strategies, S_1, at level L_8, and S_{14} and S_{15} at level L_7. The results indicate a top-level plan and leadership team (S_1) should be established to charge the overall Lean initiatives. This is essential because Lean involves a series of changes in projects' management and operational processes [50]. Moreover, this argument is supported by Netland et al. [71], who assert that Lean is a systemic project involving various aspects of man, machine, materials, method, and environment, requiring a top-level leadership group to mobilize resources for strategy implementation and process change. Additionally, fostering a lean and intelligent culture and raising employee awareness (S_{14}) is crucial, as organizations and employees may resist transitioning from traditional practices to Lean implementation [63]. Similarly, Santorella [72] emphasized that Lean tools alone are insufficient for successful Lean implementation, highlighting the critical role of Lean culture. Thereby, skills training and knowledge sharing (S_{15}) should be provided to employees to enhance employees' theoretical knowledge and practical experience in Lean [64]. However, the development of these skill sets remains limited [73].

The organizational-level strategies comprise Lean implementation strategies, S_2, S_3, and S_{13} at level L_6. The results indicate that successful Lean implementation requires a sound organizational structure, including standardized systems and incentives. This is particularly important as the Chinese construction industry and its stakeholders are in the early stage of transitioning from traditional construction to PC and have not yet developed an organizational structure suitable for this shift [52]. In this regard, Bajjou et al. [51] argued that Lean implementation necessitates organizational structure transformation, posing additional challenges for the construction industry and its stakeholders. Therefore, a comprehensive system of standards, rules, and regulations (S_{13}) needs to be constructed to support Lean transformation, which is also indicated by Demirkesen, S., and Bayhan, H.G. [74]. Secondly, a performance evaluation and incentive mechanism (S_3) should be developed to promote employee mindset transformation and motivate their participation in Lean implementation. The effectiveness of such mechanisms has been demonstrated in numerous project management studies [75–77]. Additionally, it is important to improve the decision-making efficiency and structure of the organization (S_2) to support the rapid allocation of resources during the Lean implementation [51,52].

The technical-level strategies include Lean implementation strategies S_{16}, S_{10}, and S_{12} at level L_5, S_{17} at level L_4, and S_6 and S_7 at level L_3. In fact, stakeholders in PC often lack Lean knowledge and experience, and internal and external operations are segmented, which are key factors hindering successful Lean [19]. To address this, selecting appropriate Lean tools and information technologies (S_{10}) is the most key strategy for successful Lean implementation [59]. This has been justified by Lermen et al. [78] and Deanese et al. [79]. Furthermore, the standardized system and operational processes (S_{12}) are also vital for Lean implementation [46]. Then, a continuous improvement mechanism and culture (S_{17}) should be established, as Lean is a long-term dynamic process that requires ongoing optimization of technologies and processes in response to changes in business and project environments [11,65]. In this process, external consulting firms and academic institutions (S_{16}) can provide valuable assistance in addressing challenges encountered during Lean implementation [50]. Additionally, internal and external communication and collaboration among stakeholders (S_6 and S_7), such as concurrent engineering and collaborative supply chains, need to be enhanced to further improve the effectiveness of Lean [56].

The control-level strategies include Lean implementation strategies S_8 at level L_1 and S_4, S_9, S_{11}, and S_5 at level L_2. Lean success is reflected in the performance of PC projects, which depends on meeting customer needs and values [53]. Thus, prioritizing customer needs and values (S_4) in terms of duration, cost, and quality is essential, as it directly affects project performance. Accurate project implementation and scheduling plans (S_9) are also crucial to ensuring these outcomes. However, various factors, such as environmental and technological changes, can disrupt PC processes [58]. Thereby, continuous optimization of project plans and processes (S_{11}) and a comprehensive resource management system (S_5) are necessary for managing various resources. In fact, risks related to scheduling, process management, and resource allocation, influenced by external uncertainties, can negatively impact project performance [57,80]. Therefore, a comprehensive risk management system (S_8) should be developed to identify and control risks through the whole processes of Lean in PC, as supported by Ghosh and Jason [81].

The roles of these Lean strategies have been validated by, but are not limited to, Netland et al. [71]; Anaç et al. [57]; Li et al. [80]; Demirkesen, S., and Bayhan, H.G. [74]; Mostafa et al. [53]; and Lista et al. [73]. However, previous studies have only highlighted their importance without evaluating their priorities. Compared to these studies, the current developed framework considers the interactions between Lean strategies and identifies four levels of prioritization, with emphasis placed on the top levels.

5. Conclusions

This study developed a network-based ISM-MICMAC model to analyze the planning of Lean implementation in line with strategic control and projects' performances under China's PC. This model included a qualitative analysis to identify Lean implementation strategies based on a literature review and experts' interviews and a quantitative analysis via the structural part and the importance part to analyze the interrelationships of Lean implementation. The results revealed that:

1. Lean is systematic engineering, where various implementation strategies are interconnected and mutually influenced into a complex network.
2. The planning of Lean implementation consists of foundation, organizational, technical, and control levels, reflecting the hierarchical order, priorities, and importance for the successful Lean implementation.
3. Efficient measures of Lean implementation are establishing a top-level Lean promotion group, cultivating participants Lean awareness and skills, constructing a comprehensive standard system, selecting the appropriate technologies, enhancing inter-external collaboration, continuously optimizing plans and processes, and building a risk monitoring system.

Managerial implications are summarized. This study is expected to lead stakeholders in overcoming challenges in the Lean implementation process and guide them about success parameters for strategy and performance to consider and prioritize tasks when implementing Lean in PC. Firstly, the 17 Lean implementation strategies and their interactions are revealed to successfully implement Lean for stakeholders in PC, China. Secondly, selecting the appropriate Lean tools and information technologies is crucial, which is based on the foundation of establishing a top-level management team and fostering Lean culture. Thirdly, it is important to build a standard system of processes and activities, enhance the inter-external collaboration, and continuously improve the processes in response to changes.

The contribution of this study is twofold. From a theoretical view, it has explored the effective Lean in PC by an innovative network-based analysis of the interrelationships of Lean implementation strategies to enrich the existing knowledge body of PC performance. This is not only an application of existing theories of Lean, PC, ISM-MICMAC, and complex networks but also a theoretical enrichment in the field of project management. From a practical point of view, the proposed planning could serve as a road map for stakeholders to improve strategic control and projects' performance. This study also provides clues for the advancement of digital construction, particularly through improving the processes of PC projects. In fact, Lean is the basis of transformation towards digital construction.

Admittedly, this study has limitations. Firstly, the validation of the findings in real-world projects should be considered, as this will strengthen the authority of the network-based model. Secondly, the analysis was measured from a static perspective, while Lean may evolve dynamically with the development of PC. Future studies could incorporate a longitudinal approach to capture the changes in Lean and PC development. Lastly, it should be noted that this study specifically pertained to the Lean PC projects in China, with a focus on the initial development phase. However, this network-based analysis is applicable in developed countries, making it conducive to cross-country comparisons.

Author Contributions: Conceptualization, P.D. and C.S.; Software, Methodology, Validation, P.D.; resources, Z.N.; writing—original draft preparation, P.D. and S.J.; writing—review and editing, L.G. and P.D.; Supervision, P.D. and Z.N. All authors have read and agreed to the published version of the manuscript.

Funding: This research received no external funding.

Data Availability Statement: The datasets in the current study are available from the corresponding author upon reasonable request.

Conflicts of Interest: The authors declare no conflict of interest.

Appendix A

The SSIM of Lean implementation strategies.

Strategy	S_1	S_2	S_3	S_4	S_5	S_6	S_7	S_8	S_9	S_{10}	S_{11}	S_{12}	S_{13}	S_{14}	S_{15}	S_{16}	S_{17}
S_1	Z	Z	W	W	Z	Z	Z	Z	Z	Z	Z	Z	Z	Z	Z	Z	—
S_2	Z	Z		X	Y	Z	Z	W	Z	Z	Z	Z	Z	Z	Z	—	
S_3	Z	Z	X	X	Y	Z	Z	W	Z	Z	Z	Z	Z	Z	—		
S_4	Z	Z	Z	Z	Z	Z	Z	Z	Y	Z	Z	Z	Z	—			
S_5	Z	Z	Z	Z	Z	Z	Y	Z	Z	W	Z	Z	—				
S_6	X	Z	Z	Z	Z	X	W	X	W	Z	Y	—					
S_7	X	Z	Z	Z	Z	X	W	X	W	Z	—						
S_8	Z	Z	Z	Z	Z	Z	X	Z	X	—							
S_9	Z	Z	Z	Z	Z	Z	Y	Z	—								
S_{10}	W	Y	Z	Z	X	Y	Z	—									
S_{11}	Z	Z	Z	Z	Z	Z	—										
S_{12}	W	Z	Z	Z	Z	—											
S_{13}	Z	Z	X	X	—												
S_{14}	Z	Z	Y	—													
S_{15}	Z	Z	—														
S_{16}	Z	—															
S_{17}	—																

Appendix B

The AM of Lean implementation strategies.

Strategy	S_1	S_2	S_3	S_4	S_5	S_6	S_7	S_8	S_9	S_{10}	S_{11}	S_{12}	S_{13}	S_{14}	S_{15}	S_{16}	S_{17}
S_1	0	0	0	0	0	0	0	0	0	0	0	0	0	1	1	0	0
S_2	0	0	0	0	0	0	0	0	0	1	0	0	1	0	0	0	0
S_3	0	0	0	0	0	0	0	0	0	1	0	0	1	0	0	0	0
S_4	0	0	0	0	0	0	0	0	1	0	0	0	0	0	0	0	0
S_5	0	0	0	0	0	0	0	1	0	0	1	0	0	0	0	0	0
S_6	0	0	0	0	0	0	1	0	1	0	1	0	0	0	0	0	0
S_7	0	0	0	0	0	1	0	0	1	0	1	0	0	0	0	0	0
S_8	0	0	0	0	0	0	0	0	0	0	0	0	0	0	0	0	0
S_9	0	0	0	1	0	0	0	1	0	0	1	0	0	0	0	0	0
S_{10}	0	0	0	0	0	1	1	0	0	0	0	1	0	0	0	1	1
S_{11}	0	0	0	0	1	0	0	1	1	0	0	0	0	0	0	0	0
S_{12}	0	0	0	0	0	1	1	0	0	1	0	0	0	0	0	0	1
S_{13}	0	1	1	0	0	0	0	0	0	1	0	0	0	0	0	0	0
S_{14}	0	1	1	0	0	0	0	0	0	0	0	0	1	0	1	0	0
S_{15}	0	0	1	0	0	0	0	0	0	0	0	0	1	1	0	0	0
S_{16}	0	0	0	0	0	0	0	0	0	1	0	0	0	0	0	0	0
S_{17}	0	0	0	0	0	1	1	0	0	0	0	0	0	0	0	0	0

Appendix C

The RM of Lean implementation strategies.

Strategy	S_1	S_2	S_3	S_4	S_5	S_6	S_7	S_8	S_9	S_{10}	S_{11}	S_{12}	S_{13}	S_{14}	S_{15}	S_{16}	S_{17}
S_1	1	1	1	1	1	1	1	1	1	1	1	1	1	1	1	1	1
S_2	0	1	1	1	1	1	1	1	1	1	1	1	1	0	0	1	1
S_3	0	1	1	1	1	1	1	1	1	1	1	1	1	0	0	1	1
S_4	0	0	0	1	1	0	0	1	1	0	1	0	0	0	0	0	0
S_5	0	0	0	1	1	0	0	1	1	0	1	0	0	0	0	0	0
S_6	0	0	0	1	1	1	1	1	1	0	1	0	0	0	0	0	0
S_7	0	0	0	1	1	1	1	1	1	0	1	0	0	0	0	0	0
S_8	0	0	0	0	0	0	0	1	0	0	0	0	0	0	0	0	0
S_9	0	0	0	1	1	0	0	1	1	0	1	0	0	0	0	0	0
S_{10}	0	0	0	1	1	1	1	1	1	1	1	1	0	0	0	1	1
S_{11}	0	0	0	1	1	0	0	1	1	0	1	0	0	0	0	0	0
S_{12}	0	0	0	1	1	1	1	1	1	1	1	1	0	0	0	1	1
S_{13}	0	1	1	1	1	1	1	1	1	1	1	1	1	0	0	1	1
S_{14}	0	1	1	1	1	1	1	1	1	1	1	1	1	1	1	1	1
S_{15}	0	1	1	1	1	1	1	1	1	1	1	1	1	1	1	1	1
S_{16}	0	0	0	1	1	1	1	1	1	1	1	1	0	0	0	1	1
S_{17}	0	0	0	1	1	1	1	1	1	0	1	0	0	0	0	0	1

Appendix D

The DRP and DEP values of Lean implementation strategies based on RM.

Strategy	S_1	S_2	S_3	S_4	S_5	S_6	S_7	S_8	S_9	S_{10}	S_{11}	S_{12}	S_{13}	S_{14}	S_{15}	S_{16}	S_{17}	DRP
S_1	1	1	1	1	1	1	1	1	1	1	1	1	1	1	1	1	1	17
S_2	0	1	1	1	1	1	1	1	1	1	1	1	1	0	0	1	1	14
S_3	0	1	1	1	1	1	1	1	1	1	1	1	1	0	0	1	1	14
S_4	0	0	0	1	1	0	0	1	1	0	1	0	0	0	0	0	0	5
S_5	0	0	0	1	1	0	0	1	1	0	1	0	0	0	0	0	0	5
S_6	0	0	0	1	1	1	1	1	1	0	1	0	0	0	0	0	0	7
S_7	0	0	0	1	1	1	1	1	1	0	1	0	0	0	0	0	0	7
S_8	0	0	0	0	0	0	0	1	0	0	0	0	0	0	0	0	0	1
S_9	0	0	0	1	1	0	0	1	1	0	1	0	0	0	0	0	0	5
S_{10}	0	0	0	1	1	1	1	1	1	1	1	1	0	0	0	1	1	11
S_{11}	0	0	0	1	1	0	0	1	1	0	1	0	0	0	0	0	0	5
S_{12}	0	0	0	1	1	1	1	1	1	1	1	1	0	0	0	1	1	11
S_{13}	0	1	1	1	1	1	1	1	1	1	1	1	1	0	0	1	1	14
S_{14}	0	1	1	1	1	1	1	1	1	1	1	1	1	1	1	1	1	16
S_{15}	0	1	1	1	1	1	1	1	1	1	1	1	1	1	1	1	1	16
S_{16}	0	0	0	1	1	1	1	1	1	1	1	1	0	0	0	1	1	11
S_{17}	0	0	0	1	1	1	1	1	1	0	1	0	0	0	0	0	1	8
DEP	1	6	6	16	16	12	12	17	16	9	16	9	6	3	3	9	10	

Appendix E

D among the 17 Lean implementation strategies.

d_{SiSj}	S_1	S_2	S_3	S_4	S_5	S_6	S_7	S_8	S_9	S_{10}	S_{11}	S_{12}	S_{13}	S_{14}	S_{15}	S_{16}	S_{17}
S_1	-	2	2	6	6	4	4	6	5	3	5	4	2	1	1	4	4
S_2	∞	-	2	4	4	2	2	4	3	1	3	2	1	∞	∞	2	2
S_3	∞	2	-	4	4	2	2	4	3	1	3	2	1	∞	∞	2	2
S_4	∞	∞	∞	-	3	∞	∞	2	3	∞	2	∞	∞	∞	∞	∞	∞
S_5	∞	∞	∞	3	-	∞	∞	1	2	∞	1	∞	∞	∞	∞	∞	∞
S_6	∞	∞	∞	2	2	-	1	2	1	∞	1	∞	∞	∞	∞	∞	∞

d_{SiSj}	S_1	S_2	S_3	S_4	S_5	S_6	S_7	S_8	S_9	S_{10}	S_{11}	S_{12}	S_{13}	S_{14}	S_{15}	S_{16}	S_{17}
S_7	∞	∞	∞	2	2	1	-	2	1	∞	1	∞	∞	∞	∞	∞	∞
S_8	∞	∞	∞	∞	∞	∞	∞	-	∞	∞	∞	∞	∞	∞	∞	∞	∞
S_9	∞	∞	∞	1	2	∞	∞	1	-	∞	1	∞	∞	∞	∞	∞	∞
S_{10}	∞	∞	∞	3	3	1	1	3	2	-	2	1	∞	∞	∞	1	1
S_{11}	∞	∞	∞	2	1	∞	∞	1	1	∞	-	∞	∞	∞	∞	∞	∞
S_{12}	∞	∞	∞	3	3	2	1	3	2	1	2	-	∞	∞	∞	2	1
S_{13}	∞	1	1	4	4	2	2	4	3	1	3	2	-	∞	∞	2	2
S_{14}	∞	1	1	5	5	3	3	5	4	2	4	3	1	-	1	3	3
S_{15}	∞	2	1	5	5	3	3	5	4	2	4	3	1	1	-	3	3
S_{16}	∞	∞	∞	4	4	2	2	4	3	1	3	2	∞	∞	∞	-	2
S_{17}	∞	∞	∞	3	3	1	1	3	2	∞	2	∞	∞	∞	∞	∞	-

Appendix F

The ICM for the ISM network with 17 Lean implementation strategies.

ICV_{SiSj}	S_1	S_2	S_3	S_4	S_5	S_6	S_7	S_8	S_9	S_{10}	S_{11}	S_{12}	S_{13}	S_{14}	S_{15}	S_{16}	S_{17}
S_1	1	0	0	0	0	0	0	0	0	0	0	0	0	0.329	0.263	0	0
S_2	0	1	0	0	0	0	0	0	0	0.526	0	0	0.329	0	0	0	0
S_3	0	0	1	0	0	0	0	0	0	0.526	0	0	0.329	0	0	0	0
S_4	0	0	0	1	0	0	0	0	0.329	0	0	0	0	0	0	0	0
S_5	0	0	0	0	1	0	0	0.197	0	0	0.329	0	0	0	0	0	0
S_6	0	0	0	0	0	1	0.394	0	0.329	0	0.329	0	0	0	0	0	0
S_7	0	0	0	0	0	0.394	1	0	0.329	0	0.329	0	0	0	0	0	0
S_8	0	0	0	0	0	0	0	1	0	0	0	0	0	0	0	0	0
S_9	0	0	0	0.066	0	0	0	0.197	1	0	0.329	0	0	0	0	0	0
S_{10}	0	0	0	0	0	0.394	0.394	0	0	1	0	0.263	0	0	0	0.066	0.263
S_{11}	0	0	0	0	0.131	0	0	0.197	0.329	0	1	0	0	0	0	0	0
S_{12}	0	0	0	0	0	0.394	0.394	0	0	0.526	0	1	0	0	0	0	0.263
S_{13}	0	0.197	0.197	0	0	0	0	0	0	0.526	0	0	1	0	0	0	0
S_{14}	0	0.197	0.197	0	0	0	0	0	0	0	0	0	0.329	1	0.263	0	0
S_{15}	0	0	0.197	0	0	0	0	0	0	0	0	0	0.329	0.329	1	0	0
S_{16}	0	0	0	0	0	0	0	0	0	0.526	0	0	0	0	0	1	0
S_{17}	0	0	0	0	0	0.394	0.394	0	0	0	0	0	0	0	0	0	1

Appendix G

The IEM for the ISM network with 17 Lean implementation strategies.

IEV_{SiSj}	S_1	S_2	S_3	S_4	S_5	S_6	S_7	S_8	S_9	S_{10}	S_{11}	S_{12}	S_{13}	S_{14}	S_{15}	S_{16}	S_{17}
S_1	0.374	0	0	0	0	0	0	0	0	0	0	0	0	0.149	0.111	0	0
S_2	0	0.401	0	0	0	0	0	0	0	0.230	0	0	0.142	0	0	0	0
S_3	0	0	0.401	0	0	0	0	0	0	0.230	0	0	0.142	0	0	0	0
S_4	0	0	0	0.104	0	0	0	0	0.072	0	0	0	0	0	0	0	0
S_5	0	0	0	0	0.177	0	0	0	0	0	0.072	0	0	0	0	0	0
S_6	0	0	0	0	0	0.281	0.111	0	0.072	0	0.072	0	0	0	0	0	0
S_7	0	0	0	0	0	0.111	0.281	0	0.072	0	0.072	0	0	0	0	0	0
S_8	0	0	0	0	0	0	0	0	0	0	0	0	0	0	0	0	0
S_9	0	0	0	0.007	0	0	0	0	0.219	0	0.072	0	0	0	0	0	0
S_{10}	0	0	0	0	0	0.111	0.111	0	0	0.438	0	0.099	0	0	0	0.018	0.066
S_{11}	0	0	0	0	0.023	0	0	0	0.072	0	0.219	0	0	0	0	0	0
S_{12}	0	0	0	0	0	0.111	0.111	0	0	0.230	0	0.375	0	0	0	0	0.066
S_{13}	0	0.079	0.079	0	0	0	0	0	0	0.230	0	0	0.432	0	0	0	0
S_{14}	0	0.079	0.079	0	0	0	0	0	0	0	0	0	0.142	0.454	0.111	0	0
S_{15}	0	0	0.079	0	0	0	0	0	0	0	0	0	0.142	0.149	0.423	0	0
S_{16}	0	0	0	0	0	0	0	0	0	0.230	0	0	0	0	0	0.276	0
S_{17}	0	0	0	0	0	0.111	0.111	0	0	0	0	0	0	0	0	0	0.250

References

1. Jiang, L.; Li, Z.; Li, L.; Gao, Y. Constraints on the Promotion of Prefabricated Construction in China. *Sustainability* **2018**, *10*, 2516. [CrossRef]
2. Dang, P.; Geng, L.; Niu, Z.; Chan, M.; Yang, W.; Gao, S. A Value-Based Network Analysis for Stakeholder Engagement through Prefabricated Construction Life Cycle: Evidence from China. *J. Civ. Eng. Manag.* **2024**, *30*, 49–66. [CrossRef]
3. Tan, T.; Chen, K.; Xue, F.; Lu, W. Barriers to Building Information Modeling (BIM) implementation in China's prefabricated construction: An interpretive structural modeling (ISM) approach. *J. Clean. Prod.* **2019**, *219*, 949–959. [CrossRef]
4. Binh, N.T. Applying Lean Construction to Construction Project. *Res. Mater. Manuf. Technol.* **2014**, *834–836*, 1976–1983.
5. Habibi Rad, M.; Mojtahedi, M.; Ostwald, M.J.; Wilkinson, S. A Conceptual Framework for Implementing Lean Construction in Infrastructure Recovery Projects. *Buildings* **2022**, *12*, 272. [CrossRef]
6. Koskela, L.; Ferrantelli, A.; Niiranen, J.; Pikas, E.; Dave, B. Epistemological Explanation of Lean Construction. *J. Constr. Eng. Manag.* **2018**, 145. [CrossRef]
7. Habidin, N.F.; Yusof, S.M.; Fuzi, N.M. Lean Six Sigma, strategic control systems, and organizational performance for automotive suppliers. *Int. J. Lean Six Sigma* **2016**, *7*, 110–135. [CrossRef]
8. Li, S.; Fang, Y.; Wu, X. A systematic review of lean construction in Mainland China. *J. Clean. Prod.* **2020**, *257*, 120581. [CrossRef]
9. Hector, C.; Joanna, D.; Julien, L.D.; Benoît, E. Strategic Lean Management Integration of operational Performance Indicators for strategic Lean management. *IFAC-Papers Online* **2016**, *49*, 65–70.
10. Sarhan, J.G.; Xia, B.; Fawzia, S.; Karim, A.; Olanipekun, A.O.; Coffey, V. Framework for the implementation of lean construction strategies using the interpretive structural modelling (ISM) technique. *Eng. Constr. Archit. Manag.* **2019**, *27*, 1–23. [CrossRef]
11. Hussein, M.; Zayed, T. Critical factors for successful implementation of just-in-time concept in modular integrated construction: A systematic review and meta-analysis. *J. Clean. Prod.* **2021**, *284*, 124716. [CrossRef] [PubMed]
12. Gao, S.; Low, S.P. The Toyota Way model: An alternative framework for lean construction. *Total Qual. Manag. Bus. Excell.* **2014**, *25*, 664–682. [CrossRef]
13. Yunus, R.; Noor, S.R.M.; Abdullah, A.H.; Nagapan, S.; Hamid, A.R.A.; Tajudin, S.A.A.; Jusof, S.R.M. Critical Success Factors for Lean Thinking in the Application of Industrialised Building System (IBS). *IOP Conf. Ser. Mater. Sci. Eng.* **2017**, *226*, 012045. [CrossRef]
14. Chiarini, A.; Vagnoni, E. Strategic Planning for Lean Production, Comparing Hoshin Kanri with Balanced Scorecard. In *Understanding the Lean Enterprise*; Springer: Berlin/Heidelberg, Germany, 2016; pp. 221–236.
15. Mao, C.; Liu, G.W.; Shen, L.Y.; Wang, X.Y.; Wang, J. Structural Equation Modeling to Analyze the Critical Driving Factors and Paths for Off-site Construction in China. *KSCE J. Civ. Eng.* **2018**, *22*, 2678–2690. [CrossRef]
16. Wankhade, N.; Kundu, G.K. Interpretive Structural Modelling (ISM) Methodology and its Application in Supply Chain Research. *Int. J. Innov. Technol. Explor. Eng.* **2020**, *9*, 1101–1109. [CrossRef]
17. Singh, C.; Singh, D.; Khamba, J.S. Developing a conceptual model to implement green lean practices in Indian manufacturing industries using ISM-MICMAC approach. *J. Sci. Technol. Policy Manag.* **2021**, *12*, 587–608. [CrossRef]
18. Luo, L.Z.; Shen, G.Q.; Xu, G.Y.; Liu, Y.L.; Wang, Y.J. Stakeholder-Associated Supply Chain Risks and Their Interactions in a Prefabricated Building Project in Hong Kong. *J. Manag. Eng.* **2019**, *35*, 05018015. [CrossRef]
19. Aslam, M.; Gao, Z.; Smith, G. Exploring factors for implementing lean construction for rapid initial successes in construction. *J. Clean. Prod.* **2020**, *277*, 123295. [CrossRef]
20. Ahmed, S.; Sobuz, M.H.R. Challenges of implementing lean construction in the construction industry in Bangladesh. *Smart Sustain. Built Environ.* **2020**, *9*, 174–207. [CrossRef]
21. Enshassi, A.; Saleh, N.; Mohamed, S. Barriers to the application of lean construction techniques concerning safety improvement in construction projects. *Int. J. Constr. Manag.* **2021**, *21*, 1044–1060. [CrossRef]
22. Mano, A.P.; Gouvea da Costa, S.E.; Pinheiro de Lima, E. Criticality assessment of the barriers to Lean Construction. *Int. J. Product. Perform. Manag.* **2020**, *70*, 65–86. [CrossRef]
23. Bashir, A.M.; Suresh, S.; Oloke, D.A.; Proverbs, D.G.; Gameson, R. Overcoming the Challenges facing Lean Construction Practice in the UK Contracting Organizations. *Int. J. Archit. Eng. Constr.* **2015**, *4*, 10–18. [CrossRef]
24. Ebrahimi, M.; Daneshvar, A.; Valmohammadi, C. Using a comprehensive DEMATEL-ISM-MICMAC and importance–performance analysis to study sustainable service quality features. *J. Econ. Adm. Sci.* 2024, *ahead-of-print*. [CrossRef]
25. Thakkar, J.; Deshmukh, S.G.; Gupta, A.D.; Shankar, R. Selection of Third-Party Logistics (3PL): A Hybrid Approach Using Interpretive Structural Modeling (ISM) and Analytic Network Process (ANP). *Supply Chain. Forum Int. J.* **2015**, *6*, 32–46. [CrossRef]
26. Ali, A.A.; Mahmood, A. Developing a causal framework of internet of things adoption barriers for agile manufacturing in post COVID-19. *Int. J. Eng. Bus. Manag.* **2024**, *16*, 18479790231223623. [CrossRef]
27. Taherdoost, H.; Sahibuddin, S.; Jalaliyoon, N. Exploratory factor analysis; concepts and theory. *Adv. Appl. Pure Math.* **2022**, *27*, 375–382.
28. Sun, C.; Xu, H.; Jiang, S. Understanding the risk factors of BIM technology implementation in the construction industry: An interpretive structural modeling (ISM) approach. *Eng. Constr. Archit. Manag.* **2020**, *27*, 3289–3308. [CrossRef]
29. Xu, X.; Zou, P.X. Analysis of factors and their hierarchical relationships influencing building energy performance using interpretive structural modelling (ISM) approach. *J. Clean. Prod.* **2020**, *272*, 122650. [CrossRef]

30. Gardas, B.B.; Raut, R.D.; Narkhede, B.E. A state-of the-art survey of interpretive structural modelling methodologies and applications. *Int. J. Bus. Excell.* **2017**, *11*, 505–560. [CrossRef]
31. Li, M.; Yang, J. Analysis of interrelationships between critical waste factors in office building retrofit projects using interpretive structural modelling. *Int. J. Constr. Manag.* **2014**, *14*, 15–27. [CrossRef]
32. Pimentel, M.; Arantes, A.; Cruz, C.O. Barriers to the adoption of reverse logistics in the construction industry: A combined ISM and MICMAC approach. *Sustainability* **2022**, *14*, 15786. [CrossRef]
33. Ahmad, M.; Tang, X.-W.; Qiu, J.-N.; Ahmad, F. Interpretive Structural Modeling and MICMAC Analysis for Identifying and Benchmarking Significant Factors of Seismic Soil Liquefaction. *Appl. Sci.* **2019**, *9*, 233. [CrossRef]
34. Al-Zarooni, H.; Bashir, H. An integrated ISM fuzzy MICMAC approach for modeling and analyzing electrical power system network interdependencies. *Int. J. Syst. Assur. Eng. Manag.* **2020**, *11*, 1204–1226. [CrossRef]
35. Gothwal, S.; Raj, T. Analyzing the factors affecting the flexibility in FMS using weighted interpretive structural modeling (WISM) approach. *Int. J. Syst. Assur. Eng. Manag.* **2016**, *8*, 408–422. [CrossRef]
36. Gan, X.L.; Chang, R.D.; Zuo, J.; Wen, T.; Zillante, G. Barriers to the transition towards off-site construction in China: An Interpretive structural modeling approach. *J. Clean. Prod.* **2018**, *197*, 8–18. [CrossRef]
37. Khaba, S.; Bhar, C. Modeling the key barriers to lean construction using interpretive structural modeling. *J. Model. Manag.* **2017**, *12*, 652–670. [CrossRef]
38. Singh, M.K.; Kumar, H.; Gupta, M.P.; Madaan, J. Competitiveness of Electronics manufacturing industry in India: An ISM–fuzzy MICMAC and AHP approach. *Meas. Bus. Excell.* **2018**, *22*, 88–116. [CrossRef]
39. Tavakolan, M.; Etemadinia, H. Fuzzy Weighted Interpretive Structural Modeling: Improved Method for Identification of Risk Interactions in Construction Projects. *J. Constr. Eng. Manag.* **2017**, *143*, 04017084. [CrossRef]
40. Chen, J.K. Improved DEMATEL-ISM integration approach for complex systems. *PLoS ONE* **2021**, *16*, e0254694. [CrossRef]
41. Wang, W.; Wang, Y.; Wang, G.; Li, M.; Jia, L. Identification of the critical accident causative factors in the urban rail transit system by complex network theory. *Phys. A Stat. Mech. Its Appl.* **2023**, *610*, 128404. [CrossRef]
42. Lü, L.; Chen, D.; Ren, X.L.; Zhang, Q.M.; Zhang, Y.C.; Zhou, T. Vital nodes identification in complex networks. *Phys. Rep.* **2016**, *650*, 1–63. [CrossRef]
43. Chen, B.; Wang, Z.; Luo, C. Integrated evaluation approach for node importance of complex networks based on relative entropy. *J. Syst. Eng. Electron.* **2016**, *27*, 1219–1226. [CrossRef]
44. Yang, Y.; Yu, L.; Wang, X.; Chen, S.; Chen, Y.; Zhou, Y. A novel method to identify influential nodes in complex networks. *Int. J. Mod. Phys. C* **2020**, *31*, 2050022. [CrossRef]
45. Liu, W.; Xu, L. Identification of key brittleness factor for manufacturing system based on ISM and complex network. *Comput. Integr. Manuf. Syst.* **2021**, *27*, 3076–3092. (In Chinese)
46. Evans, M.; Farrell, P.; Mashali, A.; Zewein, W. Critical success factors for adopting building information modelling (BIM) and lean construction practices on construction mega-projects: A Delphi survey. *J. Eng. Des. Technol.* **2020**, *19*, 537–556. [CrossRef]
47. Yuan, M.; Li, Z.; Li, X.; Luo, X. Managing stakeholder-associated risks and their interactions in the life cycle of prefabricated building projects: A social network analysis approach. *J. Clean. Prod.* **2021**, *323*, 129102. [CrossRef]
48. Yu, T.; Shi, Q.; Zuo, J.; Chen, R. Critical factors for implementing sustainable construction practice in HOPSCA projects: A case study in China. *Sustain. Cities Soc.* **2018**, *37*, 93–103. [CrossRef]
49. Xue, H.; Zhang, S.J.; Su, Y.K.; Wu, Z.Z.; Yang, R.J. Effect of stakeholder collaborative management on off-site construction cost performance. *J. Clean. Prod.* **2018**, *184*, 490–502. [CrossRef]
50. Xing, W.; Hao, J.L.; Qian, L.; Tam, V.W.; Sikora, K.S. Implementing lean construction techniques and management methods in Chinese projects: A case study in Suzhou, China. *J. Clean. Prod.* **2021**, *286*, 124944. [CrossRef]
51. Bajjou, M.S.; Chafi, A.; En-nadi, A. A Comparative Study between Lean Construction and the Traditional Production System. *Int. J. Eng. Res. Afr.* **2017**, *29*, 118–132. [CrossRef]
52. Liu, Y.; Chang, R.D.; Zuo, J.; Xiong, F.; Dong, N. What leads to the high capital cost of prefabricated construction in China: Perspectives of stakeholders. *Eng. Constr. Archit. Manag.* **2023**, *30*, 805–832. [CrossRef]
53. Mostafa, S.; Tam, V.W.; Dumrak, J.; Mohamed, S. Leagile strategies for optimizing the delivery of prefabricated house building projects. *Int. J. Constr. Manag.* **2020**, *20*, 867–881. [CrossRef]
54. Sarhan, J.; Xia, B.; Fawzia, S.; Karim, A.; Olanipekun, A. Barriers to implementing lean construction practices in the Kingdom of Saudi Arabia (KSA) construction industry. *Constr. Innov.* **2018**, *18*, 246–272. [CrossRef]
55. Sarhan, J.G.I.; Olanipekun, A.O.; Xia, B. Critical success factors for the implementation of lean construction in the Saudi Arabian construction industry. In Proceedings of the 2016 International Conference on Sustainable Built Environment: Actions for the Built Environment of Post-Carbon Era, Yogyakarta, Indonesia, 12–14 October 2016; CIB-International Council for Building: Ottawa, ON, Canada, 2016; pp. 338–341.
56. Du, J.; Zhang, J.; Castro-Lacouture, D.; Hu, Y. Lean manufacturing applications in prefabricated construction projects. *Autom. Constr.* **2023**, *150*, 104790. [CrossRef]
57. Anaç, M.; Gumusburun Ayalp, G.; Erdayandi, K. Prefabricated construction risks: A holistic exploration through advanced bibliometric tool and content analysis. *Sustainability* **2023**, *15*, 11916. [CrossRef]
58. Zhang, K.; Tsai, J.S. Identification of critical factors influencing prefabricated construction quality and their mutual relationship. *Sustainability* **2021**, *13*, 11081. [CrossRef]

59. Aslam, M.; Gao, Z.; Smith, G. Framework for selection of lean construction tools based on lean objectives and functionalities. *Int. J. Constr. Manag.* **2022**, *22*, 1559–1570. [CrossRef]
60. Saini, M.; Arif, M.; Kulonda, D.J. Critical factors for transferring and sharing tacit knowledge within lean and agile construction processes. *Constr. Innov.* **2018**, *18*, 64–89. [CrossRef]
61. Wu, G.B.; Yang, R.; Li, L.; Bi, X.; Liu, B.S.; Li, S.Y.; Zhou, S.X. Factors influencing the application of prefabricated construction in China: From perspectives of technology promotion and cleaner production. *J. Clean. Prod.* **2019**, *219*, 753–762. [CrossRef]
62. Bajjou, M.S.; Chafi, A.; Ennadi, A. Development of a Conceptual Framework of Lean Construction Principles: An Input-Output Model. *J. Adv. Manuf. Syst.* **2019**, *18*, 1–34. [CrossRef]
63. Al Balkhy, W.; Sweis, R.; Lafhaj, Z. Barriers to adopting lean construction in the construction industry—The case of Jordan. *Buildings* **2021**, *11*, 222. [CrossRef]
64. Parameswaran, A.; Ranadewa KJ, S.; Environment, S.B. Learning-to-learn sand cone model integrated lean learning framework for construction industry. *Smart Sustain. Built Environ.* **2024**, *13*, 856–882. [CrossRef]
65. Feldmann, F.G. Towards lean automation in construction—Exploring barriers to implementing automation in prefabrication. *Sustainability* **2022**, *14*, 12944. [CrossRef]
66. Sushil Interpreting the Interpretive Structural Model. *Glob. J. Flex. Syst. Manag.* **2012**, *13*, 87–106. [CrossRef]
67. Chandramowli, S.; Transue, M.; Felder, F.A. Analysis of barriers to development in landfill communities using interpretive structural modeling. *Habitat Int.* **2011**, *35*, 246–253. [CrossRef]
68. Congliang, T.; Minggong, W.; Xiangxi, W.; Chen, H. Key Nodes Detection of Aviation Network Based on Complex Network theory. In Proceedings of the 4th International Conference on Machinery, Materials and Information Technology Applications (ICMMITA 2016), Xi'an, China, 10–11 December 2016; Atlantis Press: Amsterdam, The Netherlands, 2016; pp. 1368–1374.
69. Huang, W.; Li, H.; Yin, Y.; Zhang, Z.; Xie, A.; Zhang, Y.; Cheng, G. Node importance identification of unweighted urban rail transit network: An Adjacency Information Entropy based approach. *Reliab. Eng. Syst. Saf.* **2024**, *242*, 109766. [CrossRef]
70. Yu, W.; Jinli, G.; Han, L. A New Evaluation Method of Node Importance in Directed Weighted Complex Networks. *J. Syst. Sci. Inf.* **2017**, *5*, 367–375.
71. Netland, T.H.; Powell, D.J.; Hines, P. Demystifying lean leadership. *Int. J. Lean Six Sigma* **2020**, *11*, 543–554. [CrossRef]
72. Santorella, G. *Lean Culture for the Construction Industry: Building Responsible and Committed Project Teams*; Productivity Press: New York, NY, USA, 2017.
73. Lista, A.P.; Tortorella, G.L.; Bouzon, M.; Thürer, M.; Jurburg, D. Soft and hard skills development in lean management trainings. *Int. J. Lean Six Sigma* **2022**, *13*, 1137–1158. [CrossRef]
74. Demirkesen, S.; Bayhan, H.G. Critical success factors of lean implementation in the construction industry. *IEEE Trans. Eng. Manag.* **2019**, *69*, 2555–2571. [CrossRef]
75. Shi, Q.; Zhu, J.; Hertogh, M.; Sheng, Z. Incentive mechanism of prefabrication in mega projects with reputational concerns. *Sustainability* **2018**, *10*, 1260. [CrossRef]
76. Zhu, J.; Hertogh, M.; Zhang, J.; Shi, Q.; Sheng, Z. Incentive mechanisms in mega project-risk management considering owner and insurance company as principals. *J. Constr. Eng. Manag.* **2020**, *146*, 04020120. [CrossRef]
77. Staedele, A.E.; Ensslin, S.R.; Forcellini, F.A. Knowledge building about performance evaluation in lean production: An investigation on international scientific research. *J. Manuf. Technol. Manag.* **2019**, *30*, 798–820. [CrossRef]
78. Lermen, F.H.; Echeveste, M.E.; Peralta, C.B.; Sonego, M.; Marcon, A. A framework for selecting lean practices in sustainable product development: The case study of a Brazilian agroindustry. *J. Clean. Prod.* **2018**, *191*, 261–272. [CrossRef]
79. Danese, P.; Manfè, V.; Romano, P. A systematic literature review on recent lean research: State-of-the-art and future directions. *Int. J. Manag. Rev.* **2018**, *20*, 579–605. [CrossRef]
80. Li, X.; Wang, C.; Kassem, M.A.; Alhajlah, H.H.; Bimenyimana, S. Evaluation method for quality risks of safety in prefabricated building construction using SEM–SDM approach. *Int. J. Environ. Res. Public Health* **2022**, *19*, 5180. [CrossRef]
81. Ghosh, S.; Burghart, J. Lean construction: Experience of us contractors. *Int. J. Constr. Educ. Res.* **2021**, *17*, 133–153. [CrossRef]

buildings

MDPI

Article

Development of an Evaluation System for Intelligent Construction Using System Dynamics Modeling

Jing Guan *, Boyang Liu and Wenxin Shen

School of Economics and Management, Beijing Jiaotong University, Beijing 100044, China; 20113060@bjtu.edu.cn (B.L.); shwenxin@bjtu.edu.cn (W.S.)
* Correspondence: 20113058@bjtu.edu.cn

Abstract: Under the new wave of scientific and technological revolution, the construction industry finds itself with a critical need to alter the traditional and outdated production mode through technological innovation in order to realize industry transformation and move towards a new era characterized by digitalization, informatization, and intelligence. As intelligent construction is the indispensable pathway for the transformation and upgrading of the construction industry, it is of great significance to conduct in-depth research on its evaluation indicators and causality. This paper adopts the system dynamics method, based on the overall structure of intelligent construction, extracts the causality chain and causal feedback loop of intelligent construction, and presents a causality diagram and system dynamics diagram to build a robust system dynamics model for intelligent construction. On this basis, an evaluation index system for intelligent construction is constructed from the five dimensions—investment, design, construction, operation, and environment—for a holistic assessment of the current state of intelligent construction. The research aims to provide a valuable reference for professionals focusing on intelligent construction and the broader development of the industry.

Keywords: intelligent construction; system dynamics; causality; evaluation index system

Citation: Guan, J.; Liu, B.; Shen, W. Development of an Evaluation System for Intelligent Construction Using System Dynamics Modeling. *Buildings* **2024**, *14*, 1489. https://doi.org/10.3390/buildings14061489

Academic Editors: Huayang Yu and Jurgita Antucheviciene

Received: 24 January 2024
Revised: 5 May 2024
Accepted: 8 May 2024
Published: 21 May 2024

1. Introduction

As a crucial pillar of the China's economy, the construction industry lends great support to the sustainable development. However, the current production mode in the construction industry remains relatively traditional, causing notable differences when compared to the requirements for high-quality development that involves efficiency improvement, structural optimization, and innovation-driven, green development [1]. With the introduction of new infrastructure policy and directives from the Ministry of Housing and Urban-Rural Development, the Ministry of Science and Technology, the National Development and Reform Commission, and other relevant departments in the issuance of guidance on "promoting the coordinated development of intelligent construction and building industrialization", digitalization in the construction industry refers to the use of digital technologies to streamline and automate processes. Informatization involves the extensive collection, processing, and use of information to enhance decision-making and productivity. Intelligence in construction incorporates artificial intelligence (AI) and smart technologies to predict outcomes, optimize operations, and improve safety and efficiency, marking a shift towards more adaptive and responsive construction practices. How to use policy dividends; seize the opportunity in informatization, digitalization, and intelligent development; and realize construction enterprise "intellectualization" transformation and upgrading have become crucial imperatives for the industry and, notably, for small- and medium-sized enterprises [2]. For a long time, the construction industry has suffered problems such as low information levels, extensive management, serious waste, and low efficiency. The leading cause of these problems lies in the lack of technological innovation.

According to McKinsey's 2019 report "China in the Digital Era: Building a New Economy with Global Competitiveness", the construction industry's level of informatization is only marginally higher than that of agriculture. Investments in new technology and scientific research and innovation accounted for less than 1% of the total revenue [3].

Intelligent construction represents the development direction of construction 4.0. It encompasses all stages of a building's life cycle, including design, construction, operation, and maintenance. It is based on civil engineering construction technology and is supported by modern information and intelligent technologies. Guided by project management theory, intelligent construction is represented by intelligent management information systems. By constructing a digital twin model and establishing bidirectional mapping between the real and virtual worlds, intelligent construction enables the perception, analysis, and control of the construction process and buildings. The result is a refined, high-quality, and efficient civil engineering construction mode. Intelligent construction has become a requirement for innovative development in the construction industry [4,5]. The emergence of "intelligent infrastructure" presents a critical opportunity for small- and medium-sized enterprises, which enhances the efficiency and precision of traditional infrastructure construction but also provides feasible solutions for the intelligence, digitalization, and industrialization of engineering projects, which greatly improves the production and management standards of engineering projects [6,7].

In recent years, many scholars have directed considerable attention to the research on intelligent construction. Štefanič et al. [8] delve into the application of intelligent construction in the construction stage. Ding [9], on the other hand, reconstructs the construction industry from the aspects of construction mode, business philosophy, product form, market form, and industry management. Tian [10] emphasizes the importance of intelligent construction. Taking the coal mining enterprises as an example, Tian illustrates that when the intelligent construction of coal mining enterprises is in the mature stage, with the improvement in the cost and benefits of intelligent construction, the "intelligent construction" strategy will become a general consensus of coal mining enterprises. Wang [11] believes that the advanced intelligent construction concept involves the adoption of the most advanced technology to replace the traditional construction means, which aims to optimize natural resource utilization, greatly improve the construction environment, and enhance project control, ultimately fostering the long-term healthy development of the construction industry. Intelligent construction, as a brand new construction model, will change the traditional construction industry and affect the entire industrial chain. This model will reshape production modes, production factors, and production relations and will require technological and managerial transformations. The various elements in the intelligent construction system are interwoven to form an organic whole characterized by dynamics, complexity, and nonlinearity. In the research on intelligent construction evaluation systems, You et al. [12] introduce the concept of "information-physics" fusion and provide a crucial framework for the development of the field of intelligent architecture. This framework explores five key features: digital twin (the process of digitizing and modeling the composition, features, functions, and performance of physical entities using information technology), data-information-driven, ubiquitous link, object-oriented work, and system autonomy. In addition, it proposes a new, sustainable, and measurable model for technology development and a complete set of competency testing criteria. Through the application of G1, the COWA empowerment method, and TOPSIS technology, Liu et al. [13] successfully developed a new, measurable, and intelligent management system for road and railway construction projects. Moreover, Chen et al. also proposed a new three-layer architecture of "capability element-capability domain-sub-capability" and its corresponding five levels of intelligent management system [14]. The application of gray system theory enables the comprehensive measurement and analysis of the level of intelligent construction within construction companies.

However, since conventional empirical approaches may fall short in the study of the links between different factors, it has become inevitable to use more complex, interdisciplinary, new approaches, such as system dynamics (SDs) [15]. SDs offer a comprehensive framework for understanding and addressing complex problems across various domains by modeling the intricate interactions and feedback loops within systems. Its holistic approach integrates diverse factors—social, economic, environmental, and technological—providing valuable insights into system behavior over time. SDs support strategic decision-making through scenario testing and sensitivity analysis, allowing for the exploration of potential policies and interventions before implementation. Moreover, its visual modeling tools enhance communication among stakeholders, facilitating a shared understanding of complex dynamics. This adaptability, combined with the ability to incorporate both quantitative and qualitative data, makes SDs an invaluable tool for forecasting, strategic planning, and fostering systemic change, thereby offering a versatile approach to tackling multifaceted challenges in a wide range of fields [16,17]. At present, the system dynamics method has been applied to the study of the productivity influence factors of green construction enterprises [18], analyses on the carbon emission process of the construction industry at different stages. [19], and research on urban solid waste management and urban earthquake disaster reduction evaluation systems [20,21]. The development of intelligent construction technology facilitates the understanding of causal relationships but also exhibits characteristics of multiple circuits, making it an indispensable component of complex social and economic management systems. Therefore, the system analysis method proves highly suitable for comprehensive analysis of the various factors of intelligent construction and further evaluation of their influence.

In summary, the transformation of intelligent construction brings opportunities and major changes to the engineering construction industry, especially for small enterprises, given their constraints in talent, scale, capital, and technology. Therefore, in the process of their own transformation, it is crucial to maximize their own advantages. Resources should be concentrated on breaking through some key links of engineering construction. With this in mind, the paper aims to study the influence mechanism among the subsystems within the intelligent construction system, explore the important factors within the system as variables, and construct a dynamic model of the intelligent construction system. This paper focuses on dynamic feedback loops and causality chains within the intelligent construction ecosystem, which divides the intelligent construction system into five subsystems: input, design, construction, operation, and environment. The overall structure is analyzed using an intelligent construction system model. According to the system dynamics model, the paper reveals the mutual relationship among elements within each system. Additionally, it constructs an evaluation index system including five second-level indexes and twenty-six three-level indexes from the five dimensions of intelligent construction input, design, construction, operation, and environment. Through evaluation and analysis of the input, design, construction, operation, and environment subsystems, it aims to provide a valuable reference for the development of the industry.

2. Connotation and Theoretical Framework of Intelligent Construction

2.1. Connotation of Intelligent Construction

The increasing integration of industrialization and information technology has brought intelligent construction into focus, including discussions on the theoretical system, various single-point application scenarios, and the overall intelligence of the construction industry. However, the current academic understanding of these concepts remains insufficient. Many scholars have proposed different interpretations of intelligent construction according to their respective research. For instance, Guo and Li [22] view the core idea of intelligent building as the organic combination of informatization, industrialization, and other fields through the introduction of advanced technology, all kinds of original labor organization, management mode, management philosophy, such as organic structure transformation, and the organic combination of various factors to improve building quality, safety, and sustainability. The goal is to achieve time savings, cost reductions, and service optimization. Dewit [23] emphasizes the transformation of construction through the integration of robots to change the construction industry, reduce project costs, improve accuracy, minimize waste, and fortify resilience and sustainability. Ding [9] introduces the concept of intelligent building as a new engineering construction innovation mode. It is the integration of engineering construction and new information technology. Through standardized modeling, high-performance computing, and intelligent decision support, it aims to achieve digitally driven engineering project planning, planning and design, construction production, operational service integration, and efficient coordination. You et al. [24] present a comprehensive theoretical framework of intelligent architecture, which consists of 11 key components. Each component is explained in terms of its scientific connotation and reveals the internal logical connection between them. This paper believes that intelligent construction optimizes the investment, design, construction, operation, and maintenance through the collaborative development of information technology, advanced equipment and industrialization, and professional and technical personnel. The ultimate goal is to establish an efficient, intelligent, and sustainable construction mode.

2.2. The Components and System Structure of Intelligent Construction

The components of intelligent building systems exhibit considerable diversity. Based on previous discussions and research on the developmental influence of the construction industry, this paper undertakes a qualitative analysis using the feedback mechanism in the system dynamics model. Key factors were selected to form the feedback loop model. The selected elements mainly cover five dimensions: input, construction, design, operation and maintenance, and the environmental subsystem. Interactions among elements in these dimensions contribute to the dynamic system of intelligent construction. Elements within intelligent buildings contain technology, personnel, capital, equipment, information, policies, laws, and regulations, etc. Funding and personnel represent the core elements of intelligent construction, while equipment and information serve as supporting elements. Policy, laws, and regulations act as guarantee elements to facilitate contact between and restrict these elements. The intelligent construction system plays a leading role in the market; the government plays a regulating and guiding role. Attention is directed towards the cultivation of an intelligent construction environment. Universities, research institutions, and science and technology intermediaries contribute to the realization of the purpose and functions of the intelligent construction system.

The overall structure of intelligent construction reflects the connection and interaction among its various elements. In recognition of many random factors involved in intelligent construction, this study simplifies the analysis by dividing the intelligent construction system into four subsystems: input, design, construction, and operation and maintenance. At the same time, the impact of the environment on the intelligent construction process is considered an important subsystem during the overall structural analysis. Among them, the input, design, construction, and operation and maintenance subsystems mutually promote and restrict each other while carrying out their own behavioral activities, forming a closed-loop operation mode. At the same time, the environmental subsystem is at the support level of intelligent construction, providing support and affecting its operation. The composition of the intelligent construction system and the interaction relationships among the subsystems are illustrated in Figure 1.

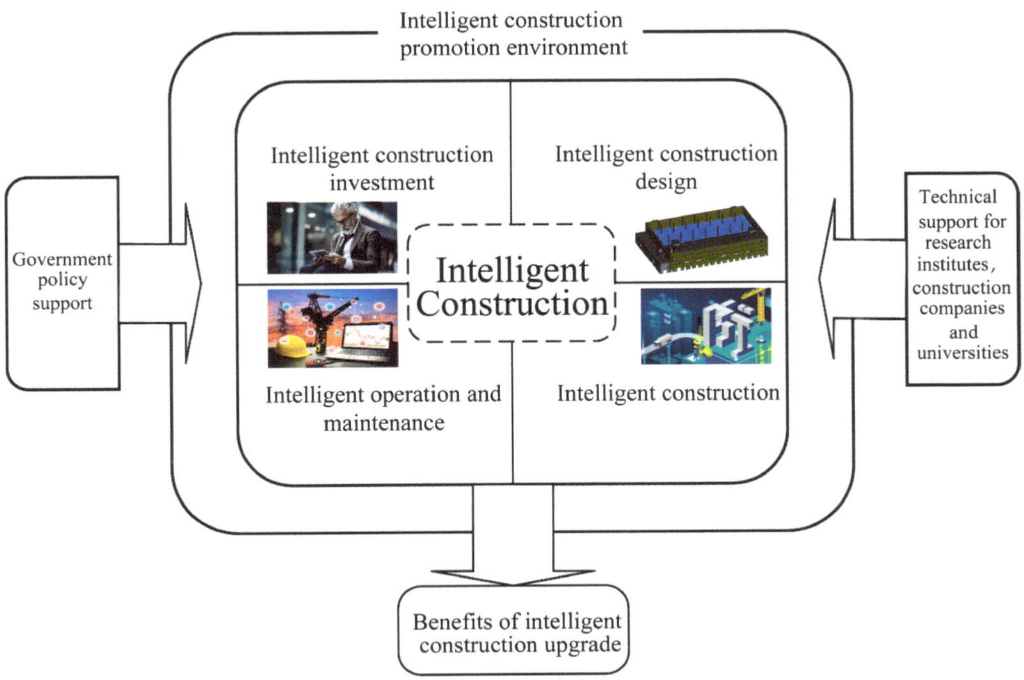

Figure 1. Structure of the intelligent construction system.

3. System Dynamics Model Construction of Intelligent Construction

3.1. Definition of the Model Boundaries

The time span under consideration and the main variables studied determine the boundaries of the system. It is a basic principle to include variables closely related to the modeling objective and ensure the closure of the system boundary. Only after satisfying these two points can the system boundary be accurately demarcated [25,26]. The principal focus of this paper is intelligent construction; consequently, the system boundary aligns with the construction industry. In this paper, the construction industry is broadly encompasses construction, material production, environmental protection, and real estate. Considering the temporal dimension and combined with the previous analysis of the overall structure of intelligent construction, attention is directed towards the five subsystems of input, design, construction, operation, and environment as research factors to further study the causal feedback relationship within and between the systems.

3.2. Basic Assumptions

To facilitate the analysis, this paper temporarily ignores the secondary factors and simplifies the intelligent construction system into five subsystems. Before the construction of a model of an intelligent building system, it is essential to establish reasonable assumptions that enhance the understanding of the complexity and details in the system [27,28], without compromising simulation accuracy. Therefore, after thorough research and analysis, several effective assumptions are put forward; based on these assumptions, a complete model of the intelligent construction system has been constructed. These assumptions include the following:

(1) The system of intelligent construction is intricate, necessitating compliance with certain rules and guidelines to guarantee its ongoing and seamless progression. Hence, the

system of intelligent construction being analyzed must demonstrate steadiness to secure its lasting growth and preclude any sporadic leaps.
(2) Considerations of the external milieu, significant shifts in governmental policies, and the influence of unforeseen events on the construction sector are momentarily set aside.
(3) The advantages of the intelligent construction system are assessable via metrics like efficiency in construction, safety, expense, consumer contentment, and acknowledgment by the sector.

3.3. Systematic Causality Diagram

The flow diagram of the system dynamics is based on the causality diagram, wherein the causal interactions among various elements form the function and behavior of the system [29]. Therefore, the formation of the system mainly depends on the establishment of causal relationships in this system. Due to the complexity of the intelligent construction system, it includes five subsystems: decision-making, design, production, construction, and operation and maintenance. These subsystems are related to each other and all have their own unique functional and structural characteristics. Therefore, based on the overall structure of the intelligent construction system, this paper establishes a causal relationship chain and a causal feedback loop to analyze the operation of the system, as illustrated in Figure 2. The subsequent analysis delves into the following causal feedback loops:

(1) Intelligent construction upgrade investment + intelligent construction design and research and development + intelligent construction technology output + intelligent construction operation and maintenance level/intelligent construction level + intelligent construction upgrade benefit + intelligent construction implementation concept + intelligent construction upgrade investment. This loop constitutes a positive feedback loop, reflecting the development of enterprise strategy, profit, industry recognition, customer satisfaction, and external environment and making decisions on the personnel investment and capital investment of intelligent construction transformation and upgrading according to the market demand.
(2) Intelligent construction design and research and development + intelligent construction technology output + intelligent construction operation and maintenance level/intelligent construction level + intelligent construction upgrade benefit + intelligent construction promotion environment + combination of industry, university, and research + intelligent construction design and research and development/intelligent construction technology output. This loop is another positive feedback loop, reflecting the impact of intelligent construction design and development on the transformation and upgrading process of construction enterprises. Collaboration with educational institutions and research entities helps overcome technical barriers, increase research and development facilities, improve the project rate and technology yield, and enhance the enterprise construction and operations level, thus increasing the enterprise benefit and promoting the construction of an environment for intelligent construction.
(3) Market demand + intelligent construction promotion environment + intelligent construction execution concept + intelligent construction upgrade investment + intelligent construction design and development + intelligent construction technology output + intelligent construction operation and maintenance level/intelligent construction level + industry recognition level + market demand. This positive feedback loop reflects the impact of market demand on the intelligent construction transformation and upgrading process of construction enterprises. Technological advancements leading to improved construction, operation, and maintenance levels will enhance industry recognition and corporate social image and stimulate user demand. Increased market demand further boosts intelligent construction to promote the transformation and upgrading of the construction industry.
(4) Government policy support + investment in intelligent construction upgrade + intelligent construction design and development + design and construction technology

output + intelligent construction operation and maintenance level/intelligent construction level + benefit of intelligent construction upgrade + intelligent construction promotion environment + government policy support. This positive feedback loop signifies the construction external environment's impact on intelligent construction, with the government as the main driver for intelligent construction upgrading. Legal, tax, and fiscal incentives guide the construction enterprise to their consciousness regarding intelligent construction transformation and upgrading, increase investment in transformation and upgrading, and play a role in its regulation and guidance.

(5) Financial sector support + capital investment for intelligent construction and upgrading + intelligent construction design and development + design and construction technology output + intelligent construction operation and maintenance level/intelligent construction level + intelligent construction upgrade benefit + intelligent construction promotion environment + support from financial sector. This positive feedback loop reflects the construction industry's reliance on the external environment, with crucial support from the financial sector to meet the capital requirements of construction enterprises during the intelligent upgrade transformation. It promotes progress in intelligent construction design research and development, improves the technical yield, and fully embodies the intelligent building upgrade transformation in enterprises and the financial sector.

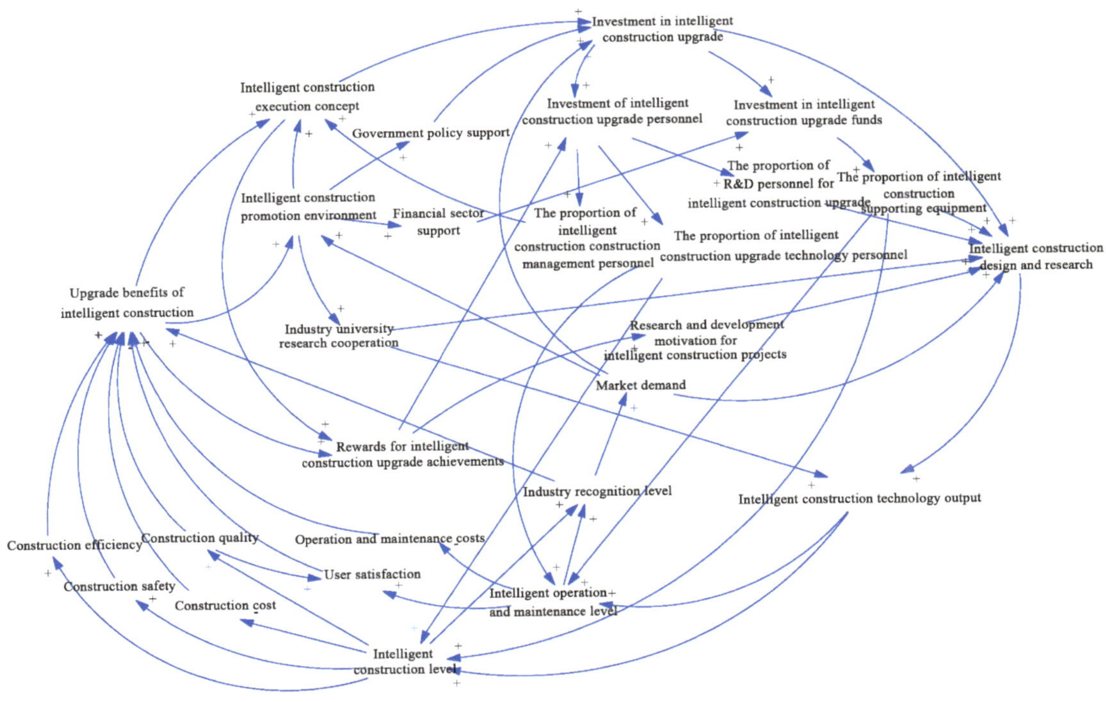

Figure 2. Intelligent manufacturing causality.

3.4. System Dynamics Model

Through the analysis of the causality in intelligent construction, this paper establishes six state variables for intelligent construction upgrade input; intelligent construction design, research, and development; intelligent construction technology output; intelligent construction operations level; and intelligent construction environment. The selected state variables include the input increment, R&D project rate, technology output, operational level increment, and environmental construction rate. The remaining variables are considered auxiliary variables or constants. According to the theory of system dynamics, the system dynamics model is constructed using Vensime 9.4.2 software, as shown in Figure 3.

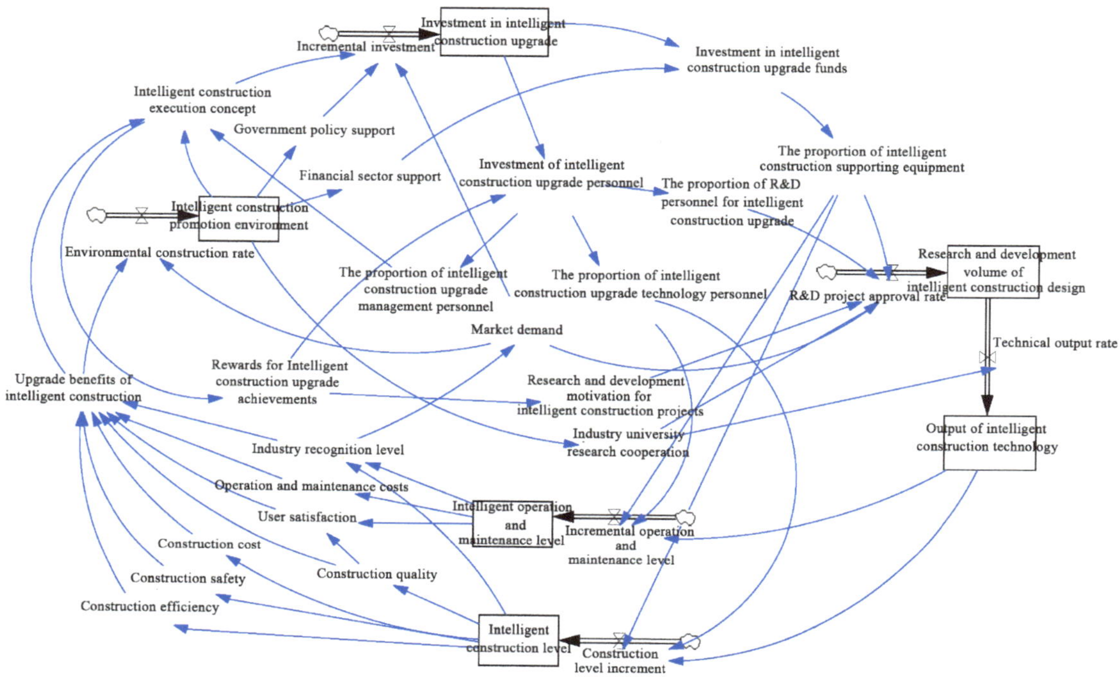

Figure 3. System dynamics model of intelligent manufacturing.

4. Intelligent Construction Evaluation Index System

According to the rules revealed by the aforementioned system dynamics model of intelligent construction, this paper will construct an evaluation index system of intelligent construction from the five dimensions: investment, design, construction, operation, and environment of intelligent construction.

4.1. *Evaluation Index of Intelligent Construction Input Subsystem*

The advent of intelligent construction has changed decision-making paradigms from the traditional "Decision-making based on experience" to "Decision-making based on data". In engineering and construction activities, a substantial volume of data is generated, which contains crucial information such as market trends and consumption patterns. By using new technologies to mine the rules of these data and to assist in the decision-making process, the utilization of data resources can be facilitated for well-informed choices. As illustrated in Figure 4, intelligent construction is initiated by investment in intelligent construction. With the enhancement of the implementation concept of intelligent construction in combination with a supportive environment for intelligent construction, including governmental and financial backing, enterprises increase their design and development of intelligent construction, as seen in the higher proportion of management personnel, technical personnel, and R&D personnel. Therefore, the fundamental investment indicators for intelligent construction are personnel and capital. Moreover, a portion of the R&D investment is allocated to non-intelligent upgrading, which also constitutes the basic elements and conditions of intelligent construction, basically reflected in the personnel and capital. In intelligent construction, the increase in economic benefits (e.g., internal rate of return) indirectly stimulates incremental investments in intelligent construction. Based on the above analysis, this paper mainly represents the status of the intelligent construction input subsystem, where personnel input is quantified by the proportion of intelligent construction R&D personnel and management personnel, while capital input is expressed through indicators of intelligent construction and upgrading equipment, technology introduction costs, and personnel training funds.

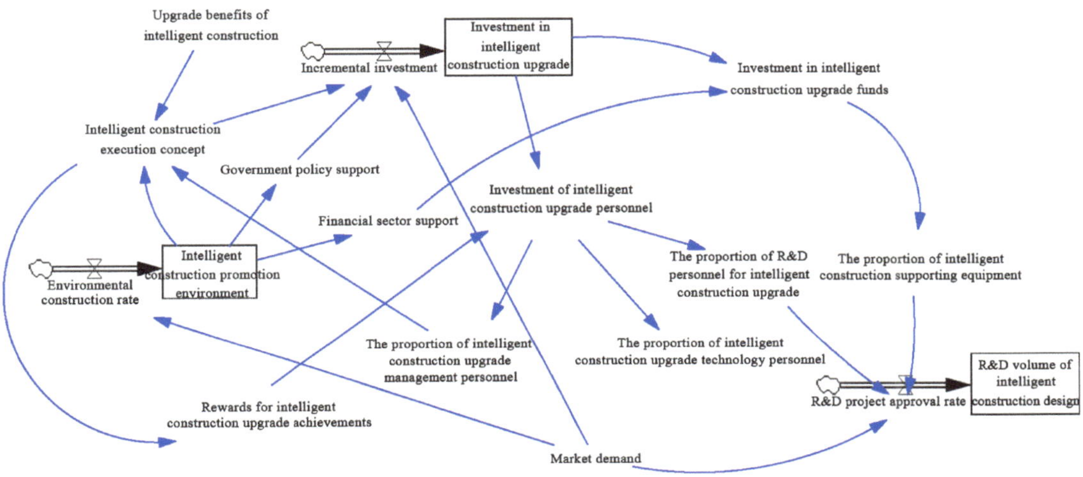

Figure 4. System dynamic flow of intelligent construction upgrade input.

4.2. Evaluation Index of Intelligent Construction Design and Development Subsystem

As seen in Figure 5, the upgrade of intelligent construction design and development mainly focuses on the transformation of research and development project results to improve the level of construction, operation, and maintenance. Therefore, this paper selects intelligent construction research and development and the technology output rate as status variables. In the design R&D subsystem, the proportion of capital and R&D talents, coupled with the incentive of intelligent construction achievements, acts as the driving force behind R&D quantity. At the same time, industry–university–research cooperation enhances the technology output rate. For example, enterprises will cooperate with universities and research institutes to solve the technical challenges in intelligent construction. In addition, the informal conduct of intelligent builders also plays an important role in the promotion of research and development, such as consulting with external experts during the tumultuous stages of intelligent construction development and informal communication and communication among intelligent construction developers within the industry. Variables related to the number of intelligent construction developments and their relationships are shown in Figure 5. Based on the above analysis, this paper believes that the intelligent construction R&D subsystem can be explained by the following indicators: intelligent construction R&D power, the output rate of intelligent construction technology, communication with external consulting experts, informal communication, intelligent construction communication and cooperation, etc.

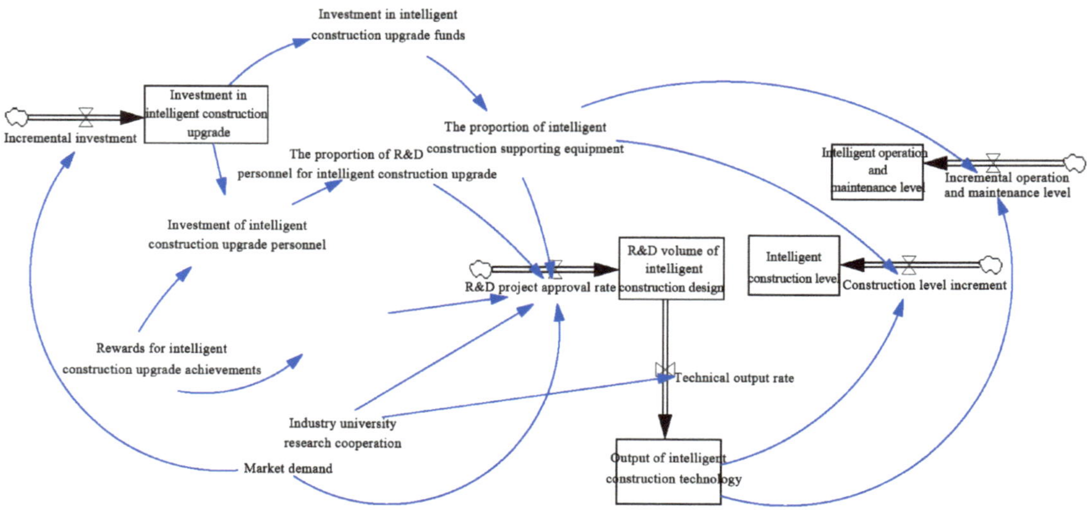

Figure 5. System dynamic flow of intelligent construction design and development.

4.3. Evaluation Index of Intelligent Construction and Construction Subsystem

The improvement of the intelligent construction level plays an important role in the output of intelligent construction technology. The improvement can reduce material consumption in the building process, lower waste rates, improve the energy utilization rate, and, thus, decrease construction costs. In addition, the improvement can also optimize construction workflows and safety monitoring, especially for high-risk construction positions. The wide application of intelligent construction supporting equipment can effectively reduce employee accidents and improve safety. The improvement substantially impacts construction quality and efficiency. In the whole process of construction, the integration and promotion of new technology (such as the Internet of Things (a network based on information carriers such as the Internet and traditional telecommunications networks, which enable all ordinary physical objects that can be independently addressed to achieve interconnection), big data, remote sensing, automation, cloud computing, artificial intelligence, block chain, etc.), as well as the research and development, manufacturing, and application of advanced manufacturing equipment, intelligent equipment, and smart site-related tools, can upgrade construction machinery and amplify data resource utilization to boost quality and efficiency. It can also directly improve consumer satisfaction through tangible feedback on construction quality. Lower construction costs and higher construction safety, quality, and efficiency generate considerable transformation and upgrading benefits for construction enterprises. These upgrading benefits further promote the transformation and upgrading environment of intelligent construction. The system dynamic flow diagram of the intelligent construction subsystem for construction enterprises is shown in Figure 6. According to the analysis, this paper mainly uses the following indicators to characterize intelligent construction levels: shortening rate of the construction cycle, construction safety, reduced building material consumption rate, lowered energy consumption rate, the share of intelligent construction supporting equipment, the quality inspection pass rate, and more.

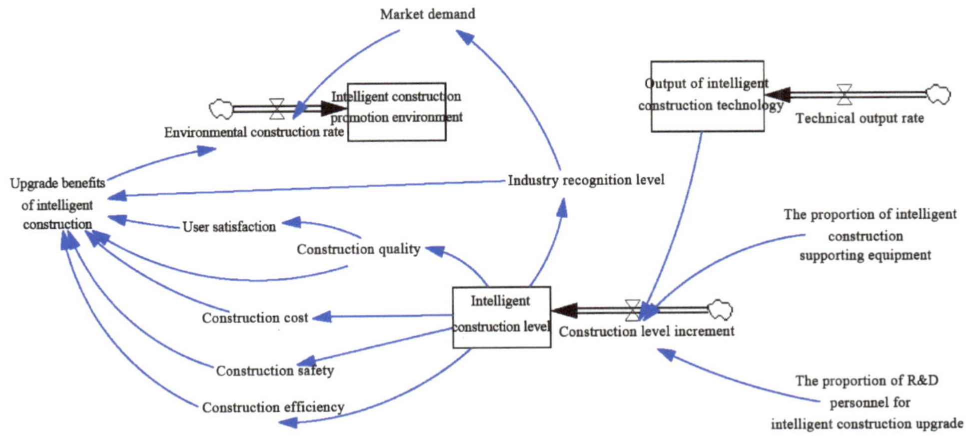

Figure 6. System dynamic flow of intelligent construction.

4.4. Evaluation Index of Intelligent Construction, Operation, and Maintenance Subsystem

The improvement of the operational and maintenance levels in intelligent construction plays a crucial role in the output of intelligent construction technology. This improvement can reduce the need for building operation and maintenance personnel, lower labor costs, and improve the energy efficiency of buildings. In addition, it contributes to higher consumer satisfaction with operation and maintenance services, enhances public safety, and reduces the failure rate of building systems, such as power supply, water supply and drainage, and fire prevention equipment. The system dynamic flow diagram for the intelligent construction operation and maintenance subsystem within construction enterprises is shown in Figure 7. This paper mainly uses the following indicators to characterize intelligent construction and operation and maintenance levels: reduction in operation and maintenance labor costs, improved energy utilization efficiency in buildings, decreased failure rates of construction equipment, and enhanced user satisfaction with property service, among others.

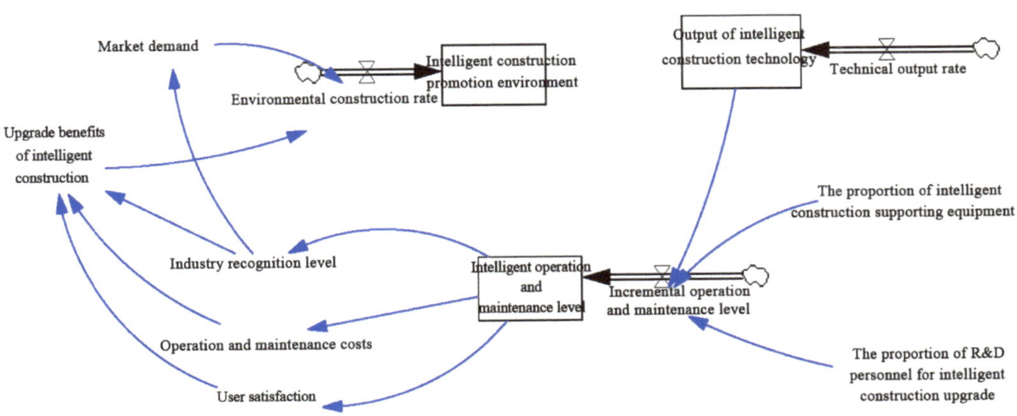

Figure 7. System dynamic flow of intelligent construction operation and maintenance.

4.5. Evaluation Index of Intelligent Construction Environment Subsystem

Whether enterprises embrace the concept of intelligent building execution, whether their employees possess an awareness of intelligent building upgrade transformation, and whether enterprise managers recognize the importance of the intelligent upgrade transformation are critical factors. The enterprise's innovation reward index for intelligent building directly impacts the enthusiasm of construction enterprises for intelligent building upgrade transformation, which, in turn, either promotes or hinders investment and implementation in the transformation and upgrading of intelligent buildings. Since these indicators revolve around construction enterprises and occur within enterprises, they can be summarized as the environment influencing the transformation and upgrading of construction enterprises. On the other hand, the external environment involves key players such as the government, financial departments, universities, and research institutes, which jointly provide economic resources and the overall economic environment and a supportive environment for scientific research. This external support system promotes intelligent construction, upgrading, and transformation initiatives within enterprises. The internal environment and the external environment jointly influence the upgrading and transformation process of intelligent construction, as shown in Figure 8. According to the analysis, this paper mainly employs the internal and the external environment to characterize the condition of the intelligent built propulsion environment subsystem. The internal environment includes aspects of construction enterprise culture, including the adoption of intelligent construction execution concepts and employee awareness. It also considers the institutional environment, such as the reward and punishment system. The external environment includes economic resources and the overall economic environment and a supportive environment for scientific research.

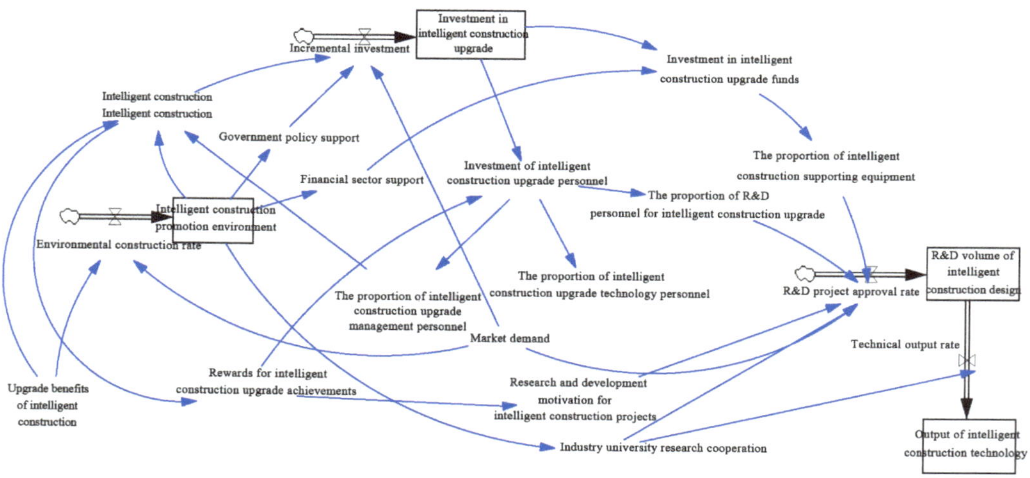

Figure 8. System dynamic flow of intelligent construction environment.

4.6. Evaluation Index System of Intelligent Construction System

The aforementioned intelligent construction input subsystem index, design subsystem index, construction subsystem index, operation and maintenance subsystem index, and environment subsystem index are organically integrated together to form a comprehensive intelligent construction evaluation index system. This system contains five secondary indicators and twenty-six third level indicators, which comprehensively reflect the operation process of intelligent construction. The detailed breakdown of these indicators is presented in Table 1 below.

Table 1. Evaluation index system of intelligent construction.

Primary Indicators	Secondary Indicators	Third Level Indicators	
Intelligent construction	Intelligent construction investment	Capital investment	Cost of intelligent construction upgrade equipment Technology introduction costs Personnel training funds
		Personnel input	The proportion of intelligent construction personnel to total employees The proportion of intelligent construction R&D personnel The proportion of management personnel The proportion of technical personnel
	Intelligent construction research and development	Intelligent construction research and development power Output rate of intelligent construction technology Communication with external consulting experts Informal communication situation Internal intelligent construction communication and cooperation within the enterprise	
	Intelligent construction	Shortening rate of construction period Construction safety Reduction rate of building materials consumption Reduction rate in energy consumption Construction quality inspection qualification rate The proportion of intelligent construction supporting equipment	
	Intelligent construction and operation maintenance	Reduce labor costs for operations and maintenance Building energy utilization efficiency Building equipment failure rate User satisfaction with property services	
	Intelligent building environment	Internal environment	Corporate culture (intelligent construction execution concept, employee awareness, etc.) Institutional environment (reward and punishment system)
		External environment	Economic resources and environment Research support environment

5. Effectiveness of the Case

This section illustrates the example of an intelligent construction and industrialization research and development base and its service project in Shaanxi Province. In its early stage, the base adhered to intelligent construction as the driving force behind the development of the construction industry and actively formulated strategies for the development of the intelligent construction industry. The base invests more than CNY 2 million annually in intelligent construction and equipment upgrades, with the technology introduction cost reaching CNY 1 million, and personnel training costs CNY 500,000. The number of management personnel exceeds 50, accounting for more than 70% of the total personnel in the base. Most operational processes are overseen by the management professionals operating the mechanical equipment. Half of the personnel in the base are technical experts, with much preliminary technical groundwork already undertaken. A total of 70% of the staff are dedicated to intelligent construction, including 30% focused on intelligent construction research and development, much higher than industry benchmarks. In terms of intelligent construction research and development, the base invests an annual investment of more than CNY 500,000, with the output of intelligent construction technology accounting for more than 80% of the total revenue. More than ten annual exchanges occur with external consulting experts. In the first five months of 2023 alone, the base conducted six informal technical exchanges and industrial collaboration. The base has effectively integrated intelligent construction research and development with on-site implementation, with widespread application of the research outcomes across various projects. In project construction services, there is an active deployment and adoption of intelligent construction equipment and systems, of which intelligent building equipment accounted for 36%. For instance, in the room installation construction process, the overall construction period has been dramatically reduced from 2 months to 0.5 months, with a 75% cycle shortening rate and enhanced construction security without any damage. At the same time, the base achieved a 35% reduction in on-site pipelines and other installed building materials, significantly lower than industry averages. Energy consumption reduction stands at 30%, and the quality qualified rate of primary installation surged to 97%.

In this study, a system dynamics analysis identified five secondary indicators and twenty-six tertiary indicators for evaluating intelligent construction. This section employs a questionnaire survey to investigate and determine the final system of impact factors by querying personnel from various sectors within the industry.

Based on the intelligent construction evaluation indicator system, which includes five secondary indicators and twenty-six tertiary indicators derived from system dynamics analysis, a survey titled "Intelligent Construction Evaluation Indicator Survey" was developed. The survey was conducted using a targeted questionnaire approach, focusing on professionals in the construction industry. The target respondents included personnel from government and public institutions, universities and research institutions, development and construction entities, survey and design firms, construction companies, consulting firms, supervisory and inspection entities, and others, totaling 174 individuals.

The survey comprised two parts. The first part collected basic information from the respondents, as shown in Table 2. The second part assessed the degree of impact of the evaluation indicators, as shown in Table 3. After collecting the surveys, data were compiled and analyzed to determine the influence of each indicator on intelligent construction evaluation.

Table 2. Statistics of survey respondent information.

Survey Questions	Options	Quantity	Percentage (%)
Age	Over 50 years old	17	9.77
	40–50 years old	41	23.56
	30–40 years old	69	39.66
	25–30 years old	34	19.54
	Under 25 years old	13	7.47
Educational background	Doctor	14	10.34
	Master	47	27.01
	Bachelor	82	47.13
	Others	31	15.52
Professional Title Status	Senior Engineer	23	13.22
	Associate Senior Engineer	41	23.56
	Engineer	71	40.80
	Associate Engineer and Others	39	22.41
Workplace	Government and public institutions	7	4.02
	Universities and research institutions	6	3.45
	Development and construction entities	16	9.20
	Survey and design firms	27	15.52
	Construction companies	66	37.93
	Consulting firms	12	6.90
	Supervisory and inspection entities	17	9.77
	Others	23	13.22

Table 3. Degree of impact of the evaluation indicators.

Secondary Indicators	Tertiary Indicators	Distribution of Choices (%)				
		Very Important	Important	Somewhat Important	Not Important	Very Unimportant
Intelligent construction investment	Intelligent construction and equipment upgrade costs	38 (21.84)	73 (41.95)	56 (32.18)	7 (4.02)	0 (0)
	Technology introduction costs	52 (29.89)	68 (39.08)	45 (25.86)	9 (5.17)	0 (0)
	Personnel training expenses	39 (29.89)	54 (31.03)	61 (35.06)	15 (8.62)	5 (2.87)
	Proportion of intelligent construction staff in total workforce	28 (16.09)	35 (20.11)	83 (47.70)	19 (10.92)	9 (5.17)
	Proportion of R&D personnel in intelligent construction	41 (23.56)	38 (21.84)	72 (41.38)	17 (9.77)	6 (3.45)
	Proportion of management personnel	26 (14.94)	36 (20.69)	61 (35.06)	35 (20.11)	16 (9.20)
	Proportion of technical staff	36 (20.69)	42 (24.14)	76 (43.68)	15 (8.62)	3 (1.72)
Intelligent construction research and development	Intelligent construction R&D motivation	38 (21.84)	46 (26.44)	73 (41.96)	12 (6.90)	5 (2.87)
	Intelligent construction technology output rate	46 (26.43)	53 (30.46)	68 (39.08)	7 (4.02)	0 (0)
	Exchange with external consulting experts	26 (14.94)	45 (25.86)	76 (43.68)	19 (10.92)	8 (4.60)
	Informal communication situations	21 (12.07)	36 (20.69)	75 (43.10)	29 (16.67)	13 (7.47)
	Internal cooperation on intelligent construction within the company	35 (20.11)	42 (24.14)	68 (39.08)	23 (13.22)	6 (3.45)
Intelligent construction implementation	Construction cycle reduction rate	42 (24.14)	56 (32.18)	67 (38.51)	9 (5.17)	0 (0)
	Construction safety	46 (26.43)	64 (36.78)	62 (35.63)	2 (1.15)	0 (0)
	Reduction rate in building materials consumption	43 (24.71)	59 (33.91)	65 (37.36)	7 (4.02)	0 (0)
	Energy consumption reduction rate	39 (22.41)	47 (27.01)	76 (43.67)	10 (5.75)	2 (1.15)
	Construction quality inspection pass rate	52 (29.89)	55 (31.61)	63 (36.21)	4 (2.30)	0 (0)
	Proportion of intelligent construction support equipment	38 (21.84)	46 (26.44)	67 (38.51)	16 (9.20)	5 (2.87)
Intelligent construction operation and maintenance	Reduction in operation and maintenance labor costs	46 (26.44)	61 (35.06)	56 (32.18)	11 (6.32)	0 (0)
	Building energy utilization efficiency	53 (30.46)	59 (33.91)	56 (32.18)	6 (3.45)	0 (0)
	Building equipment failure rate	56 (32.19)	46 (26.44)	63 (36.21)	9 (5.17)	0 (0)
	Customer satisfaction with property services	38 (21.84)	49 (28.16)	67 (29.89)	16 (13.21)	5 (6.32)

Table 3. Cont.

Secondary Indicators	Tertiary Indicators	Distribution of Choices (%)				
		Very Important	Important	Somewhat Important	Not Important	Very Unimportant
Intelligent construction environment	Corporate culture	36 (20.69)	46 (26.44)	66 (37.93)	18 (10.34)	8 (4.60)
	Institutional environment (reward and punishment system)	44 (25.29)	56 (32.18)	59 (33.91)	13 (7.47)	2 (1.14)
	Economic resource environment	36 (20.69)	42 (24.14)	55 (31.61)	26 (14.94)	15 (86.21)
	Research support environment	41 (25.56)	49 (28.16)	64 (36.78)	15 (86.21)	5 (2.87)

Combining the aforementioned data and the method of the intelligent construction conceptual evaluation system based on system dynamics proposed in this paper, the application and promotion of intelligent construction throughout the entire process is anticipated to yield impressive benefits. Projections indicate a reduction of 30% in operation and maintenance labor costs, a 15% decrease in the failure rate of construction equipment, a 20% increase in building energy utilization efficiency, and a great enhancement in customer satisfaction.

In terms of the environmental construction of intelligent construction, the Intelligent Construction and Industrialization Research and Development Base believes that intelligent construction is the indispensable path for high-quality development of the construction industry. To shape innovation as the foremost driving force for the development of the base, a multi-category research group has been instituted. At the same time, an incentive mechanism for intelligent construction is in the process of establishment, designed to reward individuals and teams who deliver outstanding contributions to research and development and implementation on an annual basis. The total annual reward allocation is set at more than CNY 150,000. The base vigorously promotes the comprehensive development of intelligent construction, provides sufficient economic resources and various support mechanisms, constantly improves the product research and development and service capabilities of the base, and promotes the transformation of intelligent construction and high-quality development of the industry.

6. Conclusions

This paper focuses on dynamic feedback loops and causality chains within the intelligent construction ecosystem, which divides the intelligent construction system into five subsystems: input, design, construction, operation, and environment. The overall structure is analyzed using an intelligent construction system model. Through an examination of the logic of system variables and consideration of the feedback mechanism in system dynamics research, it is concluded that intelligent construction itself is a complex system with multiple feedback and cyclical interactions. It is the result of the connection and interaction among input, design, construction, operation, and environment. Therefore, according to the system dynamics model, the paper reveals the mutual relationship among elements within each system. Additionally, it constructs an evaluation index system including five second-level indexes and twenty-six three-level indexes from the five dimensions of intelligent construction input, design, construction, operation, and environment. The findings indicate that in the intelligent construction system the central role of the government, universities, research institutes, and technology intermediary agencies should be clarified. While emphasizing investment, design, construction, and operation and maintenance, it is crucial to strengthen the cultivation and improvement of the internal and external environment of intelligent construction to disrupt traditional construction processes, apply advanced information technology to realize intelligent construction, and strive for an efficient, intelligent, and sustainable construction mode.

The research in this paper not only provides an systematic, comprehensive, and reasonable reference for the practical operation of intelligent construction but also establishes a theoretical basis for further exploration within the construction industry. However, the value of considering the entire lifecycle should be considered, including deconstruction, for future research to fully embrace the principles of a circular economy. Intelligent construction is a comprehensive system involving theoretical and practical factors. As the intelligent construction standard system progresses into the practical application stage, updates and improvements are required. The role and value of legacy or historic intelligence and data as well as technological change are not extensively discussed in this paper. It will be necessary to verify the evaluation method proposed in this paper through case studies or experiments.

These issues require further discussion in future research combined with the actual situation.

Author Contributions: Conceptualization, Methodology, Investigation and Data Curation, Writing—Original Draft, J.G.; Methodology, Project administration, Writing—Review and Editing, W.S.; Visualization, Software, B.L. All authors have read and agreed to the published version of the manuscript.

Funding: This research received no external funding.

Data Availability Statement: The datasets generated during and/or analyzed during the current study are available from the corresponding author on reasonable request.

Acknowledgments: The authors would like to express their sincere gratitude to the editor and the anonymous reviewers for their insightful and constructive comments.

Conflicts of Interest: The authors declare no conflicts of interest.

References

1. Pan, Y.-R.; Wang, J.; Lu, Z.; Liu, Y.-S.; Li, Y.-R. High-quality development in China: Measurement system, spatial pattern, and improvement paths. *Habitat Int.* **2021**, *118*, 102458. [CrossRef]
2. Boadu, E.-F.; Wang, C.-C.; Sunindijo, R.-Y. Characteristics of the construction industry in developing countries and its implications for health and safety: An exploratory study in ghana. *Int. J. Environ Res. Public Health* **2020**, *17*, 4110. [CrossRef]
3. Liu, Z.S.; Sun, X.T.; Shi, G.L. Summary of Application of Intelligent Construction in Civil Engineering Construction. *Archit. Technol.* **2021**, *50*, 40–53. (In Chinese)
4. Qian, Q.H. The field of engineering construction should move towards intelligent construction. *Constr. Archit.* **2020**, *18*, 17–18. (In Chinese)
5. Abd-Alhamid, F.; Kent, M.; Bennett, C.; Calautit, J.; Wu, Y. Developing an innovative method for visual perception evaluation in a physical-based virtual environment. *Build. Environ.* **2019**, *162*, 106278. [CrossRef]
6. Kazemi, M.-Z.; Elamer, A.; Theodosopoulos, G.; Khatib, S. Reinvigorating research on sustainability reporting in the construction industry: A systematic review and future research agenda. *J. Bus. Res.* **2023**, *167*, 114145. [CrossRef]
7. Cortés, D.; Traxler, A.; Greiling, D. Sustainability reporting in the construction industry—Status quo and directions of future research. *Heliyon* **2023**, *9*, e21682. [CrossRef]
8. Štefanič, M.; Stankovski, V. A review of technologies and applications for smart construction. *Proc. Inst. Civ. Eng.-Civ. Eng.* **2018**, *172*, 83–87. [CrossRef]
9. Ding, L.-Y. *Digital Construction Introduction*; China Building Industry Press: Beijing, China, 2019.
10. Tian, Y.; Yang, X.; Yang, J.; Mao, K.; Yao, Y.; Liang, H. Evolution dynamic of intelligent construction strategy of coal mine enterprises in China. *Heliyon* **2022**, *8*, e10933. [CrossRef]
11. Wang, L.; Huang, X.; Zheng, R. The application of BIM in intelligent construction. *Appl. Mech. Mater.* **2012**, *188*, 236–241. [CrossRef]
12. You, Z.-J.; Wu, C. A framework for data-driven informatization of the construction company. *Adv. Eng. Inform.* **2019**, *39*, 269–277. [CrossRef]
13. Liu, S.-Y.; Su, Y.-K. Research on intelligent construction capability evaluation of road and bridge construction enterprises. *J. Eng. Manag.* **2022**, *36*, 51–56.
14. Chen, K.; Du, P.; Duan, W.-J. Research on maturity based intelligent construction capability evaluation model and application of construction enterprises. *J. Eng. Manag.* **2022**, *36*, 37–42.
15. Zhong, Y.-G.; Jia, X.-J.; Qian, Y. *System Dynamics*; Science Press: Beijing, China, 2013.

16. Yuan, H.-P.; Yang, B.-X. System dynamics approach for evaluating the interconnection performance of cross-border transport infrastructure. *J. Manag. Eng.* **2022**, *38*, 04022008. [CrossRef]
17. Malbon, E.; Parkhurst, J. System dynamics modelling and the use of evidence to inform policymaking. *Policy Stud. J.* **2023**, *44*, 454–472. [CrossRef]
18. Sun, G.-S.; Jiang, D.-L.; Zhang, X.-L. Study on the influencing factors of green construction enterprise productivity based on system dynamics. *Constr. Econ.* **2021**, *42*, 99–104.
19. Zhang, Z.; Gao, Q.-F.; Shao, S. Carbon emission scenarios of China's construction industry using a system dynamics methodology—Based on life cycle thinking. *J. Clean. Prod.* **2024**, *435*, 140457. [CrossRef]
20. Sancheta, L.-N.; Chaves, G.L.; Siman, R.-R. The use of system dynamics on urban solid waste management: A literature analysis. *Gest. Producao* **2021**, *28*, e5336. [CrossRef]
21. Zhao, H.; Dong, H.; Qu, W.-L. Constructing a dynamic risk evaluation system for environmental protection ppp projects based on improved system dynamics. *Account. Mon.* **2017**, 65–71. (In Chinese)
22. Guo, Z.; Li, L. A conceptual framework for collaborative development of intelligent construction and building industrialization. *Front. Environ. Sci.* **2022**, *10*, 904518. [CrossRef]
23. Dewit, A. Komatsu, Smart construction, Creative destruction, and Japan's robot revolution. *Asia-Pac. J. Ophthalmol.* **2015**, *13*, 2.
24. You, Z.-J.; Wu, C.; Liu, Z.-W. Review of intelligent construction research. *J. Civ. Eng. Manag.* **2022**, *39*, 82–87+139.
25. Liu, J.P.; Zhou, X.H.; Wu, Z.; Cao, L.; Feng, L.; Li, D.S. *Intelligent Construction Basic Algorithm Tutorial*; China Architecture & Building Press: Beijing, China, 2021.
26. Bala, B.-K.; Arshad, F.-M.; Noh, K.-M. *System Dynamics: Modelling and Simulation Singapore*; Springer: Berlin/Heidelberg, Germany, 2017.
27. Kazantsev, N.; Petrovskyi, O.; Müller, J. From supply chains towards manufacturing ecosystems: A system dynamics model. *Forecast Soc. Chang.* **2023**, *197*, 122917. [CrossRef]
28. Naveed, F.; Khan, K.I.A. Investigating the influence of information complexity on construction quality: A systems thinking approach. *Eng. Constr. Archit. Manag.* **2021**, *29*, 1427–1448. [CrossRef]
29. Riaz, H.; Khan, K.; Ullah, F.; Tahir, M.; Alqurashi, M.; Alsulami, B. Key factors for implementation of total quality management in construction sector: A system dynamics approach. *Ain Shams Eng. J.* **2023**, *14*, 101903. [CrossRef]

buildings

Article

Study on Resilience Evaluation for Construction Management of Major Railway Projects

Wenchao Jiang [1], Xi Zhao [2], Chaoxun Cai [3,*], Kai Chang [3], Kai Liu [2] and Yuming Liu [2]

[1] State Key Laboratory for Track Technology of High-Speed Railway, Beijing 100081, China; 13810715250@163.com
[2] School of Economics and Management, Beijing Jiaotong University, Beijing 100044, China; 20113062@bjtu.edu.cn (X.Z.); kailiuyy@bjtu.edu.cn (K.L.); ymliu@bjtu.edu.cn (Y.L.)
[3] Railway Engineering Research Institute, China Academy of Railway Sciences Corporation Limited, Beijing 100081, China; 18513583648@163.com
* Correspondence: q15515267015@gmail.com

Abstract: The construction of major railway projects poses significant risks, which present considerable challenges to construction management. To accurately assess the level of construction management for these projects, this study incorporated resilience theory into the field. The grounded theory method was utilized to establish a resilience evaluation indicator system for managing a major railway project construction. Additionally, a resilience evaluation model based on the Analytic Hierarchy Process (AHP) and fuzzy comprehensive evaluation method was proposed. This model was applied to evaluate the construction management resilience of a major railway project located in the mountainous region of southwest China. The results indicated that the project exhibits a very high overall level of construction management resilience. Specifically, it demonstrates high levels of ability to monitor and warn, an ability to resist absorption, and an ability to respond to emergencies. Additionally, it showcases high levels of ability to recover and rebuild, and an ability to learn to adapt. The evaluation results were consistent with the actual situation and verified the correctness and reliability of the method. Based on the aforementioned research findings, this paper puts forward recommendations on material redundancy and resource security from a resource perspective, and suggestions on organizational optimization and personnel capacity improvement from a subject perspective, thus indicating directions for enhancing the management level of major engineering railway constructions.

Keywords: construction management resilience; major railway; evaluation model; empirical study

Citation: Jiang, W.; Zhao, X.; Cai, C.; Chang, K.; Liu, K.; Liu, Y. Study on Resilience Evaluation for Construction Management of Major Railway Projects. *Buildings* **2024**, *14*, 732. https://doi.org/10.3390/buildings14030732

Academic Editors: Derek Clements-Croome and Paulo Santos

Received: 27 January 2024
Revised: 29 February 2024
Accepted: 6 March 2024
Published: 8 March 2024

1. Introduction

The construction and management of major railway projects face significant challenges due to the complex environmental characteristics [1]. Firstly, the geological environment's complexity not only increases the construction difficulty but also demands higher levels of stability and safety for the project [2]. Secondly, the extreme harshness of the climate environment poses another problem in railway engineering construction [3]. For instance, under extreme climatic conditions, such as high temperatures, low temperatures, and strong winds, railway equipment and structures are susceptible to damage, compromising the safety and stability of railway operations. Thirdly, the vulnerability of the ecological environment requires sufficient attention during railway project construction. Additionally, a variety of disaster risks, including climatic disasters, geological disasters, engineering disasters, and plateau disasters, can significantly impact railway projects [4].

Currently, most studies on major railway engineering construction rely on traditional risk prevention and control methods [5–7]. These studies primarily focus on risk management, emphasizing a pre-construction analysis of explicit risk sources, predicting the probability and impact of risks, and formulating risk plans and contingency measures in advance to control and mitigate damage caused by disasters within an acceptable range. However, major railway projects differ significantly from general engineering and construction projects. The natural environment in which major railway projects are undertaken possesses multiple high-energy characteristics, and our understanding of it is still not comprehensive enough [8,9]. This implies that further improvement is needed in our comprehension of natural environmental conditions, potential risk factors, and interactions faced by the project. Particularly, interactions with construction activities may give rise to various hidden hazards that are difficult to accurately portray and predict using traditional risk management methods.

In contrast to the traditional risk management concept, crisis management under the resilience perspective emphasizes the adaptability and sustainable development of the engineering construction system during crises. The objective is to enhance the system's resilience capabilities in the face of shocks, ensuring uninterrupted and stable engineering and construction activities without breaching the system's resilience threshold [10]. As shown in Table 1. Therefore, this study aims to introduce resilience theory into the construction management of significant railway projects, facilitating the shift in project management philosophy from risk control to risk adaptation. Building upon an analysis of the composition of construction management for major railway projects and a comprehensive understanding of resilience management in existing projects, the study proposes that the core concept of resilience in construction management is the system's ability to prevent, absorb, adapt to, and recover from risk impacts. This allows the construction management system to maintain its characteristics, functions, and operational modes, even exceeding its pre-impact state. Furthermore, this paper integrates the Analytic Hierarchy Process (AHP) and the fuzzy comprehensive evaluation (FCE) method to develop a systematic and comprehensive resilience assessment system for major railway construction project management. It also proposes targeted recommendations for improvement, enhancing the adaptability and sustainability of the construction project system in dealing with explicit and implicit risk impacts. This aligns with the management needs of major railway construction project systems and bears significant theoretical research significance and practical application value.

Table 1. The difference between traditional risk management and resilience management.

Approaches	Advantages	Disadvantages
Traditional Risk Management	When engineering project uncertainty is low, engineering construction risks can often be accurately identified, controlled, and even avoided.	When engineering project uncertainty is high, engineering construction risks often exceed cognition and are difficult to accurately identify.
Resilience Management	When engineering project uncertainty is high, the improvement of system resilience can help it adapt, recover, and even surpass its own steady state during the shock of unknown risks.	When engineering project uncertainty is low, the system may not suffer shocks, and excessive resilience will cause a large waste of resources.

The remainder of this paper is organized as follows. Section 2 reviews the related literature. Section 3 proposes a resilience evaluation indicator system for the construction management of major railway projects. Section 4 constructs a resilience evaluation model for major railway construction management. Section 5 conducts an empirical analysis based on the model. Section 6 discusses the results. Finally, Section 7 presents the conclusions, recommendations for resilience enhancement, and the limitations of this study.

2. Literature Review

2.1. Engineering Resilience

The concept of engineering resilience has its origins in ecological resilience [11], and has since been adopted by scholars in the field of engineering construction management. In this context, robustness can be seen as synonymous with engineering resilience [12]. The core principle of engineering resilience is to enable major engineering or infrastructure systems to maintain their core functions in the face of internal or external shocks or conflicts, and to recover quickly from unforeseen crises [13,14]. Key elements of engineering resilience include monitoring and warning capabilities, resistance to absorption, recovery and reconstruction, and adaptive learning in major projects [15]. Bruneau et al. developed a framework for resilience capabilities that includes robustness, resourcefulness, redundancy, and rapidity [16]. Li and Lence highlighted the importance of engineering resilience in the context of water resource systems, specifically the ability to recover from failure and return to a safe state [17]. Building on the 4R framework of resilience developed by Bruneau et al., a resilience capacity triangle model was proposed that consists of absorptive capacity, adaptive capacity, and recoverability [12]. Vugrin et al. focused on the ability to restore engineering resilience after a disturbance, acknowledging that certain crises cannot be avoided in the current environment [18]. Therefore, resilience emphasizes the development of flexibility and adaptability in critical engineering and infrastructure systems in order to maintain minimal functionality during crises, and to recover as quickly as possible afterwards. The introduction of this concept has led to a shift in the critical infrastructure industry, moving from an attitude of simply protecting critical infrastructure to actively improving its resilience [19,20].

2.2. Resilience Evaluation

With the increasing research on the meaning and conceptual framework of resilience in major infrastructure engineering, quantifying engineering resilience and evaluating the level of system engineering resilience has become a focal point for researchers. Resilience evaluation research stems from well-established risk assessment research, but there are clear differences between the two. Risk assessment is a common method for quantifying potential risks and reducing their occurrence, focusing on measuring the potential losses to the system [21]. In contrast, resilience evaluation emphasizes measuring the system's ability to withstand and recover from adaptive disturbances [22–24]. Major infrastructure projects, as complex adaptive systems, are characterized by significant complexity and uncertainty. Under the influence of internal and external disturbances, system losses are difficult to avoid, which is precisely the challenge that resilience evaluation addresses. In terms of the time dimension, resilience evaluation focuses more on the timing of resilience activities rather than analyzing the entire process of risk occurrence as in risk assessment studies. For example, it involves reconfiguring the organizational structure and resources to cope with the impact of internal and external disturbances within the system. Additionally, conducting regular internal vulnerability analyses of the system is crucial for cultivating the system's adaptive capacity and enhancing its resilience [25–28].

Existing resilience evaluations of major infrastructure projects have mainly been conducted from two perspectives: the single dimension and the network dimension. Evaluations from a unidimensional perspective focus on a specific major infrastructure project, identifying weak points within the system to improve reliability and reduce vulnerability by assessing the resilience level of that particular infrastructure [29,30]. Resilience evaluations from a network dimension perspective are concentrated in the field of urban resilience. The research primarily focuses on evaluating the importance and interconnectivity of each infrastructure node within the resilience network composed of multiple lifeline infrastructure projects. This provides a basis for enhancing the resilience of the infrastructure network and proposing strategies to improve the resilience of lifeline infrastructure systems [31,32].

Regarding resilience assessment methods, existing mainstream methods can be categorized into three types: qualitative, quantitative, and a combination of both. Qualitative research mainly employs conceptual framework modeling, constructing models based on relevant literature and works in the field of resilience. This approach effectively elucidates the correlation and relationship between system resilience and key internal elements at a macro level [33]. The data involved in this method are typically textual, and are thus less costly. However, due to the lack of data information, it is challenging to accurately describe the changes in internal parameters of technical systems. Empirical analysis is another popular qualitative assessment method, constructing qualitative indicators based on the practical experience of typical engineering projects to evaluate the resilience of infrastructure [34,35]. Researchers using this method often have close cooperation with the construction and construction units of the engineering projects they study, allowing them to accurately refine resilience indicators and effectively and realistically assess the resilience level of projects. However, this method tends to focus on a specific engineering project, and its evaluation results may lack universality. Quantitative research on resilience assessment mainly utilizes tools such as Bayesian networks [36,37], Monte Carlo simulation [38,39], machine learning, and nonlinear dynamic analysis based on the OpenSees platform [40–42]. These tools are used to simulate the characteristics and behavior of the system for nonlinear dynamic analysis to assess the system resilience [43,44]. Quantitative research methods significantly improve the accuracy and reliability of resilience assessments by conducting various predictive simulation experiments based on a large dataset of infrastructure system structural properties and impact actions and selecting the most effective models. The main challenges of this approach lie in the complexity of the models, the time-consuming model construction and simulation processes, and the demand for a large amount of technical, managerial, and organizational data. Finally, the method combining qualitative and quantitative assessments of resilience has become increasingly popular in recent years. These methods involve assigning indicators representing resilience based on the nature of the system and its operating conditions and calculating the resilience index using a weighted average [45]. This approach combines a large amount of objective data and expert experience from relevant fields, yielding scientifically reasonable results with moderate implementation difficulty, but it also has its limitations. Indicator construction may be limited by historical data, and expert assessments as a significant source of data may introduce considerable subjectivity.

3. Resilience Evaluation Indicator System for Major Railway Project Construction Management

3.1. Methodology for the Selection of Evaluation Indicators

The resilience of major railway project construction management is a comprehensive and complex concept characterized by dynamism and evolution. Relying solely on statistical research techniques can lead to the exclusion of qualitative data and overlook hidden details. To address this, this study employed the semi-structured field interview method to collect qualitative data from various stakeholders involved in major railway projects, including owners, constructors, designers, and supervisors. Grounded theory research was then utilized to analyze the qualitative material, aiming to clarify, illustrate, and systematize the resilience of construction management iteratively and deductively in major railway projects. The objective was to enhance the evaluation system for resilience in major railway construction management based on the field-collected data.

The data collection process spanned 17 months and involved multiple symposiums with each respondent from major railway project owners, construction units, and scientific research units. These symposiums addressed the topics outlined in the interview guide, as well as the actual situation of the projects. Following the completion of the interviews, audio and video recordings were transcribed into written materials, resulting in a total of 40 interview memos. A random selection of 30 memos was chosen for coding analysis, while the remaining 10 were used to test theoretical saturation.

3.2. Determination of Evaluation Indicators

This research was based on the analysis of 40 interview transcripts and incorporates existing literature resources, such as journals, dissertations, newspapers, and monographs, from both domestic and international sources. The objective was to identify the key factors that influence major railway project construction management or construction management resilience. During the coding process, relevant literature was consulted to assign names and definitions to these factors [46–49]. The interview records, totaling 50,000 words, were analyzed using Nvivo 11 software, which employed open coding, axial coding, and selective coding techniques. Initially, initial concepts were extracted and refined from the organized interview records, resulting in 23 categories. Subsequently, utilizing the coding feature of Nvivo 11, these 23 concept categories were classified based on their inherent connections, leading to the identification of 12 main categories. Through repeated discussions and debates, employing the steps of selective coding, including outlining storylines, describing primary and secondary categories along with their relevant dimensional attributes, and establishing connections between core and other categories, five core categories were ultimately determined. Table 2 summarizes these core categories and their definitions regarding the influential factors of major railway project construction management resilience. Subsequently, the remaining 10 interview records underwent independent coding, which did not introduce any new concepts or categories, thereby satisfying the requirements for theoretical saturation testing.

Table 2. Encoding of factors influencing resilience of major railway construction management.

Core Categories	Concept Description	Main Category	Category
Ability to monitor early warning	During the construction of major railway projects, advanced and reliable technological methods, such as big data, cloud computing, sensors, and intelligent terminals, are employed to gather information on construction safety, ecological environment, geological hazards, and occupational health risks and hazards. Dynamic tracking and monitoring activities are conducted to proactively identify and monitor potential risks before they have a significant impact, thereby providing valuable support for risk analysis. Additionally, a well-established risk assessment system is implemented, utilizing the monitoring data collected on-site to conduct comprehensive analysis and evaluation. With the aid of a risk warning platform, modern information technology is utilized to intelligently disseminate safety warnings.	Shock Monitoring	Natural and geological Hazards monitoring
			Production safety inspection
			Occupational health and safety management
			Construction environment and water protection monitoring and inspection
			Material security monitoring and inspection
		Information Processing	Processing and analysis of various types of monitoring data
		Shock Warning	Early warning platforms and mechanisms

Table 2. *Cont.*

Core Categories	Concept Description	Main Category	Category
Ability to resist absorption	On one hand, the ability to resist absorption refers to the system's capacity to promptly develop emergency plans and issue command instructions upon the occurrence of major internal or external risks. This involves mobilizing all departments to respond to the emergency, coordinating the deployment of personnel and resources, and collaborating with external organizations to collectively address the situation. On the other hand, it pertains to the capability of ongoing construction activities, as well as the equipment and facilities on-site, to withstand risk impacts and prevent or minimize casualties and property damage. It also encompasses the system's ability to sustain its construction functions and mitigate losses in the face of regular risk events.	Shock Resistance	Rapid decision-making response and organizational mechanisms
		Shock Absorption	Occupational health protection facilities and medical personnel
			On-site redundancy of key construction equipment
			Repair and maintenance of large machinery and equipment
			Redundant stockpiles of important key materials
Ability to respond to emergencies	In order to cope with the extremely high-risk challenges faced by the construction management of major railway projects, the construction site and rescue bases are strengthened through the construction of hardware and software rescue facilities and equipment, with the ability to enter, rescue, transport, and treat after a disaster occurs.	Emergency organization	Emergency rescue leadership team and rescue team
		Emergency Management	Emergency plan and emergency response
			Three-tier emergency medical treatment and transportation systems
		Emergency Facilities and Materials	Emergency rescue equipment configuration
Ability to recover and rebuild	After the occurrence of risk impact, the ability to quickly mobilize engineering machinery and equipment and materials to arrive at the construction site, quickly clean up the construction site, and quickly restore the damaged engineering entity.	Resource Recovery	Rapid restoration of key equipment or rapid arrival at the site
			Rapid restoration of construction roads, power and communication networks
		Engineering Recovery	Rapid cleanup of the site
			Rapid resumption of normal construction activities
Ability to adapt to learning	After a shock occurs, through summarizing, refining, and reflecting on the practical experience of coping with risky shocks, and drawing on the lessons learned from other risky shock coping experiences and relevant theories, it optimizes the organizational structure and operational mechanism, and trains field staff in crisis cognition and shock-handling ability, so as to enhance the system's overall ability to cope with shocks and adapt to changes in the external environment.	Lessons Learned	Summarizing and improving experience and learning exchange
			Organization and performance of emergency drills
		Adaptation	Plateau acclimatization and technical training

3.3. Indicator System for Resilience Evaluation

By summarizing and refining the influencing factors of construction management resilience through grounded theoretical research methods, the resilience evaluation indicator system of major railway project construction management was formed as shown in Figure 1.

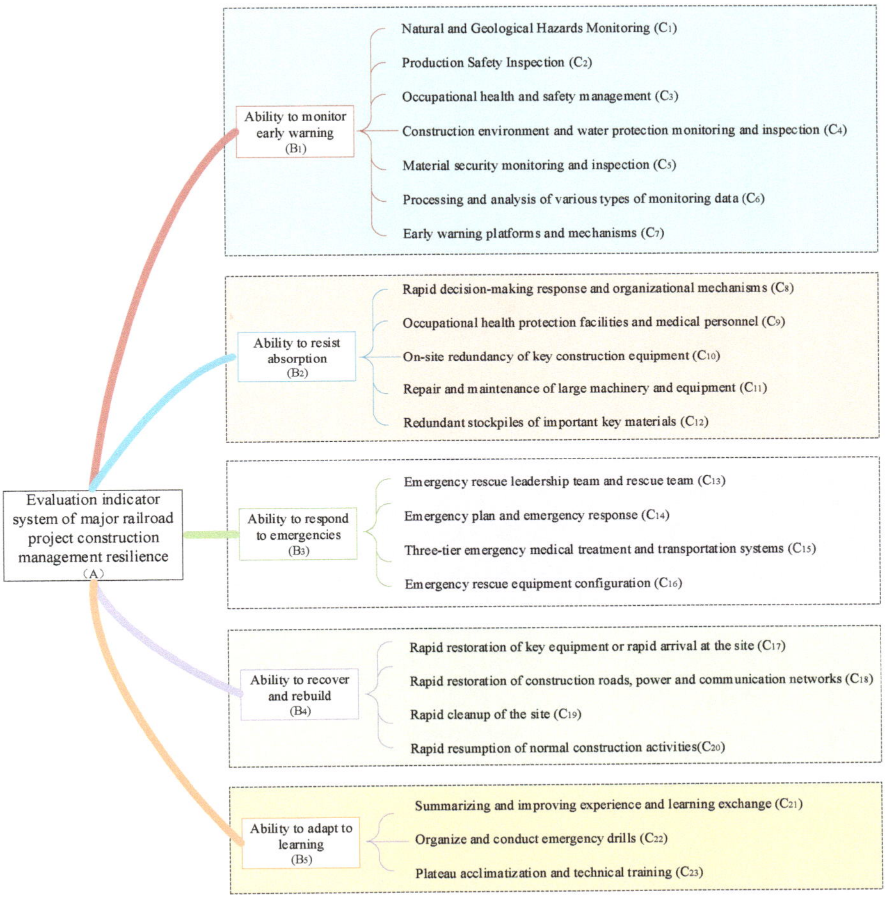

Figure 1. Resilience evaluation indicator system of major railway project construction management.

4. Resilience Evaluation Model for Major Railway Construction Management

4.1. Calculation of Evaluation Indicator Weights

This study used the AHP to determine the weights of indicators. AHP first hierarchizes a problem, decomposes it into different constituent factors according to the nature of the problem and the overall goal to be achieved, and forms different levels of aggregation combinations in accordance with the interrelated influences and affiliations among the factors, so as to constitute a multilevel analytical structural model [50,51]. The scale method of a 1 to 9 scale was used to judge the importance of two elements in the matrix for two-by-two comparison, and to assign a value to the degree of importance. From the quantitative results of the scale-constructed two-by-two comparison of the judgment matrix, the judgment matrix for single sorting calculation to determine the weight of each indicator was created. The weight calculation was firstly normalized, and the number of indicators in the group was set to be n.

Step 1: Compute the product M_i of the elements of each row of the matrix.

$$M_i = \prod_{j=1}^{n} a_{ij} \quad i = 1, 2, \cdots, n \tag{1}$$

Step 2: Calculate the nth root $\overline{W}_i$ of M_i.

$$\overline{W}_i = \sqrt[n]{M_i} \quad i = 1, 2, \cdots, n \tag{2}$$

Step 3: Normalize the vector $\overline{W} = [\overline{W}_1, \overline{W}_2, \cdots, \overline{W}_n]^T$.

$$\overline{W}_i = \frac{\overline{W}_i}{\sum\limits_{i=1}^{n} \overline{W}_i} \quad i = 1, 2, \cdots, n \tag{3}$$

$W = [W_1, W_2, \cdots, W_n]^T$ is the desired weight vector.

Step 4: In order to ensure the scientific and reliability of the calculation results, the judgment matrix must be tested for consistency.

$$CI = \frac{\lambda_{\max} - n}{n - 1} \tag{4}$$

$$CR = \frac{CI}{RI} \tag{5}$$

where $\lambda_{\max}$ is the largest eigen value of the judgment matrix, n is the judgment matrix order, and RI is the average random consistency value corresponding to n. When $CR < 0.1$, it was considered that the judgment matrix had good consistency. Otherwise, the judgment matrix elements should be adjusted.

This study utilized the expert scoring method to collect questionnaires from experienced experts in the field of construction management for major railway projects. The purpose was to determine the relative importance of each rating indicator. A total of 40 questionnaires were collected, and a rigorous screening process was employed to ensure the scientific validity and applicability of the Analytic Hierarchy Process (AHP) method, resulting in the exclusion of invalid questionnaires that exhibited inconsistencies in two-by-two comparisons, had excessively short response times, or consecutively selected the same options. Ultimately, 30 valid questionnaires were retained for the AHP analysis. By employing the AHP method, the significance of each group of indicators was compared and scored by the 30 participating experts. Calculation resulted in obtaining 30 weight vectors for each group of indicators, and the average values were utilized to determine the weights of technical indicators at all levels within the construction management resilience system for major railway projects. The significance attributed to the ability to monitor early warning (B_1) was quantified at 28.41 percent, with its subordinate indicators (C_1 through C_7) assigned weights of 4.18 percent, 5.00 percent, 3.32 percent, 3.31 percent, 3.34 percent, 3.33 percent, and 3.33 percent, respectively. The ability to resist absorption (B2) was evaluated at 24.88 percent, with its secondary indicators (C_8 through C_{12}) having weights of 8.35 percent, 3.33 percent, 3.31 percent, 3.32 percent, and 3.33 percent, respectively. The ability to respond to emergencies (B_3) held a weight of 28.41 percent, with its associated indicators (C_{13} through C_{16}) weighted at 6.67 percent, 5.83 percent, 3.33 percent, and 7.50 percent, respectively. The ability to recover and rebuild (B_4) was determined to be 11.29 percent, with the weights for its secondary indicators (C_{17} through C_{20}) set at 4.18 percent, 4.17 percent, 4.17 percent, and 4.19 percent, respectively. The ability to adapt to learning (B_5) was assessed at 7.01 percent, with each of its secondary indicators

(C_{21}, C_{22}, and C_{23}) uniformly weighted at 4.17 percent. The obtained weights are presented in Table 3.

Table 3. Weights of resilience indicators for major railway construction management.

First Level Indicator	First Level Indicator Weight	Second Level Indicator	Second Level Indicator Weight
B_1	28.41%	C_1	4.18%
		C_2	5.00%
		C_3	3.32%
		C_4	3.31%
		C_5	3.34%
		C_6	3.33%
		C_7	3.33%
B_2	24.88%	C_8	8.35%
		C_9	3.33%
		C_{10}	3.31%
		C_{11}	3.32%
		C_{12}	3.33%
B_3	28.41%	C_{13}	6.67%
		C_{14}	5.83%
		C_{15}	3.33%
		C_{16}	7.50%
B_4	11.29%	C_{17}	4.18%
		C_{18}	4.17%
		C_{19}	4.17%
		C_{20}	4.19%
B_5	7.01%	C_{21}	4.17%
		C_{22}	4.17%
		C_{23}	4.17%

4.2. Fuzzy Integrated Evaluation

Fuzzy comprehensive evaluation method was used for the resilience evaluation for construction management of major railway projects [52,53], and its main steps are shown in Figure 2.

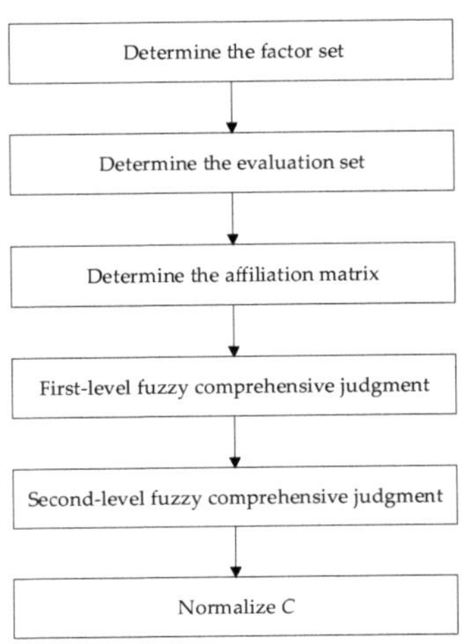

Figure 2. Step of fuzzy integrated evaluation.

Step 1: Determine the factor set. Divide the factor set of fuzzy comprehensive judgment of resilience of major railway project construction management into n sub-factor sets to obtain the resilience factor set U, $U = \{U_1, U_2, \cdots, U_n\}$.

Step 2: Determine the evaluation set. Due to the different levels of major railway project construction management resilience, the evaluation set is composed of all possible judgments. Define the evaluation level of major railway project construction management resilience as {very low, low, average, high, very high} five levels, and perform one comprehensive fuzzy judgment V for each U_i denoted as $V = \{v_1, v_2, v_3, v_4, v_5\}$. In order to facilitate the statistical calculation, we quantify the semantics scale of subjective evaluation and assign the values of 5, 4, 3, 2, and 1 in turn.

Step 3: Determine the affiliation matrix. The fuzzy relationship between the set of resilience factors U and the evaluation set V is represented, and the matrix element representing U_n evaluates its affiliation vector belonging to the mth rubric of V as $R_n = \{r_{n1}, r_{n2}, \cdots, r_{nm}\}$, to obtain the affiliation matrix R.

$$R = \begin{bmatrix} r_{11} & r_{12} & \cdots & r_{1m} \\ r_{21} & r_{22} & \cdots & r_{2m} \\ \vdots & \vdots & \cdots & \vdots \\ r_{n1} & r_{n2} & \cdots & r_{nm} \end{bmatrix} \tag{6}$$

Step 4: First-level fuzzy comprehensive judgment. Calculate the affiliation matrix R_1 of the seven influencing factors of ability to monitor early warning (B_1), the affiliation matrix R_2 of the five influencing factors of ability to resist absorption (B_2), the affiliation matrix R_3 of the four influencing factors of ability to respond to emergencies (B_3), the affiliation matrix R_4 of the four influencing factors of ability to recover and rebuild (B_4), and the affiliation matrix R_5 of the three influencing factors of ability to adapt to learning (B_5). Then, perform the synthesis operation with the relative weights of the second-level indicators W_i to obtain the first-level fuzzy comprehensive evaluation matrix.

$$B = W_i * \mathrm{R}_i \tag{7}$$

Step 5: Second-level fuzzy comprehensive judgment. According to Equation (7), the first-level fuzzy comprehensive judgment matrix B is obtained, and the importance weights of the first-level indicators W are synthesized to obtain the second-level fuzzy comprehensive judgment vector.

$$C = W * B \tag{8}$$

Step 6: Normalize C to obtain the comprehensive evaluation results of the construction management resilience of major railway projects.

5. Empirical Research

5.1. Data Collection

This study focused on a major railway project located in the mountainous region of southwest China. It involved the participation of 80 senior managers and experts directly involved in the construction of this project. Their task was to rate the resilience indicators using a 1–5 scoring method. The respondents included 50 experts in the field of engineering construction management, accounting for 62.5%, and 30 individuals in engineering risk management, representing 37.50%. Among them, there were 4 project managers (5%), 10 department heads (12.5%), 36 project engineers (45%), and 30 supervising engineers (37.5%). Over half of the respondents had a working experience of more than 5 years. For more details, see Table 4. To ensure the scientific rigor and accuracy of the fuzzy comprehensive evaluation results, stringent measures were taken to obtain high-quality data. The 80 collected questionnaires underwent a thorough screening, eliminating invalid ones that had short response times or consecutively selected the same options. Consequently, 78 valid questionnaires were retained for the subsequent fuzzy comprehensive evaluation.

The reliability test conducted on the questionnaire data yielded a result of 0.997, surpassing the required threshold with exceptional standards. This high level of reliability not only validates the research findings from a mathematical analysis perspective but also attests to the validity of the research results.

Table 4. Information on respondents.

Basic Information on Experts	Category	Number of Persons	Percentage
Years of work/research	Less than 5 years	22	27.50%
	5–10 years	28	35.00%
	More than 10 years	30	37.50%
Work unit	Project Owner	5	6.25%
	Design Unit	26	32.50%
	Construction Contractor	45	56.25%
	Research Institutes	4	5.00%
Research Field	Engineering Construction Management	50	62.50%
	Engineering Risk Management	30	37.50%
Position/Title	Project Manager	4	5.00%
	Department Heads	10	12.50%
	Project Engineer	36	45.00%
	Engineer-in-Charge	30	37.50%

5.2. Fuzzy Comprehensive Evaluation for Construction Management Resilience of Major Railway Projects

Based on the evaluation results of the single-factor indicators, the judgment matrix of the three-level indicators for ability to monitor early warning (B_1) was collated.

$$R_{B_1} = \begin{bmatrix} 0 & 0 & 0.1 & 0.5 & 0.4 \\ 0 & 0 & 0 & 0.4 & 0.6 \\ 0 & 0 & 0.2 & 0.5 & 0.3 \\ 0 & 0 & 0.2 & 0.2 & 0.6 \\ 0 & 0 & 0.2 & 0.4 & 0.4 \\ 0 & 0 & 0.2 & 0.4 & 0.4 \\ 0 & 0.2 & 0.4 & 0.4 & 0 \end{bmatrix}$$

Similarly, the same method was used to process the data for the tertiary indicators of B_2, B_3, B_4, and B_5, resulting in the following judgment matrix.

$$R_{B_2} = \begin{bmatrix} 0 & 0 & 0.1 & 0.5 & 0.4 \\ 0 & 0 & 0.4 & 0.6 & 0 \\ 0 & 0.1 & 0.4 & 0.5 & 0 \\ 0 & 0.1 & 0.3 & 0.6 & 0 \\ 0 & 0 & 0.1 & 0.3 & 0.6 \end{bmatrix}$$

$$R_{B_3} = \begin{bmatrix} 0 & 0 & 0.2 & 0.3 & 0.5 \\ 0 & 0 & 0.2 & 0.3 & 0.5 \\ 0 & 0 & 0.2 & 0.5 & 0.3 \\ 0 & 0 & 0 & 0.4 & 0.6 \end{bmatrix}$$

$$R_{B_4} = \begin{bmatrix} 0 & 0 & 0.2 & 0.6 & 0.2 \\ 0 & 0.2 & 0.2 & 0.4 & 0.2 \\ 0 & 0 & 0.2 & 0.6 & 0.2 \\ 0 & 0 & 0.2 & 0.6 & 0.2 \end{bmatrix}$$

$$R_{B_5} = \begin{bmatrix} 0 & 0.2 & 0.4 & 0.4 & 0 \\ 0 & 0.1 & 0.5 & 0.4 & 0 \\ 0 & 0.1 & 0.3 & 0.6 & 0 \end{bmatrix}$$

The fuzzy comprehensive evaluation results were obtained from the indicator weight vector A and the fuzzy relationship matrix R using the fuzzy operation, and the commonly used fuzzy operators were the Zadeh operator, weighted average type $M(\cdot, +)$, the Einstein operator, the Hamacher operator, and the Yager operator. In this study, the weighted average operator was used to calculate the results of five toughness capacity evaluations using the matrix multiplication operator in MATLAB R2023a software.

In order to presume the evaluation results more finely and intuitively, fuzzy vector singularization can be implemented, combining the value assigned to each grade rubric, $V = \{V_k\} = \{V_1, V_2, V_3, V_4, V_5\}$ = {very low, low, average, high, very high}, where the score for V_1 is 1, the score for V_2 is 2, the score for V_3 is 3, the score for V_4 is 4, and the score for V_5 has a score of 5, constructed as column vector $V^T = (1, 2, 3, 4, 5)$. Then, we calculated $G = U \times V^T$, which finally resulted in the fuzzy comprehensive evaluation scores of the five toughness abilities. The specific process was as follows.

$$B_1 = A_1 \bullet R_{B_1} = (0.0418, 0.0500, 0.0332, 0.0331, 0.0334, 0.0333, 0.0333) \bullet \begin{bmatrix} 0 & 0 & 0.1 & 0.5 & 0.4 \\ 0 & 0 & 0 & 0.4 & 0.6 \\ 0 & 0 & 0.2 & 0.5 & 0.3 \\ 0 & 0 & 0.2 & 0.2 & 0.6 \\ 0 & 0 & 0.2 & 0.4 & 0.4 \\ 0 & 0 & 0.2 & 0.4 & 0.4 \\ 0 & 0.2 & 0.4 & 0.4 & 0 \end{bmatrix}$$

$$= (0.000, 0.026, 0.171, 0.400, 0.403)$$

$$G_{B_1} = U_{B_1} \bullet V_{B_1}^T = (0.000 \quad 0.026 \quad 0.171 \quad 0.400 \quad 0.403) \bullet \begin{pmatrix} 1 \\ 2 \\ 3 \\ 4 \\ 5 \end{pmatrix} = 4.181$$

The fuzzy comprehensive evaluation was carried out for the seven resilience tertiary level 3 indicators in the ability to monitor early warning (B_1), as well as for the five comment sets, and the $M(\cdot, +)$ operator. We then calculated $G = U \times V^T$, which finally resulted in a fuzzy comprehensive evaluation score of monitoring and early warning capacity B_1 of G_{B_1} = 4.181, with an evaluation grade of "very high".

$$B_2 = A_2 \cdot R_{B_2} = (0.0835, 0.0333, 0.0331, 0.0332, 0.0333) \bullet \begin{bmatrix} 0 & 0 & 0.1 & 0.5 & 0.4 \\ 0 & 0 & 0.4 & 0.6 & 0 \\ 0 & 0.1 & 0.4 & 0.5 & 0 \\ 0 & 0.1 & 0.3 & 0.6 & 0 \\ 0 & 0 & 0.1 & 0.3 & 0.6 \end{bmatrix}$$

$$= (0.000, 0.031, 0.223, 0.247, 0.500)$$

$$G_{B_2} = U_{B_2} \bullet V_{B_2}^T = (0.000 \quad 0.031 \quad 0.223 \quad 0.247 \quad 0.500) \bullet \begin{pmatrix} 1 \\ 2 \\ 3 \\ 4 \\ 5 \end{pmatrix} = 4.216$$

$$B_3 = A_3 \cdot R_{B_3} = (0.0667, 0.0583, 0.0333, 0.0750) \bullet \begin{bmatrix} 0 & 0 & 0.2 & 0.3 & 0.5 \\ 0 & 0 & 0.2 & 0.3 & 0.5 \\ 0 & 0 & 0.2 & 0.5 & 0.3 \\ 0 & 0 & 0 & 0.4 & 0.6 \end{bmatrix}$$

$$= (0.000, 0.000, 0.136, 0.361, 0.504)$$

$$G_{B_3} = U_{B_3} \bullet V_{B_3}^T = \begin{pmatrix} 0.000 & 0.000 & 0.136 & 0.361 & 0.504 \end{pmatrix} \bullet \begin{pmatrix} 1 \\ 2 \\ 3 \\ 4 \\ 5 \end{pmatrix} = 4.368$$

$$B_4 = A_4 \cdot R_{B_4} = (0.0418, 0.0417, 0.0417, 0.0419) \bullet \begin{bmatrix} 0 & 0 & 0.2 & 0.6 & 0.2 \\ 0 & 0.2 & 0.2 & 0.4 & 0.2 \\ 0 & 0 & 0.2 & 0.6 & 0.2 \\ 0 & 0 & 0.2 & 0.6 & 0.2 \end{bmatrix}$$
$$= (0.000, 0.050, 0.200, 0.550, 0.200)$$

$$G_{B_4} = U_{B_4} \bullet V_{B_4}^T = \begin{pmatrix} 0.000 & 0.050 & 0.200 & 0.550 & 0.200 \end{pmatrix} \bullet \begin{pmatrix} 1 \\ 2 \\ 3 \\ 4 \\ 5 \end{pmatrix} = 3.900$$

$$B_5 = A_5 \cdot R_{B_5} = (0.0417, 0.0417, 0.0417) \bullet \begin{bmatrix} 0 & 0.2 & 0.4 & 0.4 & 0 \\ 0 & 0.1 & 0.5 & 0.4 & 0 \\ 0 & 0.1 & 0.3 & 0.6 & 0 \end{bmatrix}$$
$$= (0.000, 0.133, 0.400, 0.467, 0.000)$$

$$G_{B_5} = U_{B_5} \bullet V_{B_5}^T = \begin{pmatrix} 0.000 & 0.133 & 0.400 & 0.467 & 0.000 \end{pmatrix} \bullet \begin{pmatrix} 1 \\ 2 \\ 3 \\ 4 \\ 5 \end{pmatrix} = 3.333$$

Similarly, the fuzzy comprehensive evaluation score G_{B2} = 4.216 for the ability to resist absorption (B_2) had an evaluation grade of "very high". The fuzzy comprehensive evaluation score G_{B3} = 4.368 for ability to respond to emergencies (B_3) had a rating of "very high". The fuzzy composite evaluation score G_{B4} = 3.900 for ability to recover and rebuild (B_4) had an evaluation grade of "high". The fuzzy composite evaluation score G_{B5} = 3.333 for ability to adapt to learning (B_5) had an evaluation rating of "high".

$$S_1 = \omega_1^* R_1 = \begin{pmatrix} 0.2841 & 0.2488 & 0.2841 & 0.1129 & 0.0701 \end{pmatrix} * \begin{bmatrix} 0.000 & 0.026 & 0.171 & 0.400 & 0.403 \\ 0.000 & 0.031 & 0.223 & 0.247 & 0.500 \\ 0.000 & 0.000 & 0.136 & 0.361 & 0.504 \\ 0.000 & 0.050 & 0.200 & 0.550 & 0.200 \\ 0.000 & 0.133 & 0.400 & 0.467 & 0.000 \end{bmatrix}$$
$$= \begin{pmatrix} 0.000 & 0.030 & 0.193 & 0.404 & 0.372 \end{pmatrix}$$

$$G_1 = U_1 \bullet V_1^T = \begin{pmatrix} 0.000 & 0.030 & 0.193 & 0.404 & 0.372 \end{pmatrix} \bullet \begin{pmatrix} 1 \\ 2 \\ 3 \\ 4 \\ 5 \end{pmatrix} = 4.119$$

The fuzzy synthesis results of the indicators of B_1, B_2, B_3, B_4, and B_5 were summarized to obtain the evaluation matrix of the resilience management of a major railway project and we carried out a fuzzy synthesis evaluation of the same type, with the result being G_1 = 4.119, with a rating of very high. Therefore, the evaluation results of the construction management resilience of a major railway project in the mountainous area of southwest China could be obtained as shown in Table 5.

Table 5. Results of the management resilience evaluation.

Resilience System	Score	Evaluation Rating	Resilience Capability	Score	Evaluation Rating
Construction management system resilience (A_1)	4.119	very high	Ability to monitor early warning (B_1)	4.181	very high
			Ability to resist absorption (B_2)	4.216	very high
			Ability to respond to emergencies (B_3)	4.368	very high
			Ability to recover and rebuild (B_4)	3.900	high
			Ability to adapt to learning (B_5)	3.333	high

6. Discussion

(1) Ability to respond to emergencies

The ability to respond to emergencies scored the highest, with a very high rating. This result not only demonstrates the on-site managers' commitment to the safety of construction personnel, but also highlights the potential dangers of major railway construction activities. In fact, compared to other railway projects in plain and coastal areas in China, the mountainous areas in Southwest China are more prone to high-intensity earthquakes and active ruptures, rock bursts, large deformations, high geothermal temperatures and hot water, noxious gases, granite radioactivity, and sudden water and mud flows, which pose a very high risk to the safety of the project. The project established a "hole rapid self-help, effective self-help section, provincial and regional base professional rescue, local social linkage rescue" three-tier emergency rescue system, and equipped it with more emergency rescue equipment and small emergency rescue special tools than other railway projects in the plains and developed coastal areas, following the principle of "combining peace and war".

(2) Ability to resist absorption

The ability to resist absorption was rated second, with a very high rating. Resilience reflects the reliability and redundancy of the construction management system to cope with disasters, and is a core capability that site managers prioritize. At the hardware level, key equipment for tunnel construction has been over-equipped for management to prevent the high-altitude environment from severely impacting mechanical equipment. Additionally, the maintenance of mechanical equipment has exceeded the normal construction state. At the software level, the company provides occupational health training to improve the construction staff's understanding of plateau construction and dispel their fear of difficulties. Health monitoring files are established for the entire workforce, and a strict health checkup system is implemented, including regular personnel changes and rotations. Furthermore, project management personnel have significant experience in constructing major railway projects, enabling quick decision-making and proper accident handling, which are crucial to the reliability of the construction management system.

(3) Ability to monitor early warning

The ability to monitor early warning ranked third, with a very high rating. A major railway project in the mountainous region of Southwest China faces various geological challenges, including great uncertainty of geological conditions, extremely large depth of burial, and extremely high gestures environment. To predict geological conditions ahead of the project, various forecasting means are employed (micro seismic monitoring, physical exploration, and over-drilling), and rock burst occurrence patterns are investigated based on construction process records. Physical surveying (ultra-long geological drilling rig, TSP203, seismic wave method) is performed to detect large deformations of soft rock in combination with geological exploration, tunnel design, and construction palm surface sketch results. For active fracture zones, geological forecasting means such as geological radar, TSP, and over-drilling are used to prejudge surrounding rock conditions. Over-drilling explores hydrogeological conditions ahead to prevent high temperature groundwater surges in high geothermal temperature warning situations. For gas and other harmful gases, the whole section of the gas section is detected in advance and the location of harmful gases is recorded. In water-rich sand layers, overrun pipe shed-grouting or overrun grouting is adopted to carry out an overrun pretreatment of the surrounding rock, combined with the overrun through the flat guide to probe the geological conditions in advance, and overrun water discharge to improve the construction conditions of the main hole.

(4) Ability to recover and rebuild

The ability to recover and rebuild scored fourth, with a high evaluation rating. For human resources recovery, a major railway project in the mountainous region of Southwest China relies on local health resources and maintains a close linkage mechanism with local medical units, highlighting medical treatment, disease prevention, and oxygen-rich construction. It establishes a targeted three-tier medical treatment and transfer system. Local military districts have jointly established post-disaster recovery teams responsible for carrying out post-rescue cleanup of the site and assisting the project in resuming normal construction activities as soon as possible. Specialized management of the construction right-of-way is set up to restore roads, electricity, communications, and other resources. Construction teams are dispatched to carry out post-disaster corrections, while local electricity and communications units join forces with the project to set up a joint team to assist with the post-disaster restoration of key resources.

(5) Ability to adapt to learning

The ability to adapt to learning was rated fifth, with a high rating. This ability was mainly reflected in all aspects of training activities for construction personnel. The project has carried out extensive publicity and education on occupational health and safety to make all employees truly realize the importance and necessity of occupational health and safety. Regular rescue and rescue training and drills for on-site emergency rescue personnel are conducted, and construction personnel are trained in rapid self-rescue skills in cave entrances.

7. Conclusions and Recommendations

7.1. Conclusions

Based on resilience theory, this study introduced the concept of construction management resilience for major railway projects. It constructed a resilience evaluation indicator system for construction management based on five dimensions: the ability to monitor early warning, the ability to resist absorption, the ability to respond to emergencies, the ability to recover and rebuild, and the ability to adapt to learning. The study proposed a resilience evaluation method for construction management using the AHP and fuzzy comprehensive evaluation approach. An empirical study was conducted on a major railway project located in the mountainous area of southwest China as an example. The validity and reliability of the research results were verified through an empirical analysis. The research results indicated that the capabilities for emergency response and early warning monitoring are paramount, both accounting for 28.41%, and are central to the resilience of large-scale railway project construction management. The empirical analysis demonstrated that the construction projects studied exhibit high resilience levels, with scores above 4 for monitoring and early warning, resistance and absorption, and emergency response capabilities. Recovery and rebuilding, along with learning and adaptation abilities, scored between 3 and 4, indicating a relatively high level of resilience. This methodology facilitated the assessment of construction management's risk resistance capabilities in major railway projects, identifying strengths and areas for improvement, thus offering necessary references and a decision-making basis for mitigating, responding to, and adapting to uncertain risk impacts.

7.2. Recommendations

(1) Advance planning of project redundant resource reserves and strengthening of on-site resource guarantee mechanism.

The project company should verify the reserve of raw materials and components required for production in advance, based on the list of material demand, and manage them in a unified manner. A designated person should be responsible for coordinating with local meteorological and road administration departments to stay updated on road flow, control, and warning information. This will ensure that materials are prepared in advance and delivered in a staggered manner, preventing road delays from affecting normal construction. Utilizing big data, artificial intelligence, and other informational tools actively can improve and optimize the existing material management digital platform, allowing for dynamic reserve and scheduling of redundant resources. This will guarantee the smooth execution of site engineering and construction activities. It is also important to strengthen the reserves of professional highland construction personnel to ensure continuous and stable progress in the construction activities. Personnel should be allocated in an organized manner according to the work schedule and echelon. Adequate feeding security and a reasonable shift rest mechanism should be in place to alleviate the hazards associated with low oxygen in plateau areas. Implementing a robust incentive mechanism is necessary to motivate site personnel to actively engage in their work.

(2) Optimize the post-disaster organizational structure and operational mechanism, and improve personnel's crisis awareness and impact management capabilities.

Establishing and improving the operation mechanism of the special work leadership organization will facilitate compliance with relevant government emergency prevention and control requirements. Establishing a contact mechanism with the Ministry of Emergency Response, the Earthquake Bureau, and other relevant units is crucial. Continuous monitoring of disaster information and smooth information reporting should be ensured. Additionally, being prepared to support local earthquake relief preparations is important. Organizing post-disaster hidden danger inspections and establishing safety and quality inspection teams will ensure the condition of buildings and equipment at each work site. Enhancing the categorized and hierarchical training model for emergency management personnel is necessary. This includes strengthening the education and training of leaders' emergency management knowledge. Organizing sharing sessions for project managers to exchange crisis response experience will improve the crisis awareness and responsibility of the project manager team. Furthermore, enhancing the overall process of crisis prevention, response, and post-disaster emergency response capabilities is crucial. Optimizing the education and training of technical personnel on emergency management knowledge and enabling them to utilize their professional expertise and technical advantages will enhance their ability to analyze sources of danger and risks. It is essential to improve the crisis awareness of all construction personnel by disseminating relevant knowledge on emergency management, disaster prevention and relief, and self-rescue and mutual assistance. Conducting disaster defense knowledge training and organizing drills and training sessions for construction personnel to handle emergencies promptly are also important measures to be implemented.

7.3. Limitations

The data in this study were obtained from the builders and managers of major railway projects in China, which is typical of major projects in China. In the future, we will expand the data of major railroad projects in many developed and developing countries, so that the results of the study can be more generalized. In addition, the construction management resilience indicators of major railroad projects identified in this study tended to be qualitative, which is difficult to describe using objective data. In the future, we will devote ourselves to constructing an objective evaluation index system for construction management resilience and researching objective assessment methods for construction management resilience of major railroad projects.

Author Contributions: Conceptualization, X.Z. and Y.L.; methodology, X.Z.; software, X.Z.; validation, X.Z., K.L. and Y.L.; formal analysis, Y.L.; investigation, X.Z. and K.L.; resources, W.J. and X.Z.;

data curation, X.Z. and Y.L.; writing—original draft preparation, X.Z.; writing—review and editing, X.Z. and K.L.; visualization, X.Z. and K.L.; supervision, W.J., C.C. and X.Z.; project administration, W.J., C.C. and K.C.; funding acquisition, W.J. and C.C. All authors have read and agreed to the published version of the manuscript.

Funding: This research was funded by the State Key Laboratory for Track Technology of High-speed Railway (Contract No. 2021YJ111), China Academy of Railway Science.

Data Availability Statement: The original contributions presented in the study are included in the article, further inquiries can be directed to the corresponding author.

Acknowledgments: The authors express their sincere gratitude to the State Key Laboratory for Track Technology of High-speed Railway for its assistance in observation of the relevant processes and in conducting on-site interviews. Special appreciation also goes to the editors and reviewers whose constructive and invaluable comments and suggestions played a decisive role in significantly improving the quality of this article.

Conflicts of Interest: Authors Chaoxun Cai and Kai Chang were employed by the company Railway Engineering Research Institute, China Academy of Railway Sciences Corporation Limited. The remaining authors declare that the research was conducted in the absence of any commercial or financial relationships that could be construed as a potential conflict of interest. The funders had important coordination roles in the data collection.

References

1. Liu, K.; Liu, Y.M.; Kou, Y.Y.; Yang, X.X. Study on dissipative structure of mega railway infrastructure project management system. *Eng. Constr. Archit. Manag.* 2023, *ahead-of-print*. [CrossRef]
2. Wang, Y.S.; Lin, C.J.; Wang, H.P.; Wang, W.; Wang, S.; Zheng, R.J. Implementation of pollution source evaluation and treatment strategy for plateau railway construction in China: An AHP-cloud model approach. *Environ. Monit. Assess.* **2023**, *195*, 749. [CrossRef]
3. Liu, B.L.; Wagner, L.E.; Ning, D.H.; Qu, J.J. Estimation of wind erosion from construction of a railway in arid Northwest China. *Int. Soil Water Conserv. Res.* **2017**, *5*, 102–108. [CrossRef]
4. Lu, C.F.; Cai, C.X. Challenges and Countermeasures for Construction Safety during the Sichuan–Tibet Railway Project. *Engineering* **2019**, *5*, 833–838. [CrossRef]
5. Liu, K.; Liu, Y.M.; Kou, Y.Y.; Yang, X.X.; Hu, G.Z. Efficiency of risk management for tunnel security of megaprojects construction in China based on system dynamics. *J. Asian. Archit. Build.* **2023**, *23*, 712–724. [CrossRef]
6. Wu, X.; Xu, F. Detection model for unbalanced bidding in railway construction projects: Considering the risk of quantity variation. *J. Constr. Eng. Manag.* **2021**, *147*, 04021055. [CrossRef]
7. Huang, J.L.; Zeng, X.Y.; Fu, J.; Han, Y.; Chen, H.H. Safety risk evaluation using a BP Neural Network of high cutting slope construction in High-Speed Railway. *Buildings* **2022**, *12*, 598. [CrossRef]
8. Shi, J.J.; Geng, G.Q.; Ming, J. Corrosion resistance of fine-grained rebar in mortars designed for high-speed railway construction. *Eur. J. Environ. Civ. Eng.* **2018**, *22*, 562–577. [CrossRef]
9. Zhao, T.J.; Xiao, X.; Dai, Q.H. Transportation infrastructure construction and high-quality development of enterprises: Evidence from the Quasi-Natural Experiment of High-Speed Railway Opening in China. *Sustainability* **2022**, *13*, 13316. [CrossRef]
10. Wehbe, F.; Al Hattab, M.; Hamzeh, F. Exploring associations between resilience and construction safety performance in safety networks. *Saf. Sci.* **2016**, *82*, 338–351. [CrossRef]
11. Holling, C.S. Resilience and Stability of Ecological Systems. *Annu. Rev. Ecol. Evol. Syst.* **1973**, *4*, 1–23. [CrossRef]
12. Francis, R.; Bekera, B. A metric and frameworks for resilience analysis of engineered and infrastructure systems. *Reliab. Eng. Syst. Safe* **2014**, *121*, 90–103. [CrossRef]
13. Wied, M.; Oehmen, J.; Welo, T. Conceptualizing resilience in engineering systems: An analysis of the literature. *Syst. Eng.* **2020**, *23*, 3–13. [CrossRef]
14. Pawar, B.; Park, S.; Hu, P.F.; Wang, Q.S. Applications of resilience engineering principles in different fields with a focus on industrial systems: A literature review. *J. Loss Prev. Process. Ind.* **2021**, *69*, 104366. [CrossRef]
15. Liu, W.Q.; Shan, M.; Zhang, S.; Zhao, X.B.; Zhai, Z. Resilience in Infrastructure Systems: A Comprehensive Review. *Buildings* **2022**, *12*, 759. [CrossRef]
16. Bruneau, M.; Reinhorn, A. Exploring the Concept of Seismic Resilience for Acute Care Facilities. *Earthq Spectra* **2007**, *23*, 41–62. [CrossRef]
17. Li, Y.; Lence, B.J. Estimating resilience for water resources systems. *Water. Resour. Res.* **2007**, *43*, W07422. [CrossRef]
18. Vugrin, E.D.; Warren, D.E.; Ehlen, M.A. A resilience evaluation framework for infrastructure and economic systems: Quantitative and qualitative resilience analysis of petrochemical supply chains to a hurricane. *Process. Saf. Prog.* **2011**, *30*, 280–290. [CrossRef]
19. Hosseini, S.; Barker, K. Modeling infrastructure resilience using Bayesian networks: A case study of inland waterway ports. *Comput. Ind. Eng.* **2016**, *93*, 252–266. [CrossRef]

20. Wu, C.W.; Cenci, J.; Wang, W.; Zhang, J.Z. Resilient city: Characterization, challenges and outlooks. *Buildings* **2022**, *12*, 516. [CrossRef]
21. Haimes, Y.Y. On Some Recent Definitions and Analysis Frameworks for Risk, Vulnerability, and Resilience. *Risk Anal.* **2011**, *31*, 698. [CrossRef]
22. Sweetapple, C.; Fu, G.T.; Farmani, R.; Butler, D. General resilience: Conceptual formulation and quantitative evaluation for intervention development in the urban wastewater system. *Water Res.* **2022**, *211*, 118108. [CrossRef]
23. Serdar, M.Z.; Koç, M.; Al-Ghamdi, S.G. Urban Transportation Networks Resilience: Indicators, Disturbances, and Evaluation Methods. *Sustain. Cities Soc.* **2022**, *76*, 103452. [CrossRef]
24. Cardoni, A.; Borlera, S.L.; Malandrino, F.; Cimellaro, G.P. Seismic vulnerability and resilience evaluation of urban telecommunication networks. *Sustain. Cities Soc.* **2022**, *77*, 103540. [CrossRef]
25. Argyroudis, S.A.; Mitoulis, S.A.; Hofer, L.; Zanini, M.A.; Tubaldi, E.; Frangopol, D.M. Resilience evaluation framework for critical infrastructure in a multi-hazard environment: Case study on transport assets. *Sci. Total Environ.* **2020**, *714*, 136854. [CrossRef] [PubMed]
26. Franchin, P.; Cavalieri, F. Probabilistic evaluation of civil infrastructure resilience to earthquakes. *Comput. Civ. Infrastruct. Eng.* **2015**, *30*, 583–600. [CrossRef]
27. Babar, A.H.K.; Ali, Y. Framework construction for augmentation of resilience in critical infrastructure: Developing countries a case in point. *Technol. Soc.* **2022**, *68*, 101809. [CrossRef]
28. Labaka, L.; Hernantes, J.; Sarriegi, J.M. Resilience framework for critical infrastructures: An empirical study in a nuclear plant. *Reliab. Eng. Syst. Safe* **2015**, *141*, 92–105. [CrossRef]
29. Huang, C.N.; Liou, J.J.H.; Lo, H.W.; Chang, F.J. Building an evaluation model for measuring airport resilience. *J. Air. Transp. Manag.* **2021**, *95*, 102101. [CrossRef]
30. Dvorak, Z.; Chovancikova, N.; Bruk, J.; Hromada, M. Methodological Framework for Resilience Evaluation of Electricity Infrastructure in Conditions of Slovak Republic. *Int. J. Environ. Res. Public Health* **2021**, *18*, 8286. [CrossRef]
31. Zhao, T.Y.; Sun, L. Seismic resilience evaluation of critical infrastructure-community systems considering looped interdependences. *Int. J. Disaster Risk Reduct.* **2021**, *95*, 103856. [CrossRef]
32. Kong, J.J.; Simonovic, S.P.; Zhang, C. Resilience evaluation of interdependent infrastructure systems: A case study based on different response strategies. *Sustainability* **2019**, *11*, 6552. [CrossRef]
33. Bruneau, M.; Chang, S.E.; Eguchi, R.T.; Lee, G.C.; O'Rourke, T.D.; Reinhorn, A.M.; Shinozuka, M.; Tierney, K.; Wallace, W.A.; von Winterfeldt, D.A. Framework to quantitatively assess and enhance the seismic resilience of communities. *Earthq. Spectra* **2003**, *19*, 733–752. [CrossRef]
34. Ghouchani, M.; Taji, M.; Roshan, A.Y.; Chehr, M.S. Identification and evaluation of hidden capacities of urban resilience. *Int. J. Proj. Manag.* **2021**, *23*, 3966–3993. [CrossRef]
35. Murdock, H.J.; de Bruijn, K.M.; Gersonius, B. Evaluation of critical infrastructure resilience to flooding using a response curve approach. *Sustainability* **2018**, *10*, 3470. [CrossRef]
36. Mustafa, A.M.; Barabadi, A. Resilience evaluation of wind farms in the arctic with the application of Bayesian Networks. *Energies* **2021**, *14*, 4439. [CrossRef]
37. Sen, M.K.; Dutta, S. A Bayesian Network Modeling approach for time-varying flood resilience evaluation of housing infrastructure system. *Nat. Hazards Rev.* **2022**, *23*, 04022006. [CrossRef]
38. Panteli, M.; Pickering, C.; Wilkinson, S.; Dawson, R.; Mancarella, P. Power system resilience to extreme weather: Fragility modeling, probabilistic impact evaluation, and adaptation measures. *IEEE Trans. Power Syst.* **2017**, *32*, 3747–3757. [CrossRef]
39. Praks, P.; Kopustinskas, V.; Masera, M. Probabilistic modelling of security of supply in gas networks and evaluation of new infrastructure. *Reliab. Eng. Syst. Safe* **2015**, *144*, 254–264. [CrossRef]
40. Kourehpaz, P.; Molina Hutt, C. Machine learning for enhanced regional seismic risk assessments. *J. Struct. Eng.* **2022**, *148*, 04022126. [CrossRef]
41. Forcellini, D. An expeditious framework for assessing the seismic resilience (SR) of structural configurations. In *Structures*; Elsevier: Amsterdam, The Netherlands, 2023; Volume 56, p. 105015.
42. Zhu, Z.; Yuan, J.F.; Shao, Q.H.; Zhang, L.; Wang, G.Q.; Li, X.W. Developing key safety management factors for construction projects in China: A resilience perspective. *Int. J. Environ. Res. Public Health* **2020**, *17*, 6167. [CrossRef] [PubMed]
43. Mostafizi, A.; Wang, H.Z.; Cox, D.; Cramer, L.A.; Dong, S.J. Agent-based tsunami evacuation modeling of unplanned network disruptions for evidence-driven resource allocation and retrofitting strategies. *Nat. Hazards* **2017**, *88*, 1347–1372. [CrossRef]
44. Ouyang, M. Review on modeling and simulation of interdependent critical infrastructure systems. *Reliab. Eng. Syst. Safe* **2014**, *121*, 43–60. [CrossRef]
45. Pumpuni-Lenss, G.; Blackburn, T.; Garstenauer, A. Resilience in complex systems: An agent-based approach. *Syst. Eng.* **2017**, *20*, 158–172. [CrossRef]
46. Liang, Y.; Liu, Q.X. Early warning and real-time control of construction safety risk of underground engineering based on building information modeling and internet of things. *Neural. Comput. Appl.* **2021**, *34*, 3433–3442. [CrossRef]
47. Zhang, J.X.; Zha, G.Q.; Pan, X.; Zuo, D.J.; Xu, Q.X.; Wang, H.X. Community centered public safety resilience under public emergencies: A case study of COVID-19. *Risk Anal.* **2022**, *43*, 114–128. [CrossRef] [PubMed]

48. Harvey, E.J.; Waterson, P.; Dainty, A.R.J. Applying HRO and resilience engineering to construction: Barriers and opportunities. *Saf. Sci.* **2019**, *117*, 253–533. [CrossRef]
49. Saldanha, M.C.W.; Araújo, L.L.F.; Arcuri, R.; Vidal, M.C.R.; de Carvalho, P.V.R.; de Carvalho, R.J.M. Identifying routes and organizational practices for resilient performance: A study in the construction industry. *Cogn. Technol. Work.* **2022**, *24*, 521–535. [CrossRef]
50. Saaty, T.L. Modeling unstructured decision problems—The theory of analytical hierarchies. *Math. Comput. Simul.* **1978**, *20*, 147–158. [CrossRef]
51. Zhang, H. Fuzzy comprehensive evaluation and quantitative weight analysis in structure management of human resources. *PLoS ONE* **2023**, *18*, e0288795. [CrossRef] [PubMed]
52. Zadeh, L.A. Fuzzy sets. *Inf. Control* **1965**, *8*, 338–353. [CrossRef]
53. Cai, F.; Yang, L.; Yuan, Y.; Taghizadeh-Hesary, F. The application of an improved fuzzy comprehensive evaluation in coal quality rating: The case study of China. *Front. Energy Res.* **2022**, *9*, 752472. [CrossRef]

buildings

MDPI

Article

Intelligent Analysis of Construction Safety of Large Underground Space Based on Digital Twin

Caizhao Yu [1], Zhansheng Liu [2,*], Haitao Wang [1], Guoliang Shi [2] and Tianshuai Song [1]

1 Shanghai Baoye Group Corp., Ltd., Shanghai 201900, China; 18651820197@163.com (C.Y.); 18516238259@163.com (H.W.); sts909@163.com (T.S.)
2 Faculty of Architecture, Civil and Transportation Engineering, Beijing University of Technology, Beijing 100124, China; guoliangshi@emails.bjut.edu.cn
* Correspondence: liuzhansheng@bjut.edu.cn

Abstract: With the rapid development of underground space, the issue of safety in construction processes is becoming more and more significant. The purpose of this paper is to solve the problem of the existing underground space monitoring technology not being centralized and unified. In view of the problems related to large underground spaces in the process of constructing complex structures, with the introduction of Internet of Things technology and digital twins, we put forward an application of an intelligent safety-monitoring digital twin system in the construction of a large underground space structure, and at the same time, explore the Internet and digital integration mechanism of the twin system. The twin system uses BIM technology to establish the corresponding physical construction model, and collects multi-source heterogeneous monitoring data in real time through Internet of Things technology to achieve the exchange of information between the virtual construction model and the physical construction model. The twin system uses the multi-source heterogeneous data for real-time security analysis, and obtains the security status of the structure and feeds it back to the application service layer. The effectiveness and practicability of the twin system in large underground spaces are verified by an example project. Aiming at the safe performance of the orthogonal arch, the mapping relationship of various parameter indexes is obtained, and reasonable control measures are given. This study provides a new solution for improving the safety of construction projects and risk prevention and control, and has important theoretical and practical value for the safety management of underground space construction processes.

Keywords: large underground space; digital twin; internet of things technology; safety monitoring; parameter analysis

Citation: Yu, C.; Liu, Z.; Wang, H.; Shi, G.; Song, T. Intelligent Analysis of Construction Safety of Large Underground Space Based on Digital Twin. *Buildings* **2024**, *14*, 1551. https://doi.org/10.3390/buildings14061551

Academic Editor: Saeed Banihashemi

Received: 31 January 2024
Revised: 19 March 2024
Accepted: 26 March 2024
Published: 27 May 2024

1. Introduction

With the acceleration of modern urbanization, the demand for large-scale buildings and underground space projects is increasing, which also promotes the continuous growth of the number and scale of large-scale foundation pit projects. However, the foundation pit is usually underground, the construction period is long, the construction is difficult, and the influencing factors are complex and changeable; therefore, the risk associated with foundation pit construction is relatively high [1,2]. Therefore, the safety monitoring of foundation pits is one of the most important means of ensuring the safety of foundation pit construction and the stability of the foundation pit structure.

In traditional foundation pit monitoring, measurements are performed manually on a regular basis through instruments, which not only is associated with a large workload, but also is easily affected by external factors such as the weather and the site itself. In the construction process of very large underground space structures, due to the large scale of the space, there are many problems, such as its complex structure, strong disturbances, the changeable geological environment, and multiple risk factors and difficult risk prevention and control, which cannot meet the requirements of high-precision and real-time safety analysis [3]. With the continuous development of electronic technology and computer technology, more and more new technologies are applied to the monitoring of foundation pits [4–6]. The application of these technologies can enable the measurement and monitoring of parameters such as the surrounding environment, soil deformation and displacement, and the water level of the foundation pit, so as to achieve the comprehensive monitoring of foundation pit construction safety.

Xu et al. [7] designed an automatic monitoring system for foundation pits based on vibrating wire sensors, which can monitor the whole process of deep foundation pit construction and provide early warnings of structural safety in real time. Zhang et al. [8] developed a foundation pit monitoring system based on the Internet of Things and WebGIS. The wireless monitoring system is composed of sensors, data collectors, data transmission equipment, and an information management module, which can achieve the on-line monitoring and early warning of indicators such as the horizontal displacement and settlement of the foundation pit. Hashash et al. [9] captured the horizontal displacement, settlement, and other parameters of a foundation pit through sensors during the excavation of an urban underground space, and achieved the real-time monitoring of ground deformation during construction. In addition, monitoring technologies based on microwaves, lasers, and drones have also been gradually applied to the field of foundation pit monitoring. Through the three-dimensional scanning and high-precision measurement of the soil around the foundation pit, the deformation of the foundation pit can be more accurately reflected. Han et al. [10] introduced a method of using BIM and 3D laser scanning technology to monitor the deformation of a foundation pit, and took a large foundation pit as an example to verify that the method can monitor the three-dimensional overall deformation of foundation pits efficiently and accurately. In the prediction of surface subsidence caused by mining, Lee et al. [11] proposed two models based on a long short-term memory network to capture the characteristics of surface subsidence caused by mining and predict subsidence. The proposed deep learning model can accurately predict the subsidence of the training set and the test set. Li et al. [12] introduced an improved support vector machine algorithm based on multi-point measurement technology to monitor and predict the deformation of deep foundation pits. These studies provide effective technical support for the monitoring and prediction of ground deformation caused by excavation engineering.

Many researchers have studied the monitoring of foundation pits from different angles and different perspectives, with the important aim of providing a guarantee of the safety of foundation pit construction. However, at present, current research is unable to integrate various types of monitoring data into a centralized and unified system. Therefore, this paper proposes a twin system for the safety monitoring of super-large foundation pits based on the Internet of Things, addressing the above issues, which innovatively achieves the centralized processing of multi-dimensional multi-source data and the twin synchronization of underground space construction. The application layer provides decision makers with the safety status of the foundation pit, so that they can take corresponding measures in real-time according to this information to ensure the safety of personnel during the construction of the foundation pit.

2. Establishment of Digital Twin System Framework

Digital twin systems can play an important role in solving structural safety problems during construction and can reduce construction safety risks. For example, Wang et al. [13] proposed a digital twin-driven structural safety control method for a cable network considering spatio-temporal variations, which was able to compare and analyze the geometric information of the construction site with the real-time finite element simulation results to ensure the structural safety during construction. Liu et al. [14] carried out an intelligent safety assessment of the steel structure based on digital twins, and achieved the analysis of the safety performance of the structure by constructing a digital twin framework for multidimensional spatial and temporal information fusion. In view of the complex characteristics of the construction of super-large underground space structures, the concept of digital twins is introduced, combined with Internet of Things technology, and a twin system architecture that can meet the actual needs of the project is proposed to improve the level of intelligent structural construction safety monitoring and ensure that the underground space structure is in a safe state during construction.

2.1. Construction Safety Risk of Large Underground Spaces

Construction safety risk data are the basis of driving the digital twin framework. Therefore, the construction risk index system of large underground spaces is first constructed through a literature research and expert interviews. The construction risk of urban underground large space engineering is mainly divided into [15–18] retaining structure instability, support system instability, pit bottom deformation and failure, soil collapse, surface subsidence, building (structure) damage, road and bridge damage, and underground pipeline damage, etc., as shown in Figure 1.

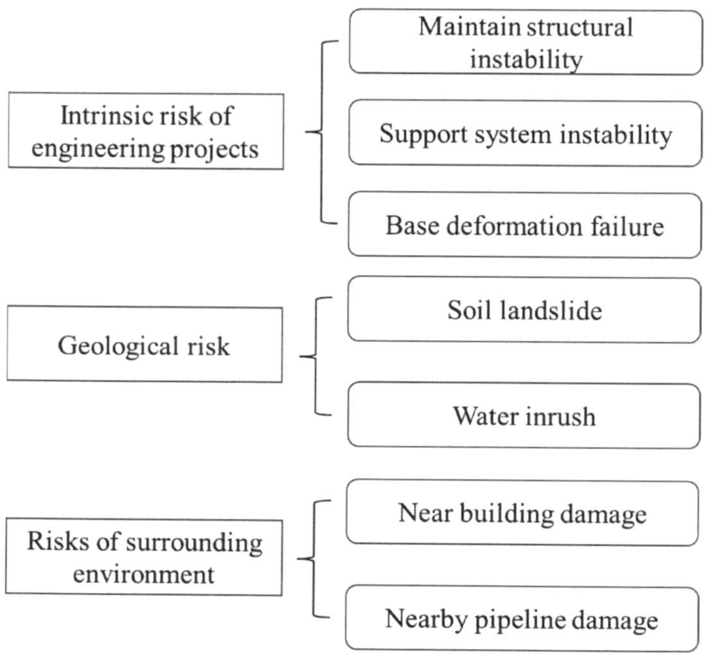

Figure 1. Construction risk sources of large-scale underground space projects.

2.2. Digital Twin System Framework

In order to meet the requirements of the high-precision and real-time construction safety monitoring of large underground spaces, considering the factors of time and space in multiple dimensions, the concept of digital twins is integrated into the construction safety monitoring system of underground structure engineering. Digital twin technology can achieve real-time information interactions between the physical space and virtual space, and map the behavior changes of entities in the real environment in the virtual space [19,20].

In the construction process of underground space structures, a large amount of measured data will be generated, which will interact with the data of the virtual simulation to form the twin data. Through the information mining of twin data, the dynamic prediction of the construction process is realized. The first and most important aim in the safety monitoring of large underground space construction is to capture the data of key safety risk factors. Through the analysis of the safety risk of large underground space construction, it can be seen that the information that needs to be captured in the construction process is mainly divided into the project's own risk, O; geological risk, G; surrounding environment risk, E; and construction personnel management, Pi. The established twin model framework can truly reflect the safety state of construction.

In the process of large underground space construction, there are many causes of safety problems. Combining the effects of various influencing factors, focusing on the risk of the project itself; the retaining structure, E; the support system, S1; the bottom deformation, F; the geological risk soil slump, S2; the mud inrush water inrush, W; the surrounding environmental risk of the adjacent building damage, A; the adjacent road and railway damage, R; the adjacent bridge damage, B; and the adjacent pipeline damage, P, they are closely related to the construction safety of the underground space structure. According to the information that needs to be captured in the construction safety analysis, the twin information is divided into three categories, which are represented by Formulas (1)–(3), respectively.

$$O = \{E, S_1, F\} \tag{1}$$

$$G = \{S_2, W\} \tag{2}$$

$$E = \{A, R, B, P\} \tag{3}$$

For each construction risk factor, the finite element model and BIM model corresponding to the entity are established, respectively, and the virtual model is uploaded to the twin system platform. According to the specification requirements and engineering experience, the management standards are set up for each key risk factor, and the intelligent auxiliary decision module of the twin system is established to ensure the scientificity of safety monitoring throughout the entire process of construction. According to the information required for the safety risk management of underground space structure construction, combined with the concept of digital twins, a twin system framework is proposed, as shown in Figure 2.

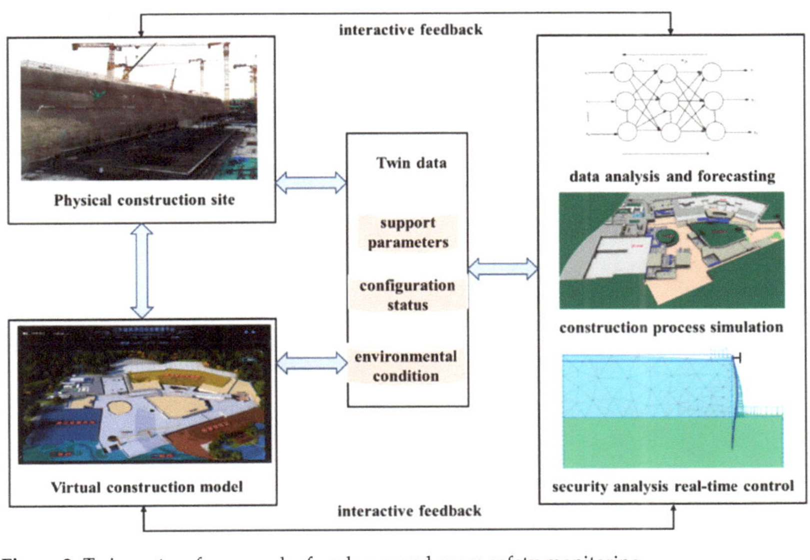

Figure 2. Twin system framework of underground space safety monitoring.

In this framework, the physical construction entity and the virtual simulation model achieve virtual and real-world interaction through the Internet of Things technology. With the help of real-time multi-factor data acquisition, they are transmitted to the cloud platform for cloud computing. Through the analysis, mining, and prediction of historical accumulated data by an intelligent algorithm, the safety state of the structure is obtained, and the information is fed back in the application layer to achieve the automatic decision-making and self-scheduling of the construction process.

3. Integration of the Internet of Things

The twin system framework of safety monitoring in the underground space construction process provides the basis of the theoretical framework for safety management in construction processes. In order to achieve the automatic decision-making and self-scheduling of the construction process of underground space structures, the most important thing is data collection. Internet of Things technology, as the supposed eye of data, can collect a large amount of monitoring data needed by twins in real time. The fusion mechanism of Internet of Things technology and twin systems is studied, and a twin system for large underground space safety monitoring based on Internet of Things technology is obtained.

3.1. The Fusion Mechanism of Internet of Things and Digital Twin Systems

The Internet of Things (IoT) refers to the use of sensing devices, such as radio frequency identification technology (RFID), global positioning systems, and other information sensing devices, on the basis of Internet technology, to connect the material itself with virtual information data to achieve the intelligent identification, positioning, and tracking of items, and to achieve information technology and remote monitoring and management [21,22].

The Internet of Things has the characteristics of convenient configuration, high security and stability, and has a simple and easy operation interface. However, the construction process of the structure is a complex task. On the one hand, the structure itself will change over time; on the other hand, the construction environment is often complex and changeable. In order to solve these specific problems, Internet of Things technology is used to transmit the construction monitoring data of the structure more appropriately and effectively, so as to monitor the state change of the structure during the construction period and for use as the control information of the construction procedure to ensure its safety.

The principle of digital twin technology is to realize the simulation, monitoring, and optimization of physical entities through the interaction of key steps such as data acquisition, virtual modeling, real-time monitoring and feedback, simulation and optimization, control and decision-making, and state diagnosis and maintenance. The data of the physical entity are collected by sensors and monitoring equipment and synchronized with the digital twin model to ensure that the model is consistent with the state of the actual entity. Based on the collected data, a virtual model of the physical entity is constructed, including detailed information such as structure, attributes, and behavior rules. By synchronizing with real-time data, the state, performance, and behavior of physical entities are monitored in real time, and real-time feedback signals are provided to control and adjust the operation of entities. The digital twin model is used for simulation and optimization analysis to predict the state and performance of the structure, and provide optimization suggestions and decision support. Through the linkage of the digital twin model and real-time data, the remote control and operation of physical entities are achieved, and intelligent decision support is provided for decision makers. Finally, based on the digital twin model and real-time data, the state diagnosis is carried out, the causes of structural safety risks are identified, and corresponding maintenance suggestions and strategies are provided.

From the summary of the principles of Internet of Things technology and the twin system, the commonality of the two can be obtained. Therefore, on the basis of the twin system framework, the Internet of Things technology is integrated to explore the following five fusion mechanisms, and the mechanism fusion framework is given, as shown in Figure 3.

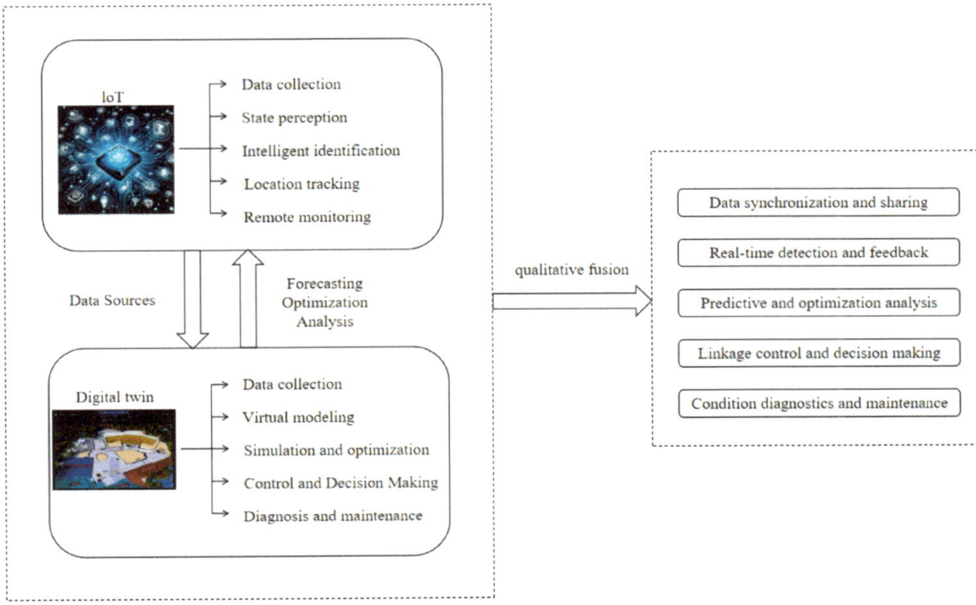

Figure 3. Fusion mechanism of IoT and digital twin.

(1) Internet of Things technology achieves real-time data acquisition and sharing by connecting and transmitting data between devices and sensors. These data can be used to update and maintain the model of the twin system and keep the model in sync with the actual physical world. Internet of Things technology provides data sources and real-time, high-quality data input for the twin system.

(2) Internet of Things technology can monitor various parameters and states in the physical world in real time, including equipment operating status, environmental conditions, energy consumption, and so on. Through combination with the twin system, the real-time monitoring data of the Internet of Things technology can be compared and analyzed with the twin model to find anomalies and provide real-time feedback. This real-time monitoring and feedback mechanism helps users to accurately understand the operating status of the physical system and perform timely adjustments and optimizations.

(3) A large amount of data generated by Internet of Things technology can be used for the prediction and optimization analysis of twin systems. By analyzing the historical records and real-time monitoring data of the Internet of Things data, the twin system can predict the future behavior and performance of the physical system and make optimization suggestions. This prediction and optimization analysis can be used for equipment maintenance, resource utilization optimization, energy efficiency improvement, etc., to achieve more efficient and sustainable operations.

(4) The integration of IoT technology and a Siamese system can enable the realization of linkage control and intelligent decision support. Through the real-time data collected by Internet of Things technology, the twin system can generate a virtual model that matches the physical system, and perform simulation and analysis. Based on the analysis results of the twin model, the remote control and operation of the physical system can be realized, and intelligent decision support can be provided for decision makers to optimize operation and management.

(5) The combination of the real-time monitoring of IoT technology and modeling analysis of a Siamese system can enable the state diagnosis and maintenance support of

physical system. Through monitoring data and model analysis, the safety status of the structure can be accurately identified, and corresponding maintenance suggestions and schemes can be provided. This helps to improve the reliability and maintenance efficiency of the structure and reduce the impact of accidents on construction.

3.2. Establishment of Mathematical Model of Fusion Mechanism

Through the analysis of the basic principles of Internet of Things technology, combined with the characteristics of digital twin technology, the fusion mechanism of Internet of Things technology and the twin system is explored. Because the safety monitoring of the underground space construction process is guided by the time dimension, multi-dimensional and multi-source data are integrated into the cloud platform for calculation and analysis. Taking the fusion mechanism as the core, a mathematical model is established to provide a reference for the development and application of the subsequent twin system.

In the mathematical model, M represents the model of the twin system. D (t) represents the real-time data collected by the Internet of Things technology, where t represents time. S (t) denotes the state of the physical system at time t.

(1) Data sharing and synchronization:

$$M(t) = f(D(t)) \tag{4}$$

The real-time data D (t) collected by the Internet of Things technology is mapped to the twin system model M (t) by function f to ensure the synchronization between the model and the actual physical system.

(2) Real-time monitoring and feedback:

$$M(t) \approx S(t) \tag{5}$$

Through the synchronization of real-time monitoring data, the twin system model M (t) is approximated to the physical system state S (t).

Feedback mechanism: According to the difference between M (t) and S (t), a feedback signal is generated to control and adjust the behavior of the physical system.

(3) Prediction and optimization analysis:

$$P(t) = g(M(t)) \tag{6}$$

The twin system model M (t) is predicted and optimized by the function g, and the prediction result P (t) is obtained.

Optimization decision: According to the prediction result P (t), the optimization decision is made, and the operating state and parameters of the physical system are adjusted.

(4) Linkage control and decision-making:

Control mechanism: According to the twin system model M (t) and real-time data D (t), the control instruction C (t) is generated to control the behavior of the physical system.

Decision support: Based on model analysis and real-time data, decision support is provided, such as intelligent decision-making based on M (t) and D (t).

(5) Status diagnosis and maintenance:

State diagnosis: According to the twin system model M (t) and real-time data D (t), the state simulation analysis is carried out to identify the causes of risk in the physical system.

Maintenance support: Provide maintenance support based on status diagnosis results, including maintenance recommendations and strategies.

According to the fusion mechanism, the following key points can be summarized. Firstly, the real-time data D (t) are mapped to the twin system model M (t) by function f to ensure the synchronization between the model and the actual physical system. Secondly, through the synchronization of real-time monitoring data, the twin system model M (t) is approximated to the physical system state S (t), so as to realize real-time monitoring and feedback. The behavior of the physical system is controlled and adjusted by generating feedback signals. In addition, the function g is used to predict and optimize the twin system model M (t), and the prediction result P (t) is obtained. According to the prediction result P (t), the optimization decision is made to adjust the operating state and parameters of the physical system. At the same time, according to the twin system model M (t) and the real-time data D (t), the control instruction C (t) is generated to control the behavior of the physical system. Based on model analysis and real-time data, decision support can be provided, such as intelligent decision-making based on M (t) and D (t). Finally, the twin system model M (t) and real-time data D (t) are used to diagnose the state to identify the causes of risk in the physical system.

In summary, the integration mechanism of Internet of Things technology and the digital twin system provides a powerful tool and decision support for the safety monitoring of the underground space construction process. Through real-time data acquisition, model analysis, and prediction optimization, the physical system can be monitored, controlled, and maintained to ensure the safety and efficiency of the construction process.

4. Case Verification

4.1. Project Case

The effectiveness of this research method is verified by taking the foundation pit construction of three major buildings' shared facilities in the urban sub-center as an example. The shared support facilities of the project are mainly located on the first and second floors of the underground, covering an area of about 154,000 square meters, with a total construction scale of about 255,000 square meters. The main functions are comprehensive support for parking, parking for three major buildings, and comprehensive support for service facilities. A plane schematic diagram of the foundation pit project for three major buildings' shared support facilities is shown in Figure 4.

Figure 4. Plane schematic diagram of foundation pit project of three major buildings' shared support facilities.

The project is based on the largest commercial underground center in China. The project has a huge volume, many work surfaces, a complex structural system, high construction accuracy requirements, many professional interfaces, and great difficulty in professional integration and coordination. Near the North Canal, the groundwater level is high; at the same time, there are two canal sites in the project that need to be supported separately.

The targets being monitored in the project mainly include horizontal displacement monitoring, settlement monitoring, deep horizontal displacement monitoring, groundwater level monitoring, anchor cable internal force monitoring, concrete support axial force monitoring, etc. The data of the foundation pit are collected in real time by arranged sensors and monitoring equipment at the monitoring points, and synchronized with the twin model to ensure that the twin model is consistent with the state of the underground space.

4.2. Construction Safety Monitoring on Twin System Platform

BIM technology is used to establish the geometric and physical models corresponding to the foundation pit, and the virtual model is uploaded to the twin system platform. At the same time, considering the dynamics of the construction process, the 4D construction simulation of the BIM technology is used to visually simulate the construction sequence and the overall construction period plan. Collection of risk indicator data is carried out by installing corresponding sensors at the construction site. The targets of safety monitoring in the foundation pit mainly include horizontal displacement, settlement, groundwater level, anchor cable internal force, etc. The data from the foundation pit are collected in real time by installing sensors and monitoring equipment at the monitoring points. The specific monitoring types and sensor models are shown in Table 1. The sensor monitoring layout requirements include the following: (1) The layout of the foundation pit engineering sensors should be able to reflect the actual state of the monitored object and its changes, such as the layout of the internal force and deformation key feature points, and should meet the monitoring needs. (2) The layout of the sensors in the foundation pit should not hinder the normal work of the monitored targets, should reduce adverse impacts on construction operations, and mark the location of the layout to help remind construction personnel and monitoring personnel how to find the sensor. (3) A regular inspection and calibration plan should be established to ensure the normal operation of the sensor, and regular maintenance measures should be taken if the sensor equipment fails or functions abnormally.

Table 1. Foundation pit safety monitoring type and sensor name.

Serial Number	Monitoring Type	Sensor Name	Sensor Type
1	Concrete supporting stress	Embedded concrete strain gauge	YB-MR01
2	Anchor cable axial force	Anchor cable axial force meter	YB-MS300
3	Deep horizontal displacement	Automatic inclinometer	LRK-CX06
4	Settlement around foundation pit	Static level	LRK-J112
5	Settlement and horizontal displacement around foundation pit	Two-dimensional laser displacement meter	LRK-DL630
6	Groundwater table monitoring	Wireless water level gauge	RYY-SW01
7	Incline monitoring of surrounding buildings	Wireless high-precision biaxial inclinometer	LRK-RG911

Through the comprehensive utilization of different transmission methods, a variety of on-site monitoring instruments, detection equipment, and wireless sensors are connected through Internet of Things technology, and data are transmitted to the twin platform through Zigbee or the 4G network, so as to realize the automatic collection and real-time transmission of monitoring data, ensure the authenticity, integrity, and real-time performance of data, and ensure that the twin model of foundation pit safety monitoring can reflect the underground space entity in physical space accurately and in real time. The twin model of the underground space of the three major buildings' shared support facilities is shown in Figure 5.

Figure 5. Twin model of underground space of three major buildings' shared support facilities.

Through the integration of the above twin model and monitoring equipment, a twin system platform for foundation pit safety monitoring is constructed, as shown in Figure 6. The system can process the original monitoring data in real time, and use mathematical models and mathematical methods such as regression analysis and difference analysis to digitally model and analyze all kinds of collected data to form various change curves, graphs, and charts. Through various forms of real-time alarm functions, the hidden dangers of engineering and the surrounding buildings and pipelines can be found in time.

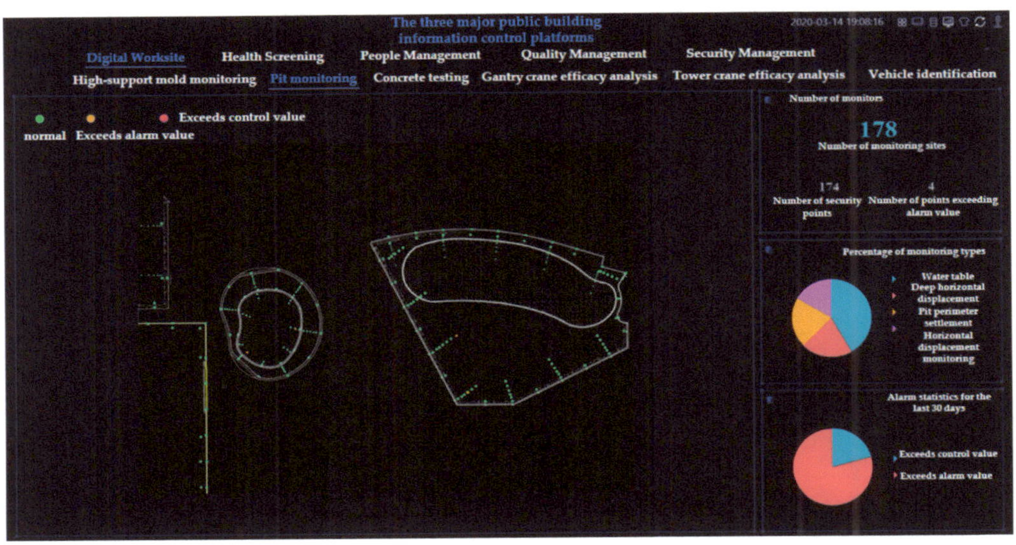

Figure 6. Twin platform for foundation pit safety monitoring.

The twin platform for pit safety monitoring, under the digital construction site board in the information control platform of the three major public buildings project, is able to display in real time whether the monitoring points have exceeded the alarm value, and to count the number of safe points and the number of points exceeding the alarm value. At the same time, the system platform can count the proportion of the monitoring types; display the monitoring data of the groundwater level, deep horizontal displacement, settlement around the foundation pit, and horizontal displacement in real time; and count the number of alarm situations occurring over a period of 30 days. Through the integration of Internet of Things technology and digital twin system, the real-time monitoring of foundation pit data can be realized, and the change in the safety state of foundation pit in the whole construction period can be measured. Through the analysis of monitoring data, alarms, and the real-time reflection of the state of foundation pit engineering, the occurrence of major accidents is avoided.

4.3. Parametric Analysis of Construction Safety

During the construction process, the displacement of the orthogonal arch is an important monitoring index. The orthogonal arches of the project are longitudinally primary arches, transversely connected beams, and flanked by secondary arches. The structure of the orthogonal arch is shown in Figure 7. Based on the twin platform, the parameter analysis between the section size and vertical displacement of the main arch, secondary arch, and coupling beam is carried out. On the basis of the parameter analysis, the control measures to ensure the structural safety are obtained, which provides an important reference for the construction safety of the underground space.

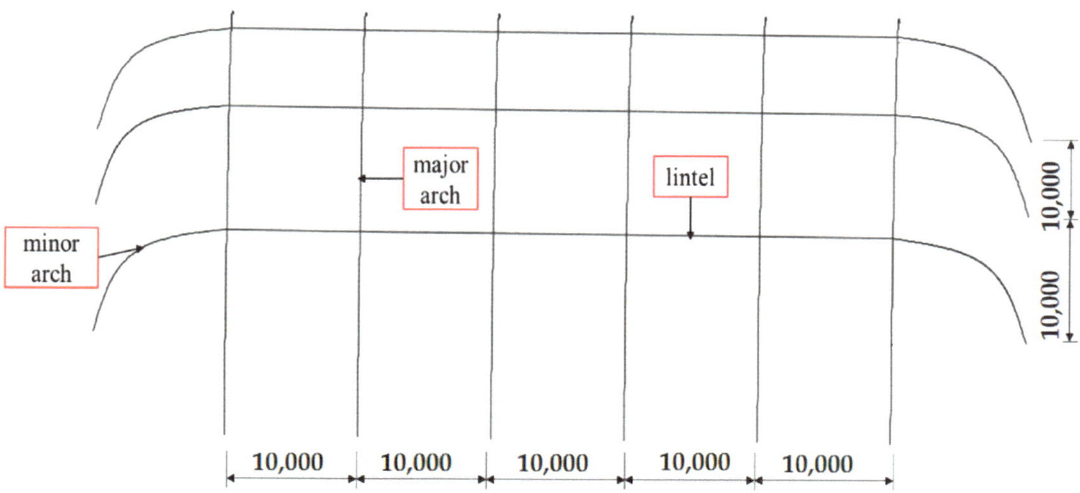

Figure 7. Construction of orthogonal arch.

In the twin platform, the cross-sectional dimensions of the three types of components are modified, respectively. The maximum vertical displacement of the structure corresponding to different types of cross-sectional dimensions is shown in Figure 8. The section size of the main arch has the greatest influence on the structural displacement, and the section size of the secondary arch has the least influence on the structural displacement.

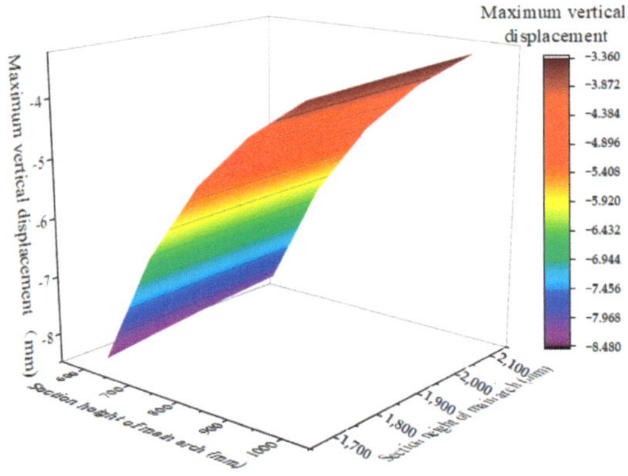

(**a**) The influence of main arch size on displacement.

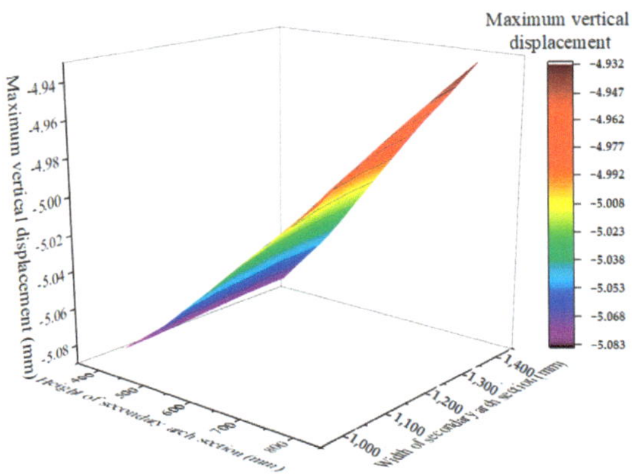

(**b**) The influence of secondary arch size on displacement.

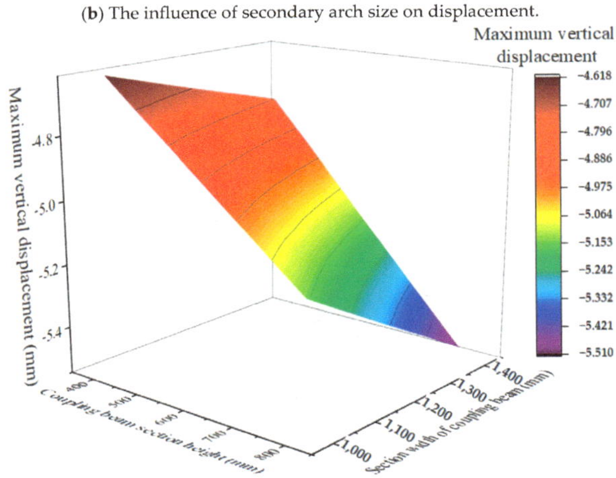

(c) The influence of coupling beam size on displacement.

Figure 8. The influence of different types of section size on the maximum vertical displacement of the structure.

According to Figure 8, the vertical displacement of the structure decreases with the increase in the section size of the main arch and the secondary arch. With the increase in the section size of the coupling beam, the vertical displacement of the structure increases. The section height of the main arch has the most significant effect on the vertical displacement. Therefore, the best measure to ensure the safety of the structure is to increase the section height of the main arch.

4.4. Discussion of Results

(1) Effectiveness

The pit safety monitoring twin platform achieves the expected effects of synchronization between the model mentioned in the theoretical research and the actual physical system, real-time monitoring and feedback, the control and adjustment of the behavior of the physical system, prediction and optimization analysis, the optimization of decision-making, state diagnosis, etc. It meets the requirements of high-precision and real-time safety analysis for the construction of the structure of a super-large underground space, and it has been verified by application case studies to more effectively solve structural safety problems during the construction process.

(2) Limitations

The scope of applications and application case studies for the fusion mechanism of Internet of Things technology and a digital twin proposed in this paper includes the construction process of large underground space structures. In the section discussing the parametric analysis of construction safety, the displacement of the orthogonal arch is analyzed only as an example, which means this theoretical research has certain limitations despite accurately solving the construction safety risk problem of a large underground space.

(3) Prospects

At present, the world is in an era of informationization, and Internet of Things technology and digital twin technology, as representatives of information technology, have a huge scope of possible applications, and can be more widely used in engineering construction. The fusion mechanism of IoT and digital twins and the digital twin system framework proposed in this paper provide a model for similar projects in the future and can be further developed to solve other problems in the construction process.

5. Conclusions

Aiming at the complex challenges faced in the construction of super-large underground space structures, an intelligent monitoring system architecture suitable for the construction of super-large underground space structures is proposed by introducing the concept of digital twins and Internet of Things technology, and the fusion mechanism of Internet of Things technology and digital twins is explored. The system can collect a large amount of monitoring data in real time, and carry out real-time safety analysis with the help of twin technology. The effectiveness and practicability of the twin system in the construction of super-large foundation pits are proven by the example project.

(1) Aiming at the complex characteristics of the construction of super-large underground space structures, the concept of digital twins is introduced. Combined with Internet of Things technology, a twin system architecture that can meet the actual needs of the project is proposed to improve the level of intelligent structural construction safety monitoring and ensure that the underground space structure is in a safe state during construction.
(2) The fusion mechanism of Internet of Things technology and the twin system is studied, and a twin system for large underground space safety monitoring based on Internet of Things technology is obtained.
(3) For construction in practical engineering, a twin system platform is developed. Based on the twin platform, a parametric analysis of construction safety is carried out, and

the mapping relationship between the section size and structural displacement is obtained. The parametric analysis effectively assists the formulation of structural safety control decisions.

This study provides a new solution for improving the construction safety level and risk prevention and control, and has important theoretical and practical value for the safety management of underground space construction processes. The safety monitoring of foundation pit construction is the core function of the twin system. The construction of the twin system for safety monitoring needs to integrate various elements such as the safety monitoring, topography, geology, and soil quality of the foundation pit, and needs to integrate multidisciplinary knowledge and industry experience to enable accurate analyses and build a decision-making knowledge base.

Author Contributions: Conceptualization, C.Y.; methodology, C.Y.; software, C.Y.; validation, C.Y., H.W., T.S., G.S. and Z.L., writing—original draft preparation, C.Y.; writing—review and editing, Z.L.; project administration, C.Y.; funding acquisition, Z.L. All authors have read and agreed to the published version of the manuscript.

Funding: The research was funded by the Beijing Science and Technology plan project (Z211100004321010).

Data Availability Statement: The data presented in this study are available on request from the corresponding author. The data are not publicly available due to confidentiality.

Acknowledgments: The authors would like to thank Shanghai Baoye Group Corp., Ltd., and China University of Mining and Technology (Beijing) for their support throughout the research project.

Conflicts of Interest: Authors Caizhao Yu, Haitao Wang, Tianshuai Song, and Kai Yang were employed by the company Shanghai Baoye Group Corp., Ltd. The remaining authors declare that the research was conducted in the absence of any commercial or financial relationships that could be construed as a potential conflict of interest.

References

1. Zhang, W.; Huang, Z.; Zhang, J.; Zhang, R.; Ma, S. Multifactor Uncertainty Analysis of Construction Risk for Deep Foundation Pits. *Appl. Sci.* **2022**, *12*, 8122. [CrossRef]
2. Chen, D.W.; Zhou, J.L.; Duan, P.S.; Zhang, J.Q. Integrating knowledge management and BIM for safety risk identification of deep foundation pit construction. *Eng. Constr. Archit. Manag.* **2023**, *30*, 3242–3258. [CrossRef]
3. Wu, J.J.; Peng, L.M.; Li, J.W.; Zhou, X.Y.; Zhong, J.B.; Wang, C.T.; Sun, J. Rapid safety monitoring and analysis of foundation pit construction using unmanned aerial vehicle images. *Autom. Constr.* **2021**, *128*, 103706. [CrossRef]
4. Schaumann, M.; Gamba, D.; Morales, H.G.; Corsini, R.; Guinchard, M.; Scislo, L.; Wenninger, J. The effect of ground motion on the lhc and hl-lhc beam orbit. *Nucl. Instrum. Methods Phys. Res. Sect. A Accel. Spectrometers Detect. Assoc. Equip.* **2023**, *1055*, 168495. [CrossRef]
5. Guinchard, M.; Develle, K.; Osborne, J.; Scislo, L.; Wenninger, J. JACoW: Investigation and Estimation of the LHC Magnet Vibrations Induced by HL-LHC Civil Engineering Activities. In Proceedings of the 9th International Particle Accelerator Conference, Vancouver, BC, Canada, 29 April–4 May 2018.
6. Shen, Y.S.; Wang, P.; Li, M.P.; Mei, Q.W. Application of Subway Foundation Pit Engineering Risk Assessment: A Case Study of Qingdao Rock Area, China. *KSCE J. Civ. Eng.* **2023**, *23*, 4621–4630. [CrossRef]
7. Xu, Z.; Wang, J. Design of deep foundation pit monitoring system based on vibrating wire sensor. *Transducer Microsyst. Technol.* **2015**, *34*, 79–81.
8. Zhang, Y.; Zhang, G.; Li, J.; Cao, Y.; Hao, B. Study on Online Monitoring Platform of Deep Foundation Pit Based on Internet of Things and WebGIS. *Chin. J. Undergr. Space Eng.* **2021**, *17*, 328–336.
9. Hashash, Y.M.; Jammoul, M.; Fleming, K.L. Construction monitoring for the transbay transit center excavation in San Francisco, California. In Proceedings of the IFCEE 2015, San Antonio, TX, USA, 17–21 March 2015; pp. 2502–2511.
10. Han, D.; Qin, G.; Zhou, Y.; Wang, D.; Yang, Y. Application of BIM and 3D Laser Scanning in Foundation Pit Monitorino. *J. Chongqing Jiaotong Univ. Nat. Sci.* **2019**, *38*, 72–76+102.
11. Lee, H.K.; Song, M.K.; Lee, S.S. Prediction of Subsidence during TBM Operation in Mixed-Face Ground Conditions from Realtime Monitoring Data. *Appl. Sci.* **2021**, *11*, 12130. [CrossRef]
12. Li, X.; Liu, X.; Li, C.Z.; Hu, Z.M.; Shen, G.Q.; Huang, Z.Y. Foundation pit displacement monitoring and prediction using least squares support vector machines based on multi-point measurement. *Struct. Health Monit.* **2018**, *18*, 715–724. [CrossRef]
13. Wang, Z.; Zhang, Z.; Liu, Z.; Dezhkam, M.; Zhao, Y. Methods and Applications of Safety Control for Cable Net Structure Considering Spatiotemporal Changes. *Sustainability* **2023**, *15*, 13922. [CrossRef]

14. Liu, Z.; Shi, G.; Jiao, Z.; Zhao, L. Intelligent safety assessment of prestressed steel structures based on digital twins. *Symmetry* **2021**, *13*, 1927. [CrossRef]
15. Zhang, L.; Li, H. Construction Risk Assessment of Deep Foundation Pit Projects Based on the Projection Pursuit Method and Improved Set Pair Analysis. *Appl. Sci.* **2022**, *12*, 1922. [CrossRef]
16. Chen, G.; Zhang, X.; Zhang, S.; Huang, F.; Xiao, H.; Ma, H.; Luo, L.; Bao, H. Response Monitoring and Analysis in Deep Foundation Pit Excavation: A Case Study in Soft Soil at Subway Tunnel Intersections. *Buildings* **2023**, *13*, 1286. [CrossRef]
17. Yuan, M.H.; Yuan, C.F.; Chen, F.; Li, L.; Hong, Y.; Yu, G.M.; Lei, J. BIM Digital Shadow Technology and Risk Assessment Method of the Deep Foundation Pit's Behavior for Zibo Light Rail. *Front. Earth Sci.* **2022**, *10*, 908032. [CrossRef]
18. Sun, Z.; Li, H.; Bao, Y.; Meng, X.; Zhang, D. Intelligent Risk Prognosis and Control of Foundation Pit Excavation Based on Digital Twin. *Buildings* **2023**, *13*, 247. [CrossRef]
19. Wu, B.; Wan, Y.; Xu, S.; Zhao, C.; Liu, Y.; Zhang, K. A Multi-Source Intelligent Fusion Assessment Method for Dynamic Construction Risk of Subway Deep Foundation Pit: A Case Study. *Sustainability* **2023**, *15*, 10162. [CrossRef]
20. Shi, G.; Liu, Z.; Meng, X.; Wang, Z. Intelligent Health Monitoring of Cable Network Structures Based on Fusion of Twin Simulation and Sensory Data. *Symmetry* **2023**, *15*, 425. [CrossRef]
21. Liu, Z.; Shi, G.; Meng, X.; Sun, Z. Intelligent Control of Building Operation and Maintenance Processes Based on Global Navigation Satellite System and Digital Twins. *Remote Sens.* **2022**, *14*, 1387. [CrossRef]
22. Ishaq, I.; Hoebeke, J.; Van den Abeele, F.; Rossey, J.; Moerman, I.; Demeester, P. Flexible Unicast-Based Group Communication for CoAP-Enabled Devices. *Sensors* **2014**, *14*, 9833–9877. [CrossRef] [PubMed]

buildings

MDPI

Article

Boosting Coordination and Employee Motivation in Mega-Project Sustainable Performance Through Quality Relationships: The Key Role of Quality Management System

Li Ma [1], Azhar Ali [2,*], Muhammad Shoaib Farooq [3] and Jonathan Musonda [2]

1 International School, Hainan Tropical Ocean University, Sanya 572022, China; cm_mali@dlut.edu.cn
2 Department of Construction Management, Dalian University of Technology, Dalian 116024, China; jonathan123@mail.dlut.edu.cn
3 Institute of Business and Management, University of Engineering and Technology, Lahore 54890, Pakistan; sshoaibfarooq@uet.edu.pk
* Correspondence: aliazhar@mail.dlut.edu.cn

Abstract: Coordination and employee motivation play vital roles in a mega-project's sustainable performance, particularly in the construction industry. With increasing demands from stakeholders and employees in the construction sector, sustainable performance has emerged as a priority. However, the importance of quality relationships and quality management systems is often overlooked. This research addresses this gap by establishing a predictive model for sustainable performance. Data from 261 respondents in Pakistan's construction industry were collected, and hypotheses were tested using partial least squares structural equation modeling. The results indicate that coordination and employee motivation exert a positive impact on sustainable performance. Furthermore, quality relationships partially mediate the relationship between coordination and sustainable performance and between employee motivation and sustainable performance. Additionally, quality management systems significantly moderate the relationship between coordination and sustainable performance, whereas the relationship between employee motivation and sustainable performance is insignificant. This study provides valuable insights for project coordinators, project managers, and policymakers on enhancing the stability of construction projects in emerging economies through quality relationships and quality management systems.

Keywords: mega-project; sustainable performance; coordination; employee motivation; quality relationship; quality management system

Academic Editors: Ahmed Senouci, Wenxin Shen, Wenzhe Tang, Jin Xue and Antonio Caggiano

Received: 1 February 2025
Revised: 24 February 2025
Accepted: 25 February 2025
Published: 16 March 2025

Citation: Ma, L.; Ali, A.; Farooq, M.S.; Musonda, J. Boosting Coordination and Employee Motivation in Mega-Project Sustainable Performance Through Quality Relationships: The Key Role of Quality Management System. *Buildings* **2025**, *15*, 936. https://doi.org/10.3390/buildings15060936

1. Introduction

Mega-projects in the construction sector face particular challenges [1] and involve many professionals, such as consultants, construction managers, contractors, designers, subcontractors, and specialists. Therefore, sustainability plays an essential role in mega-projects [2]. Mega-projects are major infrastructure projects that provide crucial public services, facilitating the production of everyday goods, social goods, and economic growth; they constitute the backbone of modern society [3,4].

However, industrial activities involve waste generation, environmental contamination, resource scarcity, and worsening environmental conditions [5]. These challenges complicate project management and require collaboration and coordination (COR) until the project is accomplished [6]. Mega-projects pose several difficulties and barriers to project management [7]. First, significant financial outlays, protracted construction, and uncontrollable events may occur throughout execution. Second, lengthy designed lifecycles and intricate construction processes demand high-level construction methods and quality. Due to the large number of parties involved and the requirement for cross-professional work, significantly higher levels of cross-functional coordination and collaboration efforts are required within mega-projects compared to typical projects. Therefore, the sustainable performance of mega-projects is a key concern for governments, users, and communities.

Due to their complexity, mega-projects provide valuable opportunities for applying theories and approaches to sustainable innovation. However, as the concept of performance is intricately linked to project stakeholders and remains unclear, determining the failure or success of the project remains challenging [8]. Therefore, mega-project sustainable performance (MSF) includes social, environmental, and economic aspects [4]. Effective COR between multiple contractual partners plays a determining role in the success of construction projects [9]. COR refers to the alignment, convergence, and harmonization of various stakeholders in any sector with numerous objectives [9]. Therefore, efficient and cohesive COR among agencies, consultants, owners, contractors, project management teams, suppliers, and users during the project implementation is essential to avoid issues leading to project failure [10].

The COR process seeks to enhance project delivery and increase efficiency by addressing the interconnectedness of project tasks and parties involved [11]. Uncertainties in the mega-project lifecycle impede teamwork, influencing the MSF. Moreover, employee motivation (EM) refers to the level of enthusiasm and commitment that the employees bear toward the project. To successfully coordinate a project, employee motivation should be upkept. Hence, predicting and managing EM is a key aspect of project COR [12]. Good coordination facilitates smooth project execution and ensures that all teams can work together effectively [13].

However, predicting EM and performance involves several difficulties, such as defining the characteristics of employee performance and identifying the elements affecting employee performance. Motivation in the workplace is defined as the direction of focus, mobilization of effort, and persistence of effort through time [14]. Most studies investigating motivation in the construction sector are flawed; for instance, most motivation theories focus only on an individual's motivation and overlook the social context in which behaviors are carried out, limiting the generalizations that can be established [15]. While COR and EM are crucial factors in project performance, studies exploring how these elements interact to influence the long-term sustainability of mega-projects are scarce. This suggests a need for further research on the efficacy of COR and EM in contributing to MSF, which represent complex and resource-intensive endeavors with significant economic, social, and environmental impacts.

In recent years, numerous studies have focused on EM and COR separately by using different methods [16–18]. Surprisingly, the relationship between COR, EM, and their relationship with MSF by analyzing quality relationship (QR) in construction projects has largely been overlooked. To the best of our knowledge, little research has been conducted on this complex phenomenon, and no studies have examined how the quality management system (QMS) moderates the influences of COR and EM on MSF in the context of construction industries. To address this important research gap, the current study aims to improve sustainable performance by analyzing the effects of COR and EM on QR and MSF.

The current study's framework was developed to address these initiatives. Therefore, this study aims to determine the following: (i) effective methods for managers to proactively address the COR and EM issues in mega-projects; (ii) whether QMS moderates the relationships between COR, EM, and MSF; and (iii) whether QR plays a mediating role in the connection between COR, EM, and MSF. This study examines the hypothesized relationship between COR, EM, and QR and aims to measure MSF using the suggested indicators. Moreover, QMS has been introduced as a novel moderating component in the interplay between COR, EM, and MSF, offering novel insight into a previously unexplored research area.

2. Materials and Methods, Literature Review, and Hypotheses Development

2.1. Coordination and Mega-Project Sustainable Performance

Mega-projects require the coordination of a variety of activities, involving continuous COR throughout the implementation processes. Most of these activities require support, such as frequent meetings among different stakeholders, to enhance the progress of a project with better satisfaction [8]. Therefore, a relational corporation is suggested to mean that COR is carried out by sharing purposes, mutual respect, and information. Moreover, COR factors in construction projects can be identified as part of a process, including resource priorities for essential tasks, comprehensive procurement planning, and identifying task components and dependencies like plans [19]. The COR processes include official COR, plans, meetings, procedures, rules, schedules, informal contacts, telephone, relational COR, and in-person interactions [20]. Construction projects are usually characterized by high ambiguity, inter-organizational, and complex interdependence of activities, highlighting the importance of communication [21]. Based on the debate above, our first hypothesis is as follows:

H1. *Coordination has a significant and positive effect on the mega-project's sustainable performance.*

2.2. Employee Motivation and Mega-Project Sustainable Performance

In construction subcontractor employees, target setting, workforce needs, and incentives and rewards were identified as factors promoting constructive motivated behavior [22]. Notably, EM can enhance the sustainable performance of mega-projects. A comprehensive evaluation of the literature from disciplines other than construction revealed that efficacy, commitment, identification, and cohesion are effective at both levels [23]. These emotional attachments have been correlated with different motivational states, such as emotional connection to the organization [24]. Therefore, EM can enhance the sustainable performance of mega-projects. According to the above discussion, a second hypothesis is proposed:

H2. *Employee motivation has a significant and positive effect on mega-project sustainable performance.*

2.3. Coordination and Quality Relationship

COR is derived from effective communication, shared goals, common knowledge, and respect among team members [25]. The benefits of performing tasks and the inherent advantages of encouraging positive behaviors were identified as sources of increased job satisfaction [25]. QR here refers to the trust between the developer and the customer, as well as the dedication to upholding their working relationship over the long term [26]. Most successful collaborative partnerships provide compelling evidence of the crucial role of QR in upholding long-lasting, high-quality alliances. For example, Toyota's ability to use cutting-edge technology and Chrysler's ability to survive can be attributed to the strong bonds with their business partners [27]. QR is a crucial relational quality that enables partners to forge and grow normative links, which can lessen uncertainty in a close relationship and its unfavorable impacts [28]. According to the above discussion, a third hypothesis is proposed:

H3. *Coordination has a significant and positive effect on quality relationship.*

2.4. Employee Motivation and Quality Relationship

An individual's response to various conditions determines motivation, which manifests in many ways [14]. Therefore, manager exchange illustrates the impact of a dyadic leader–follower connection on an employee's motivation [29]. QR affects a result; relationships with business partners are crucial for enhancing performance [30]. The resulting balance of capital and skills decreases transaction costs dramatically and increases efficiency. Leonidou et al. [31] proposed that strong relationships are built by reducing opportunistic actions and conflicts, improving communication efficiency, adaptation, and managing cultural distance. Therefore, the workplace should be managed to improve the quality of partnerships. According to the above discussion, a fourth hypothesis is proposed:

H4. *Employee motivation exerts a significant and positive effect on quality relationship.*

2.5. The Mediating Role of Quality Relationship

This study presents three relationships to examine the mediating function of QR: (I) the effects of QR on COR and employee motivation; (ii) the effects of QR on the sustainable performance of mega-projects; and (iii) the mechanism of QR in converting COR and EM to MSF. Achieving these sustainability goals requires coordination among multiple stakeholders, such as suppliers, contractors, and project managers. However, determining the mechanism underlying the effects of COR on sustainable performance remains challenging. A key factor is the caliber of the connections between these stakeholders. Due to its partial win–lose interest and the short-term structure of corporate relationships, construction is a harsh atmosphere [32]. An ideal relationship between participants in mega-projects is rare. According to Pryke [33], relationship management is a core competency in the construction industry, and the consistency of relationships is a critical component of sustainable performance. Regarding how QR affects a result, relationships with business partners play an essential role in enhancing performance [30]. The resulting balance of capital and skills decreases transaction costs dramatically and increases efficiency. When a firm's allies become a better basis for learning, the influence of quality collaborations in new environments leads to a positive result [34]. Thus, the quality of partnerships should be fostered by efficient workplace management. According to the above discussion, the following hypotheses are proposed:

H5. *Quality relationship has a significant and positive effect on construction mega-projects' sustainable performance.*

H6a. *Quality relationship mediates the association between coordination and mega-project sustainable performance.*

H6b. *Quality relationship mediates the association between employee motivation and mega-project sustainable performance.*

2.6. The Moderating Role of the Quality Management System

QMS might affect the association between COR, EM, and MSF. Therefore, management responsibilities and motivational aspects should be considered when studying the effect of COR and EM on MSF. Poor management is common in the construction industry. Systematic management is also essential for the successful implementation of construction projects [35]. To avoid interruptions in activities, the roles and duties of each mega-project team participant should be well defined. A complex temporary multi-organizational structure is frequently established during the mega-project development process that constantly faces differences between two layers of goals: the participating organizations' long-term aims and the project's operational phase, and the construction project's temporary purposes [36]. Dynamic management systems must be developed by main contractors to promote the COR of activities and monitor the actions of their representatives [37]. Therefore, the QMS facilitates the mega-project's success. QMS was initially designed to fulfill the needs of stakeholders and is necessary to ensure that the project outcomes follow the corresponding requirements. Desmond [38] reported that poor staff does not lead to poor quality, but bad management does. Considering the importance of QMS on sustainable performance, the following hypothesis was proposed:

H7a. *A quality management system moderates the relationship between coordination and mega-project sustainable performance.*

H7b. *A quality management system moderates the relationship between employee motivation and mega-project sustainable performance.*

3. Research Methodology

3.1. Measures and Validation

The survey data from practitioners in the construction industry were gathered using questionnaires. In the first stage, a questionnaire was prepared to identify factors from the related literature. The final questionnaire was distributed among the construction workers in Pakistan. The comprehensive questionnaire comprised two sections. Demographic data about the respondents, including their age, degree of education, job title, and work experience, were included in the questionnaire's first section. The second component of the questionnaire was further divided into five subsections, as shown in Appendix A. The first subsection of the questionnaire focused on the questions related to factors affecting project COR, including planning (PF), resource handling and record documentation (RDF), teamwork and leadership (TLF), value engineering and facilitating (EFF), and communication (CF). The constructs for this section were based on previous research by Alaloul et al. [39].

In the second section, the measures defined by Raoufi and Fayek [23] were used to determine the association between different components of EM, such as efficacy (EF), commitment (CM), identification (ID), and cohesion (CO). The third section included four items related to QMS, which were based on studies from Yen et al. [40]. In the fourth section, QR was calculated using four items, according to the methods of Sharma [41]. Finally, MSF was calculated using three dimensions: environmental performance (ENP), social performance (SCP), and economic performance (ECP). The environmental MSF was measured using three items; social MSF was measured using four items, and economical MSF was measured using three items. The constructs for this part were derived from Ali et al. [4]. The questionnaire used closed-end questions to obtain the respondents' opinions using a 5-point Likert scale from 1 (strongly disagree) to 5 (strongly agree). The questionnaire and construct item details are provided below in Appendix A.

3.2. Data Collection and Sampling

The identification of factors from the literature provides the basis for the preliminary questionnaire. The questionnaire was designed to facilitate understanding and feasibility. Four construction professionals and three university professors examined the questionnaire's design and structure to ensure that the questionnaire was easy to comprehend. Moreover, the importance of each factor in the context of Pakistan's construction industry was evaluated. Eventually, a final questionnaire was developed with the updated list of factors, which was used to collect information from the respondents. The results were analyzed to determine the fundamental relationships and relative importance among the elements, supporting the subsequent analysis of COR, EM, and MSF.

The final survey questionnaire was distributed by email, personal visits, and other social media applications among 750 randomly selected construction experts. The sample included workers in Pakistan's construction industry. Data were retrieved from government archives, online databases, and organizational records, which were publicly accessible. Help from the parties involved regular updates in construction and follow-up emails and phone calls in case of insufficient response time. The data collection process took five months to complete, starting in October 2023 and ending in March 2024. Assuming that every participant in a study fits inside a narrow enough sample insight, Elton and Yamane [42] presented the method below to determine the proper sample size.

$$\mathrm{n} = \frac{N}{1 + Ne^2}$$

where n = sample size, N = population size, and e = margin of error. Academics have repeatedly applied this formula, demonstrating its usefulness and validity [43]. A total of 750 workers satisfied the research inclusion criteria (relevant employment experience). Therefore, the margin of error affects the findings' dependability, which in turn affects how widely the research's conclusions may be implemented. As the margin of error grows, the reliability of the results declines. Nevertheless, this study's 5% margin of error was acknowledged by earlier researchers [43]. The sample size was as follows:

$$\mathrm{n} = \frac{750}{1 + 750(0.05)^2} = \frac{750}{1 + 750(0.0025)} = 260.86 \cong 261$$

Therefore, to bolster the previously mentioned circumstances, a representative sample size of 261 was utilized in this study. Considering the lack of significant difference between the two approaches, selecting a larger sample size would seem desirable [44]. Finally, 261 responses were received, representing a 37.28% response rate, exceeding the 20% response rate, which is considered good [45]. The demographic information is shown in Figure 1. The distribution of respondents based on education, organization position, and experience is displayed in Figure 1. Respondents were required to have more than ten years of experience in the construction industry. In addition, most respondents were engineers who held master's and Ph.D. degrees in the construction field. Therefore, the collected data were deemed adequate for perception analysis. Furthermore, the survey questionnaire included a cover letter outlining the study purpose and ensuring the privacy of the respondents. In this research, data were collected from different institutions such as PWD (Public Work Department), NESPAK (National Engineering Service Pakistan), NHA (National Highway Authority), and some other construction companies in Pakistan. Most of the respondents were civil engineers. The respondents included chief executive officers (CEOs), project managers (PMs), site engineers (SEs), designer engineers (DEs), project coordinators (PCs), planning engineers (PEs), and quality surveyors (QSs) in the construction industry. The participants had adequate expertise and skills to clarify the relationships in this research.

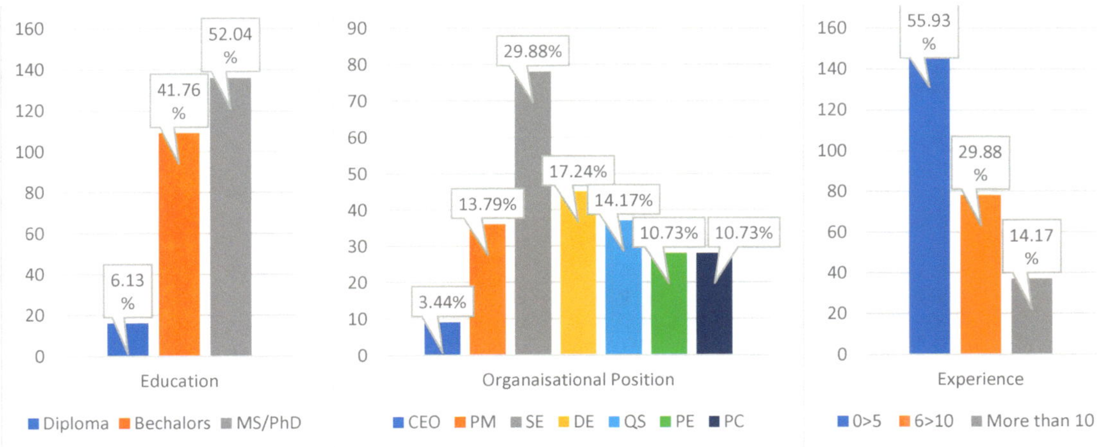

Figure 1. Demographic details of respondents.

4. Results\Data Analysis

In this work, smart-PLS version 4 and partial least squares structural equation modeling (PLS-SEM) were employed for analysis. PLS-SEM is an effective technique for analyzing complex models, particularly with complex interactions between the variables. This method facilitates the identification and modeling of the direct and indirect impacts of variables. The primary goal of this study is to identify endogenous variables, those influenced by other variables in the model contribute the least to the overall variance. By doing so, the researchers aim to improve the model's predictive accuracy and reduce noise, enhancing the clarity and reliability of the results. This approach is particularly useful in structural equation modeling with complex relationships. PLS-SEM can simultaneously handle structural and measurement models [46]. Thus, PLS-SEM was used to test the proposed model. As described by Hair et al. [46], the structural and measurement models were explicitly approximated to ensure accurate results.

4.1. Measurement Model

The investigation revealed that all constructs satisfied the standards for dependability (composite reliability (CR) and Cronbach's alpha), which have been advocated by multiple scholars working in a variety of fields [47]. The average variance extracted (AVE) and factor loading values were used to determine convergent validity; the results indicated that all items had factor loading values exceeding 0.7 on their respective constructs. However, as previously demonstrated in survey-based studies, AVE values were higher than 0.5 [46]. Therefore, convergent validity was established in our research. The authors employed the HTMT and Fornell–Larcker approach to examine discriminant validity. The findings demonstrated that both constructs' inter-correlation values were lower than those of the AVE square root. Additionally, the HTMT values were less than 0.9, as suggested by Hair et al. [46]. Tables 1 and 2 provide the specifics of the findings as mentioned.

Table 1. Factor loading, CR, AVE, HTMT, standard deviation.

Construct	Factor Loading Range	Cronbach's Alpha	CR	AVE	Mean	HTMT Below 0.9	Std. Deviation
PF	0.846–0.905	0.890	0.923	0.971	3.675	yes	1.217
RDF	0.750–0.869	0.767	0.866	0.683	4.035	yes	1.094
TLF	0.737–0.899	0.866	0.910	0.717	4.266	yes	0.825
EFF	0.833–0.899	0.897	0.928	0.764	4.256	yes	0.874
CF	0.834–0.896	0.892	0.925	0.756	4.268	yes	0.786
EF	0.741–0.887	O.801	0.874	0.699	4.186	yes	0.894
CM	0.812–0.850	0.783	0.874	0.698	4.275	yes	0.830
ID	0.811–0.844	0.773	0.869	0.688	3.778	yes	1.181
CO	0.808–0.883	0.792	0.879	0.707	4.296	yes	0.788
QR	0.785–0.877	0.861	0.906	0.706	4.235	yes	0.775
QMS	0.807–0.875	0.865	0.908	0.712	4.246	yes	0.783
ENP	0.760–0.868	0.855	0.902	0.698	4.367	yes	0.691
SCP	0.820–0.881	0.783	0.839	0.691	4.362	yes	0.866
ECP	0.743–0.853	0.840	0.812	0.732	3.886	yes	0.978

Table 2. Fornell–Larcker criterion test.

	CF	CM	CO	EF	EFF	ID	PF	ENP	QMS	QR	SCP	ECP	RDF	TLF
CF	**0.870**													
CM	0.669	**0.835**												
CO	0.748	0.701	**0.841**											
EF	0.621	0.657	0.603	**0.836**										
EFF	0.687	0.754	0.765	0.664	**0.874**									
ID	0.727	0.762	0.754	0.672	0.687	**0.829**								
PF	0.665	0.760	0.663	0.812	0.836	0.710	**0.869**							
ENP	0.736	0.622	0.745	0.549	0.789	0.730	0.700	**0.835**						
QMS	1.00	0.669	0.748	0.621	0.745	0.727	0.665	0.736	**0.844**					
QR	0.833	0.692	0.780	0.639	0.633	0.720	0.681	0.739	0.834	**0.840**				
SCP	0.625	0.753	0.688	0.653	0.754	0.822	0.702	0.620	0.652	0.672	**0.826**			
ECP	0.792	0.744	0.812	0.650	0.798	0.799	0.696	0.761	0.792	0.782	0.741	**0.847**		
RDF	0.725	0.653	0.788	0.753	0.654	0.722	0.802	0.720	0.552	0.572	0.800	0.745	**0.812**	
TLF	0.692	0.544	0.712	0.550	0.698	0.799	0.796	0.661	0.592	0.582	0.641	0.737	0.711	**0.832**

However, common method bias (CMB) is a serious problem that may impact assessments of construct reliability and validity. Researchers claim that gathering information for exogenous and endogenous components from diverse sources may significantly reduce the risk of process bias [48]. Consequently, the authors gathered data from different sources over a long period. Harman's single factor test was employed to verify the absence of CMB [49].

4.2. Assessment of Structural Model

Collinearity statistics is essential for formative outer and inner models. The variance inflation factor (VIF) among exogenous constructs is a severe issue in Smart-PLS utilizing inner VIF values [46]. This research revealed that all of the exogenous latent constructs' inner VIFs were less than 3.3. As a result, no collinearity was found in this study. Moreover, R^2 is a function of the variance described in the endogenous variables and reflects the model's predictive power. An R^2 value of 0.19 is considered low, 0.33 is moderate, and 0.67 is considered substantial. In this study, the R^2 of MSF was 0.627, reflecting a moderate variance in the exogenous latent construct [46]. The model is expected to forecast by using the Q^2 Stone-Geisser value determined by blindfolding procedures [50]. The criterion of the

Stone-Geissor Q^2 indicates that the model must be capable of predicting endogenous latent variable indicators. The outcome indicates a Q^2 value of 0.546 for the model, indicating predictive validity, as evidenced by the value exceeding zero.

The proposed study model contained second-order formative constructs, COR, and EM. Weights of first-order reflective constructs were estimated by Petter et al. [51], revealing their significance as CF = 0.367, EFF = 0.305, PF = 0.132, RDF = 0.196, TLF = 0.343, CM = 0.348, CO = 0.421, EF = 0.374 and ID = 0.204, $p < 0.001$. The VIF was also examined for multicollinearity. The lower VIF values, CF = 2.217, EFF = 1.351, PF = 1.090, RDF = 1.427, TLF = 2.160, CM = 1.293, CO = 1.560, EF = 1.617, and ID = 1.230 of all first-order constructs for COR and EM confirmed its validity [46]. Finally, the goodness of fit (GOF) was determined using the following formula:

$$\text{GOF} = \sqrt{R^2 \times AVE}$$

The PLS models' global validation cutoff values range from 0 to 1, resulting in GOF large 0.36, medium 0.25 and small 0.1 [52]. The model's GOF was determined to be 0.646, indicating a high capacity for prediction and good data fit; furthermore, our SRMR value was also in a reasonable range. Moreover, as shown in Figure 2, QR mediates the link between the exogenous constructs COR, EM, and MSF. Additionally, the authors tested the moderating variable QMS, the relationship between COR, EM, and MSF. COR and EM have a more significant impact on MSF (β = 0.240), (β = 0.037) in the construction industry. Thus, the findings support H1 and H2. The research also showed that COR and EM significantly affect QR (β = 0.460 and β = 0.298, respectively), which supports hypotheses H3 and H4. Moreover, QR also significantly impacted MSF (β = 0.330), supporting hypothesis H5. Table 3 and Figure 2 display the results of all hypotheses.

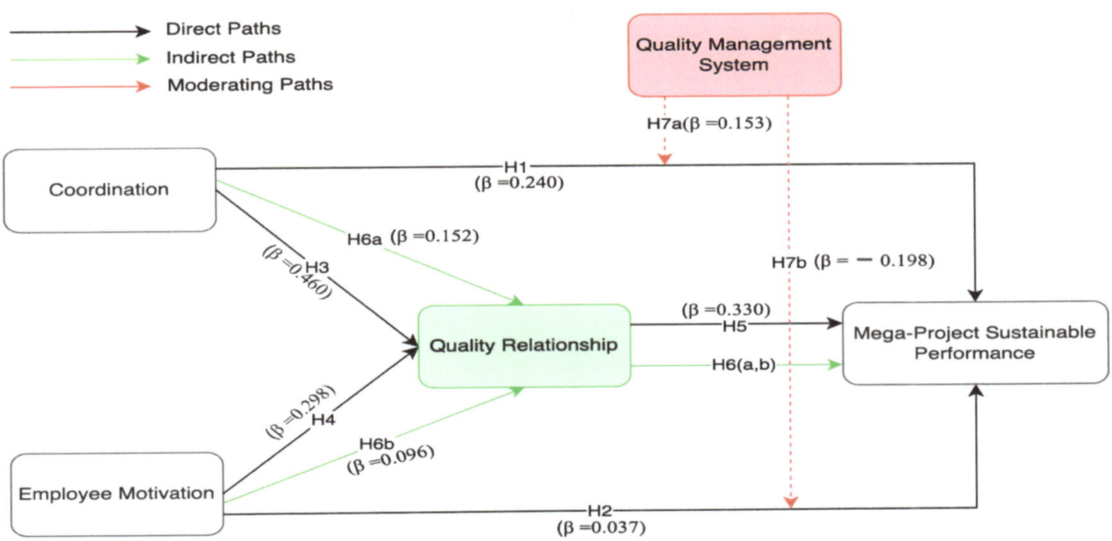

Figure 2. Estimation of the SEM.

Table 3. Hypothesis assessment.

	Path Relationship	β Value	T Statistics	*p* Values	Decision
H1	COR -> MSF	0.240	2.639	0.003	Supported
H2	EM -> MSF	0.037	2.764	0.000	Supported
H3	COR -> QR	0.460	5.763	0.000	Supported
H4	EM -> QR	0.298	3.842	0.000	Supported
H5	QR -> MSF	0.330	4.643	0.000	Supported
Mediation Analysis					
H6a	COR -> MSF	0.240	2.639	0.009	Partial
	COR -> QR -> MSF	0.152	3.899	0.000	Mediation
H6b	EM -> MSF	0.037	2.764	0.000	Partial
	EM -> QR -> MSF	0.096	3.182	0.002	Mediation
Moderation Analysis					
H7a	QMS*COR -> MSF	0.153	2.583	0.010	Supported
H7b	QMS*EM -> MSF	−0.198	2.907	0.004	Not Supported

4.3. Importance–Performance Map Analysis (IPMA)

The importance–performance map analysis (IPMA) is a useful systematic tool that expands the traditional route coefficient estimates in a more diagnostic way and graphically illustrates the discrepancy between the variables' performance and importance. The primary objective of IPMA is to determine which antecedents work well but are not very important and vice versa [47]. In our model, MSF is a dependent construct that is predicted by the following four antecedents: COR, CM, QR, and QMS. COR achieved an importance value of 0.385 and a performance value of 80.098, as shown in Figure 3 and Table 4. Similarly, the importance values of QR, QMS, and CM were 0.328, 0.229, and 0.147, with performance values of 81.451, 81.632, and 79.854, respectively. Comparing COR with QR, QMS, and CM, COR exhibited a higher importance value, while QMS achieved a higher performance value than QR, COR, and CM. In the ceteris paribus situation, an increase in QR, QMS, and CM performance by one unit leads to a 0.328-, 0.229-, and 0.147-unit improvement in MSF performance.

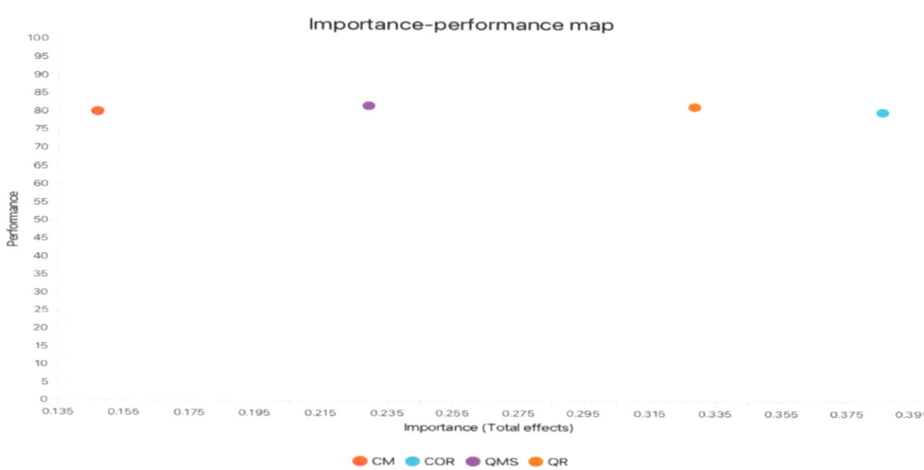

Figure 3. Importance–performance map.

Table 4. Importance–performance results.

Construct	Importance	Performance
COR	0.385	80.098
CM	0.147	79.854
QR	0.328	81.451
QMS	0.229	81.632

Similarly, a one-unit increase in COR performance was found to lead to a 0.385-unit improvement in MSF performance. Therefore, the construction industry in Pakistan should focus on QR, QMS, and CM, along with COR. To produce more observable and quantifiable results, QMS first ensures consistent procedures, quality control, and ongoing improvements. Second, QMS frequently includes more standardized frameworks that are directly measurable and amenable to improvement. However, despite its importance, coordination (COR) is frequently impacted by softer factors like interpersonal dynamics and communication, which are more difficult to quantify and control.

4.4. Mediation and Moderation Analysis

Mediation refers to the process in which the effects of an antecedent construct are transmitted to an outcome construct through a mediating construct [53]. In this study, the mediating construct QR was hypothesized to mediate the effects of COR and EM on MSF. First, COR and EM were confirmed to exert direct and significant effects on MSF, and QR was hypothesized to act as a mediator (H6a, H6b). Constructive relationships that are marked by mutual respect, trust, and good communication can facilitate COR efforts and produce longer-term results. Consequently, QR plays a mediating role in the relationship between COR and MSF. Accordingly, enhancing the QR can amplify the benefits of COR, ultimately assisting in the timely and sustainable completion of mega-projects. Moreover, good connections based on mutual respect, trust, and efficient communication improve EM over time. This suggests that the quality of these connections acts as a mediator, enhancing the relationship between EM and the MSF. As a result, cultivating strong bonds within the team can enhance EM, improving mega-project sustainability.

Construction companies may motivate their workforce to meet sustainability targets and ensure that projects are executed successfully, ethically, and efficiently by placing a high priority on QR. The authors found that QR played a significant mediating role between COR, EM, and MSF, as shown in Table 3. Thus, this study also supports hypotheses 6a and 6b of mediation. Considering that both direct and indirect effects pointed in the same direction, the mediation effects were found to be complementary [53]. Finally, the authors assessed the moderating effect of QMS on MSF (H7a, H7b). Data showed that QMS significantly strengthens the link between COR and MSF ($\beta = 0.153$, $p < 0.010$). Similarly, the effect of QMS on the association between EM and MSF was found to be insignificant ($\beta = -0.198$, $p < 0.004$). Figure 4 demonstrates that QMS enhances MSF in the construction sector when stakeholders or managers find higher QMS rather than low QMS. The moderating effect test is shown in Figures 4 and 5.

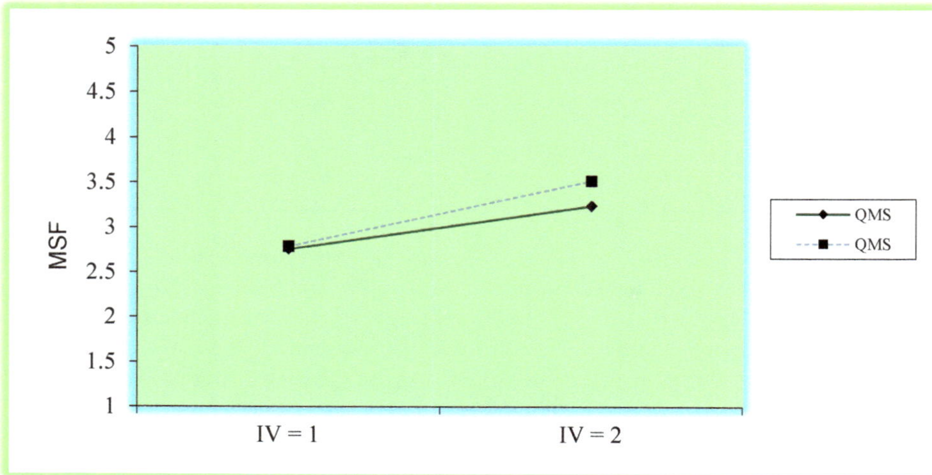

Figure 4. Moderating effect of QMS between COR and MSF.

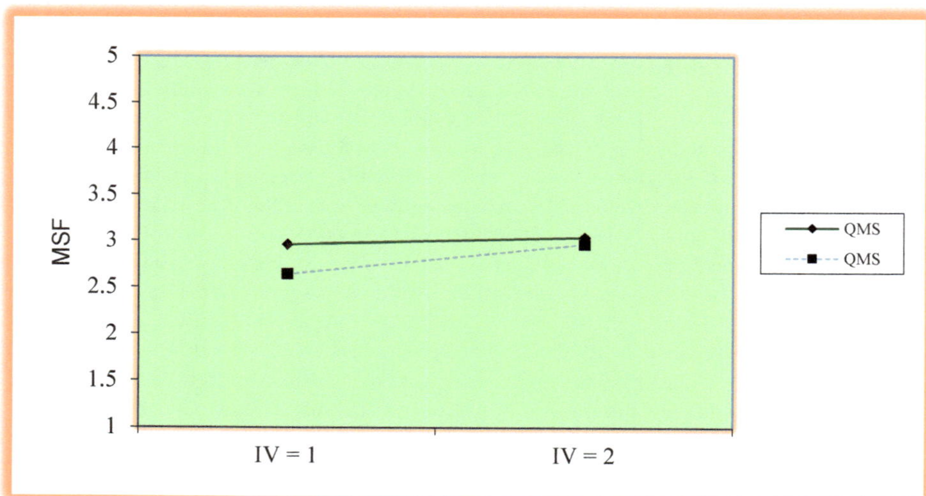

Figure 5. Moderating effect of QMS between EM and MSF.

5. Discussion

This study was conducted to explore the impact of COR and EM on MSF by using the PLS-SEM method. The introduction of QMS between COR and MSF was positive (0.153), indicating that QMS plays a significant role in this relationship. Additionally, the introduction of QMS between EM and MSF was unfavorable (−0.198), indicating that QMS did not play a significant role in this relationship. EM is an individual trait that enhances an individual's motivation level. In contrast, QMS is related to managing the quality of construction work and represents a different phenomenon, explaining the lack of positive moderation. Therefore, while examining the effect of COR on MSF, managerial duties were found to be just as essential as motivating considerations. The mediation analysis procedure developed by Baron and Kenny [54] has been used by researchers. Table 3 illustrates a significant direct and indirect impact between EM and MSF, indicating that QR exerts a partial mediation. Moreover, the relationship between COR and MSF is significant directly or indirectly, suggesting that QR exhibits a partial mediation effect.

The finding sheds light on COR and EM and is significant for MSF. Effective COR involves seamless teamwork and organized efforts, along with high EM, and significantly enhances the long-term success and sustainability of mega-projects. Good teamwork and employee motivation lead to better project outcomes, such as meeting environmental, social, and economic goals. This positive and significant relationship highlights the importance of fostering collaboration and motivation within teams to achieve sustainable performance in complex, resource-intensive mega-projects. Furthermore, QR is a key mediating variable that enhances the COR and EM for enhancing MSF. The cohesion-related factors (beta 0.421) have the most significant influence on EM regarding the relative importance of the four EM factors. Moreover, the communication-related factors (beta 0.367) have the most significant influence on COR regarding the relative importance of other COR factors. It can be regarded as the most crucial indicator affecting MSF. These factors should be analyzed to evaluate their impact on MSF to achieve better COR and EM in construction projects. Project coordinators and project managers should be involved in the project's planning to overcome this problem [55].

Project professionals have the expertise, qualification, experience, and knowledge of construction techniques. Their participation in the mega-project in the pre-construction phase could improve project COR and EM, thereby boosting the possibility of sustainable performance of mega-projects. In the pre-construction phase, effective planning, clear communication, and strategic decision-making by these key stakeholders can establish a strong foundation for COR and EM by setting clear goals and fostering a collaborative environment. A streamlined COR and motivated employees improve resource utilization, reduce delays, and enhance adherence to sustainability goals, such as minimizing environmental impact and ensuring long-term economic and social benefits. Ultimately, the proactive involvement of project managers and policymakers in this early stage creates a positive ripple effect, driving the project toward successful and sustainable outcomes. The communication-related factors and cohesion-related factors are crucial elements for all participants and are pivotal for the success of the mega-project. In addition to cohesion-related factors and communication-related factors, COR and EM are also influenced by the other model factors. This effect has been verified by evaluating relationships.

6. Conclusions and Implication

6.1. Conclusions

Previous studies have confirmed the importance of COR and EM in the construction industry and other fields. The study's findings from the conceptual model's calculation support the hypothesis. The findings of the PLS-SEM study revealed that COR and EM directly impacted mega-projects and suggested that the four most important factors were communication and cohesion, teamwork and leadership, efficacy, and commitment-related factors with maximum standardized path coefficient. The conceptual model's GOF index was calculated at 0.646, indicating that the conceptual model has sufficient validity and reliability, matching the data supported by the PLS-SEM. However, the study also indicated that these factors impact MSF and influence COR and EM in construction mega-projects. The findings of the standardized conceptual model assist project coordinators and project managers in identifying areas to improve for enhanced COR and EM.

6.2. Theoretical Implications

This study offers important new insights into a number of crucial areas related to mega-projects. Firstly, the study clarifies how efficient coordination between various stakeholders improves project outcomes, reduces delays, and maximizes resource utilization, thereby enriching our understanding of COR processes. Secondly, this research also clarifies the importance of EM in boosting output and achieving project objectives. This research offers useful techniques to improve labor productivity by identifying elements that drive motivation, such as clear communication, acknowledgment, and career growth possibilities. Moreover, this study also investigates the sustainability of mega-projects, emphasizing the importance of incorporating social, economic, and environmental factors into project development and implementation. This strategy supports global sustainability objectives while ensuring sustainability for future generations.

The study also explores the mediating effect of QR, demonstrating how robust, trust-based interactions among project participants can lead to high-quality project performance. These connections promote cooperation, lessen friction, and ease the exchange of information, all of which promote project success in its final stages. Finally, the moderating function of QMS is investigated, demonstrating how strong QMS frameworks amplified the benefits of CM and COR on project results. The QMS improved project performance by ensuring that quality requirements were regularly met by implementing defined processes and continuous improvement activities. These theoretical contributions provide construction industry workers a direction and a comprehensive understanding of the dynamics involved in managing mega-projects.

6.3. Practical Implications

This research provides solutions to problems in the construction industry. Firstly, the results emphasize the importance of setting up robust COR mechanisms. To reduce inefficiencies and prevent delays, project managers and leaders can use advanced project management tools and communication platforms to enable smooth COR among different stakeholders. The results imply that investments in motivational techniques can lead to appreciable gains in project performance regarding EM. Construction companies require proactive efforts in rewarding and acknowledging worker achievements and should offer opportunities for professional growth while promoting a positive work environment. Such initiatives can increase productivity and job satisfaction, ultimately leading to project success. Multiple teams and subcontractors are involved in mega-projects, and recognizing and rewarding key contributions can improve morale and productivity. For example, acknowledging employees who excel in safety practices or those who contribute innovative solutions to coordination issues can boost team collaboration and project success.

Secondly, the construction companies are encouraged to incorporate sustainability ideas into their project planning and execution processes by focusing on sustainable performance. Moreover, blockchain and AI can enhance transparency, security, and accountability within mega-projects by enabling immutable records of transactions, contracts, and project milestones. Such measures can reduce fraud, improve stakeholder collaboration, and streamline project management processes. AI-powered tools can also improve risk management by identifying potential issues in real time, thereby improving project outcomes and sustainability. Furthermore, by integrating BIM, project teams can streamline coordination efforts, improve accuracy, and minimize errors, making it an essential tool for managing mega-projects. Using sustainable materials, prioritizing energy efficiency first, and implementing green building methods will not only improve environmental results but also boost the company's reputation and draw in eco-aware clients. Utilizing sustainable materials such as low-carbon cement could enhance energy efficiency. Moreover, LEED-certified projects are designed to be energy-efficient, use sustainable materials, and reduce waste. These could improve project sustainable outcomes.

The study also emphasizes the importance of QR in the accomplishment of project success. Practitioners should concentrate on developing solid, trustworthy relationships with suppliers, subcontractors, and clients, among other project participants. Transparent communication, cooperative problem-solving, and consistent involvement of stakeholders can help achieve this. Finally, the moderating role of QMS implies that QMS frameworks should be adopted and strictly implemented by construction companies. Organizations can ensure that project outcomes meet or surpass quality standards by following established quality practices and always seeking improvement. This will increase client satisfaction and provide a competitive advantage.

6.4. Limitations and Future Research

Nevertheless, the limitations of the present study should be acknowledged. First, this study's conclusions are based on a questionnaire survey conducted in Pakistan, and the results cannot be extended to other countries. Researchers from other countries may conduct a similar study to examine the contribution of QR and QMS to the relationship between COR, CM, and MSF and to compare the contributions of QR and QMS in various nations. Further research is needed to explore the complex relationship between COR, EM, and sustainable performance in mega-projects, thereby identifying complexities specific to the construction sector. Examining how modern technologies like blockchain and artificial intelligence might improve stakeholder COR and communication is an interesting direction to pursue. Mega-project dependability and efficiency may be revolutionized by applying these technologies to project management procedures. Finally, analyzing the efficacy of diverse QMSs in diverse mega-project categories and cultural settings can provide a more comprehensive understanding of appropriate methodologies. The most flexible and effective quality management strategies can be found through comparative studies that examine the performance of QMS in various organizational and geographic contexts. These studies also offer a comprehensive framework for improving mega-project outcomes internationally.

Author Contributions: L.M.: Supervision, formal analysis, review and editing. A.A.: Development of theoretical framework, methodology, writing—original. M.S.F.: Data collection, conceptualization. J.M.: Software, visualization. All authors have read and agreed to the published version of the manuscript.

Funding: This research was funded by Hainan Province Social Science Planning Research funding Project No. HNSK(YB)24-14.

Institutional Review Board Statement: Not applicable.

Informed Consent Statement: Not applicable.

Data Availability Statement: The corresponding author will supply data supporting the study's conclusions upon justifiable request.

Acknowledgments: The authors thank the EIC, the AE and the reviewers for constructive suggestions.

Conflicts of Interest: The authors declare no conflicts of interest.

Abbreviations

The following abbreviations are used in this manuscript:

COR	Coordination
EM	Employee motivation
MSF	Mega-project sustainable performance
QR	Quality relationship
QMS	Quality management system

Appendix A

It is measured to what extent do you agree or disagree with the following statements on a 5-point Likert scale (5—Strongly Agree, 1—Strongly Disagree).

Section 1

Demographic Details
Education ☐ Bachelor's ☐ Master's ☐ PhD
Organizational Position ☐ Chief executive officer ☐ Project manager ☐ Site engineer ☐ Designer engineer ☐ Project coordinator ☐ Planning engineer ☐ Quality surveyor
Experience ☐ 0>5 years ☐ 6>10 years ☐ More than 10 years

Section 2

Factors affecting project coordination
Planning-related factors (PFs)
PF1 quality assurance plan PF1 Better Execution of a project Plan PF3 All Parties' participation in plan PF4 Identification of appropriate resources
Resource handling and record documentation-related factors (RDFs)
RDF1 Controlling project finances RDF2 Record maintenance RDF3 Ensuring the timeliness of all work carried
Teamwork and Leadership-related factors (TLFs)
TLF1 Joint site visit TLF2 Meetings TLF3 Managing contractual issue TLF4 Maintain proper relationships with all parties
Value engineering and facilitating-related factors (EFFs)
EFF1 Design and specification clarity EFF2 Gathering and compiling information EFF3 Identifying potential delays and strategic activities EFF4 Work integration
Communication-related factors (CFs)
CF1 Open a wide and fast communication channels CF2 Maintaining effective organizational structural and communication channels CF3 Liaison with the client and consultant CF4 Communicate instances of poor quality, unsafe or adverse situations to relevant staff
Factors affecting Employee motivation
Efficacy
EF1 Crew confidence in ability to perform tasks effectively EF2 crew confidence in ability to perform difficult tasks EF3 crew ability to concentrate on performing tasks

Commitment
CM1 Crew members very happy to spend the rest of career with the organization CM2 crew members to see the organization's problems as own CM3 crew's emotional attachment to the organization
Identification
ID1 Crew members to feel proud to be part of the crew ID2 crew members' identification with the other members of the crew ID3 crew members to like to continue working with the crew
Cohesion
CO1 Crew members get along well together CO2 defending each other from criticism CO3 crew being a close one
Factors affecting quality management system
QMS1 Effective utilization of technology and resources QMS2 the goal of the project was clear QMS3 The implementation project phase was kept on time QMS4 Team members had a expertise about the process of the organization
Factors affecting quality relationship
QR1 Respect for the local firm partners QR2 Overall partner satisfaction QR3 Long-term relationship QR4 Quick problem-solving
Factors affecting mega-project sustainable performance
Environmental performance
EP1 Our mega-project successfully reduced energy consumption. EP2 Our mega-project successfully reduced construction wastes EP3 Our project successfully decreased the frequency of environmental accidents
Social performance
SP 1 Our mega-project successfully satisfied the client's needs SP 2 Our mega-project successfully satisfied the users' needs SP 3 Our mega-project successfully satisfied the government's needs. SP 4 Our mega-project successfully satisfied the public's needs
Economic performance
EP 1 Our mega-project successfully met the budget goals. EP 2 Our mega-project successfully achieved the organization's financial objectives EP 3 Our mega-project successfully satisfied project investors' objectives

References

1. Ahmed, R.; Jawad, M. Avoiding or Disregarding: Exploring the Relationship between Scope Creep, Project Complexity, and the Success of Construction Projects. *Proj. Leadersh. Soc.* **2022**, *3*, 100064. [CrossRef]
2. Lin, X.; McKenna, B.; Ho, C.M.F.; Shen, G.Q.p. Stakeholders' Influence Strategies on Social Responsibility Implementation in Construction Projects. *J. Clean. Prod.* **2019**, *235*, 348–358. [CrossRef]
3. Alotaibi, S.; Martinez-Vazquez, P.; Baniotopoulos, C. Mega-Projects in Construction: Barriers in the Implementation of Circular Economy Concepts in the Kingdom of Saudi Arabia. *Buildings* **2024**, *14*, 1298. [CrossRef]

4. Ali, A.; Li, M.; Shahzad, M.; Musonda, J.; Hussain, S. How Various Stakeholder Pressure Influences Mega-Project Sustainable Performance through Corporate Social Responsibility and Green Competitive Advantage. *Environ. Sci. Pollut. Res.* **2024**, *31*, 67244–67258. [CrossRef] [PubMed]
5. Turan, V. Calcite in Combination with Olive Pulp Biochar Reduces Ni Mobility in Soil and Its Distribution in Chili Plant. *Int. J. Phytoremediation* **2022**, *24*, 166–176. [CrossRef] [PubMed]
6. Yang, J.B.; Peng, S.C. Development of a Customer Satisfaction Evaluation Model for Construction Project Management. *Build. Environ.* **2008**, *43*, 458–468. [CrossRef]
7. Ma, L.; Musonda, J.; Ali, A. MSR Influence on Environmental & Ecological Balance: Mediating Effect of Environmental Regulations & Strategies. *J. Clean. Prod.* **2023**, *386*, 135817.
8. Ali, A.; Ma, L.; Shahzad, M.; Hussain, S. Managing Stakeholder Pressure for Megaproject Success and Green Innovation: The Key Role of Social Responsibility. *EMJ—Eng. Manag. J.* **2024**, *36*, 366–377. [CrossRef]
9. Shen, W.; Ying, W. Large-Scale Construction Programme Resilience against Creeping Disruptions: Towards Inter-Project Coordination. *Int. J. Proj. Manag.* **2022**, *40*, 671–684. [CrossRef]
10. Titarenko, B.; Hasnaoui, A.; Titarenko, R.; Buzuk, L. Project Risk Management in the Construction of High-Rise Buildings. *E3S Web Conf.* **2018**, *33*, 03074. [CrossRef]
11. Khanzode, A.; Fischer, M.; Reed, D. Benefits and Lessons Learned of Implementing Building Virtual Design and Construction (VDC) Technologies for Coordination of Mechanical, Electrical, and Plumbing (MEP) Systems on a Large Healthcare Project. *J. Inf. Technol. Constr.* **2008**, *13*, 324–342.
12. Raoufi, M.; Robinson Fayek, A. Fuzzy Agent-Based Modeling of Construction Crew Motivation and Performance. *J. Comput. Civ. Eng.* **2018**, *32*, 04018035. [CrossRef]
13. Love, P.E.D.; Irani, Z.; Edwards, D.J. A Seamless Supply Chain Management Model for Construction. *Supply Chain Manag.* **2004**, *9*, 43–56. [CrossRef]
14. Latham, G.P.; Pinder, C.C. Work Motivation Theory and Research at the Dawn of the Twenty-First Century. *Annu. Rev. Psychol.* **2005**, *56*, 485–516. [CrossRef] [PubMed]
15. Raoufi, M.; Fayek, A.R. *Identifying Factors Affecting Motivation of Construction Crew Workers*; University of British Columbia Library: Vancouver, BC, Canada, 2015.
16. Amin, M.; Shamim, A.; Ghazali, Z.; Khan, I. Employee Motivation to Co-Create Value (EMCCV): Construction and Validation of Scale. *J. Retail. Consum. Serv.* **2021**, *58*, 102334. [CrossRef]
17. Li, R.; Liu, H.; Chen, Z.; Wang, Y. Dynamic and Cyclic Relationships between Employees' Intrinsic and Extrinsic Motivation: Evidence from Dynamic Multilevel Modeling Analysis. *J. Vocat. Behav.* **2023**, *140*, 103813. [CrossRef]
18. Parker, S.L.; Dawson, N.; Van den Broeck, A.; Sonnentag, S.; Neal, A. Employee Motivation Profiles, Energy Levels, and Approaches to Sustaining Energy: A Two-Wave Latent-Profile Analysis. *J. Vocat. Behav.* **2021**, *131*, 103659. [CrossRef]
19. Andy, K.W.; Price, D.F. Causes Leading to Poor Site Coordination in Building Projects. *Organ. Technol. Manag. Constr. Int. J.* **2010**, *2*, 167–172.
20. Daft, R.L. *Organization Theory and Design*, 8th ed.; South-Western Publ. Co.: Cincinatti, OH, USA, 2004.
21. Zhang, Y.; Wang, Y.; Yao, H. How Does the Embeddedness of Relational Behaviours in Contractual Relations Influence Inter-Organisational Trust in Construction Projects? *Eng. Constr. Archit. Manag.* **2022**, *29*, 222–244. [CrossRef]
22. Cox, R.F.; Issa, R.R.; Frey, A. Proposed Subcontractor-Based Employee Motivational Model. *J. Constr. Eng. Manag.* **2006**, *132*, 152–163. [CrossRef]
23. Raoufi, M.; Fayek, A.R. Framework for Identification of Factors Affecting Construction Crew Motivation and Performance. *J. Constr. Eng. Manag.* **2018**, *144*, 04018080. [CrossRef]
24. Johnson, R.E.; Chang, C.H.; Yang, L.Q. Commitment and Motivation at Work: The Relevance of Employee Identity and Regulatory Focus. *Acad. Manag. Rev.* **2010**, *35*, 226–245.
25. Gallego Sánchez, M.d.C.; De-Pablos-Heredero, C.; Medina-Merodio, J.A.; Robina-Ramírez, R.; Fernandez-Sanz, L. Relationships among Relational Coordination Dimensions: Impact on the Quality of Education Online with a Structural Equations Model. *Technol. Forecast. Soc. Change* **2021**, *166*, 120608. [CrossRef]
26. Athanasopoulou, P. Relationship Quality: A Critical Literature Review and Research Agenda. *Eur. J. Mark.* **2009**, *43*, 583–610. [CrossRef]
27. Chang, M.L.; Cheng, C.F.; Wu, W.Y. How Buyer-Seller Relationship Quality Influences Adaptation and Innovation by Foreign MNCs' Subsidiaries. *Ind. Mark. Manag.* **2012**, *41*, 1047–1057. [CrossRef]
28. Jiang, Z.; Henneberg, S.C.; Naudé, P. Supplier Relationship Management in the Construction Industry: The Effects of Trust and Dependence. *J. Bus. Ind. Mark.* **2011**, *27*, 3–15. [CrossRef]
29. Wang, C.-J. Does Leader-Member Exchange Enhance Performance in the Hospitality Industry? *Int. J. Contemp. Hosp. Manag.* **2016**, *28*, 969–987. [CrossRef]

30. Sharma, R.R.; Chadee, D.; Roxas, B. Effects of Knowledge Management on Client-Vendor Relationship Quality: The Mediating Role of Global Mindset. *J. Knowl. Manag.* **2016**, *20*, 1268–1281. [CrossRef]
31. Leonidou, L.C.; Samiee, S.; Aykol, B.; Talias, M.A. Antecedents and Outcomes of Exporter-Importer Relationship Quality: Synthesis, Meta-Analysis, and Directions for Further Research. *J. Int. Mark.* **2014**, *22*, 21–46. [CrossRef]
32. Chuah, C.C. Supply Chain Management through Partnering in Malaysian Construction Industry. Ph.D. Thesis, Universiti Teknologi Malaysia, Johor Bahru, Malaysia, 2003.
33. Pryke, S. *Construction Supply Chain Management: Concepts and Case Studies*; John Wiley & Sons: Hoboken, NJ, USA, 2009. ISBN 9781405158442.
34. Ma, L.; Ali, A.; Shahzad, M.; Khan, A. Factors of Green Innovation: The Role of Dynamic Capabilities and Knowledge Sharing through Green Creativity. *Kybernetes* **2025**, *54*, 54–70. [CrossRef]
35. de Freitas Filho, L.H.; Neves, C.d.C.S.; Silva, N.P.; Corsi, C.A.C.; Cardoso, E.M.; de Miranda, J.B.; de Campos, G.C. Challenges of Implementing a Human Multi-Tissue Bank in a Public Hospital in the Interior of São Paulo: Under the Light of the Quality Management System. *Transplant. Proc.* **2024**, *56*, 1041–1047. [CrossRef]
36. Mohsini, R.A.; Davidson, C.H. Determinants of Performance in the Traditional Building Process. *Constr. Manag. Econ.* **1992**, *10*, 343–359. [CrossRef]
37. Kong, I.H.; Ng, K.W.A.; Price, A.D.F. Essential Causes of the Critical Site Coordination Problems in Building Projects: A Hong Kong Study Keywords. *Organ. Technol. Manag. Constr.* **2014**, *6*. [CrossRef]
38. Desmond, C. Project management tools-beyond the basics. *IEEE Engineering Management Review* **2017**, *45*, 3, 25–26 (. [CrossRef]
39. Alaloul, W.S.; Liew, M.S.; Zawawi, N.A.W.A. Identification of Coordination Factors Affecting Building Projects Performance. *Alex. Eng. J.* **2016**, *55*, 2689–2698. [CrossRef]
40. Yen, H.J.R.; Li, E.Y.; Niehoff, B.P. Do Organizational Citizenship Behaviors Lead to Information System Success? Testing the Mediation Effects of Integration Climate and Project Management. *Inf. Manag.* **2008**, *45*, 394–402. [CrossRef]
41. Sharma, R.R. Cultural Intelligence and Institutional Success: The Mediating Role of Relationship Quality. *J. Int. Manag.* **2019**, *25*, 100665. [CrossRef]
42. Elton, H.D.; Yamane, T. Elementary Sampling Theory. *Stat.* **1968**, *18*, !65–166. [CrossRef]
43. Daniel, E.C.; Eze, O.L. The Role Of Formal And Informal Communication In Determining Employee Affective And Continuance Commitment In Oil And Gas Companies. *Int. J. Adv. Acad. Res. Manag. Sci.* **2016**, *2*, 33.
44. Hair, J.F.; Risher, J.J.; Sarstedt, M.; Ringle, C.M. When to Use and How to Report the Results of PLS-SEM. *Eur. Bus. Rev.* **2019**, *31*, 2–24. [CrossRef]
45. Dillman, D.A.; Smyth, J.D.; Christian, L.M. *Internet, Phone, Mail, and Mixed-Mode Surveys: The Tailored Design Method*, 4th ed.; John Wiley Sons, Inc.: Hoboken, NJ, USA, 2014; pp. 1–530.
46. Hair, J.F., Jr.; Hult, G.T.M.; Ringle, C.M.; Sarstedt, M. *A Primer on Partial Least Squares Structural Equation Modeling (PLS-SEM)*, 2nd ed.; Sage: Thousand Oaks, CA, USA, 2021. ISBN 9781483377445.
47. Sarstedt, M.; Ringle, C.M.; Hair, J.F. Partial Least Squares Structural Equation Modeling. In *Handbook of Market Research*; Springer International Publishing: Cham, Switzerland, 2021.
48. Podsakoff, P.M.; MacKenzie, S.B.; Podsakoff, N.P. Sources of Method Bias in Social Science Research and Recommendations on How to Control It. *Annu. Rev. Psychol.* **2012**, *63*, 539–569. [CrossRef] [PubMed]
49. Harman, H. *Modern Factor Analysis*, 3rd ed.; University of Chicago Press: Chicago, IL, USA, 1976.
50. Tenenhaus, M.; Vinzi, V.E.; Chatelin, Y.M.; Lauro, C. PLS Path Modeling. *Comput. Stat. Data Anal.* **2005**, *48*, 159–205. [CrossRef]
51. Petter, S.; Straub, D.; Rai, A. Specifying Formative Constructs in Information Systems Research. *MIS Q.* **2007**, *31*, 623–656. [CrossRef]
52. Akter, S.; D'Ambra, J.; Ray, P. Trustworthiness in MHealth Information Services: An Assessment of a Hierarchical Model with Mediating and Moderating Effects Using Partial Least Squares (PLS). *J. Am. Soc. Inf. Sci. Technol.* **2011**, *62*, 100–116. [CrossRef]
53. Carrión, G.C.; Nitzl, C.; Roldán, J.L. Mediation Analyses in Partial Least Squares Structural Equation Modeling: Guidelines and Empirical Examples. In *Partial Least Squares Path Modeling*; Springer: Cham, Switzerland, 2017. ISBN 9783319640693.
54. Baron, R.M.; Kenny, D.A. The Moderator-Mediator Variable Distinction in Social Psychological Research. Conceptual, Strategic, and Statistical Considerations. *J. Pers. Soc. Psychol.* **1986**, *51*, 1173–1182. [CrossRef]
55. Daoud, A.O.; El Hefnawy, M.; Wefki, H. Investigation of Critical Factors Affecting Cost Overruns and Delays in Egyptian Mega Construction Projects. *Alex. Eng. J.* **2023**, *83*, 326–334. [CrossRef]

buildings

Article

The Supplier Selection of Prefabricated Component Production Line: A Lean-Based AHP–Improved VIKOR Framework

Pei Dang [1], Hui Gao [2], Zhanwen Niu [3], Linna Geng [4,*], Felix Kin Peng Hui [5] and Chao Sun [1]

1 School of Economics and Management, Tianjin Chengjian University, Tianjin 300384, China
2 School of Civil Engineering, Liaoning Technical University, Fuxin 123000, China
3 College of Management and Economics, Tianjin University, Tianjin, 300072, China
4 Centre for Smart Modern Construction, School of Engineering, Design & Built Environment, Western Sydney University, Parramatta, NSW 2116, Australia
5 Department of Infrastructure Engineering, The University of Melbourne, Parkville, VIC 3052, Australia
* Correspondence: l.geng@westernsydney.edu.au

Abstract: Prefabrication is increasingly recognized as a sustainable construction practice, with the efficiency of prefabricated component (PC) production lines playing a critical role in its success. However, supplier selection for PC production lines has become more complex due to evolving industrial demands, uncertain supply chain conditions, and operational complexities. This study addresses this gap by developing a lean-based AHP–improved VIKOR decision-making framework to enhance the supplier selection for PC production lines. The framework integrates advanced lean principles with universal and specific evaluation criteria, identified through a comprehensive literature review and expert interviews. Its validity was tested via a real-world case study with Yizhong Construction Co., Ltd., Tianjin, China. The results show that the three suppliers are ranked as Zhongjian > Tianyi > Xindadi, where Zhongjian is the best supplier in this case study, with a VIKOR index of 0.156. The findings show that the developed framework can improve the supplier selection efficiency by aligning with lean principles and enhancing the performance of PC production lines. By addressing the challenges of PC supplier selection, this study provides a practical tool to advance the adoption of prefabrication in construction. Furthermore, it contributes to the development of the PC industry by offering a robust method for the selection of suitable suppliers, which can help to optimize the production efficiency and support sustainable practices in construction.

Keywords: prefabricated components (PCs); supplier selection; VIKOR; lean principle

Citation: Dang, P.; Gao, H.; Niu, Z.; Geng, L.; Hui, F.K.P.; Sun, C. The Supplier Selection of Prefabricated Component Production Line: A Lean-Based AHP–Improved VIKOR Framework. *Buildings* **2024**, *14*, 4018. https://doi.org/10.3390/buildings14124018

Academic Editor: Jorge Lopes

Received: 7 November 2024
Revised: 5 December 2024
Accepted: 14 December 2024
Published: 18 December 2024

1. Introduction

The construction industry is considered a major driver of economic and social development in our society [1,2]. However, conventional construction has faced criticism in recent years for its high energy consumption, excessive resource waste, and substantial carbon emissions [3]. In this context, prefabricated construction has emerged as an alternative, shifting from traditional to sustainable construction practice [4–6]. Potential advantages, like fast installation, high quality, reduced labor, and enhanced environmental protection, have been proven in numerous studies, like Mao, et al. [7], Wu, et al. [8], Chippagiri, et al. [9], Gao and Tian [10], Liu, et al. [11], and Luo, et al. [12]. Hence, this technology has been adopted in many developed countries and areas, e.g., Australia, the U.K., Singapore, and Hong Kong [7]. In the US, for instance, Plant Prefab has announced that the firm raised up to USD 42 million to boost manufacturing capabilities [13]. Moreover, developing countries, like China and Malaysia, have embraced prefabrication as an efficient vehicle for achieving sustainability in the construction industry [6,14].

Along with its growing popularity, the evaluation and selection of suppliers for prefabricated component (PC) production lines require significant attention. Evidence for this can be derived from Zhao, et al. [15], who stated that supplier selection directly influences the performance of prefabricated projects, as the purchase of PCs accounts for nearly 70% of the total cost. Similarly, Liang, et al. [16] argue that the selection of a supplier is critical to the success of prefabrication. This is because PC production lines act as the cornerstone of prefabrication, with its unique requirements and production rules, such as the mold arrangement, fixed-size platforms, and other facets, significantly influencing the efficiency of the prefabrication implementation [17–19]. However, selecting suppliers has become increasingly challenging due to the evolving, uncertain demands and complexity in production processes, especially in developing countries, where there is an anticipated surge in PC facilities. In fact, many PC producers are still facing the dilemma of how to evaluate and select suitable suppliers. Especially, according to the report by the China Construction Association Certification Center [20], most PC production lines are established by construction firms that lack experience in PC production. Therefore, the evaluation and selection of suppliers to address the client's demands for PC production lines has become more pressing.

Satisfactory supplier selection makes a significant difference to a firm's future, which can reduce the operational costs and improve the quality of products, and can allow for rapid responses to the customers' demands [16,21,22]. One of the most important components of supplier evaluation and selection is criteria formulation. In such a process for selection, [23] stated that firms should have pursued the "lean" thinking paradigm to construct lean evaluation criteria, where the supplier attributes involve low cost, high quality, efficiency, and flexibility. This is more important for firms without professional experience in solving PC production problems, particularly during the development stage of prefabrication. Notably, the lean principle is effective for improving the operational management and boosting the quality and efficiency of production processes in prefabrication [24–26]. When evaluating PC production lines, it is necessary to implement all-round, multi-level lean principles [27]. This includes the process flow, equipment performance, personnel arrangement, production line layout, and platform configuration [28]. Several studies have concentrated on supplier selection based on lean criteria. For example, Rashidi, et al. [29] determined seven lean manufacturing evaluation criteria for supplier selection. Although these studies have provided valuable support for lean-based supplier selection, there is limited focus on supplier selection in the PC context. Additionally, lean supplier evaluation metrics should be selectively tailored from a wide range of criteria to align with the uniqueness of PC production and industry demands. Thus, lean evaluation criteria should consist not only of universal standards to reflect suppliers' common features, but also specific standards, reflecting their own unique attributes.

Considering that PC supplier evaluation and selection is a multi-criteria decision-making (MCDM) problem [30], various quantitative models have been adopted in the MCDM area, including the Analytic Hierarchy Process (AHP), Analytic Network Process, Linear Programming, TOPSIS, Data Envelopment Analysis, and VIKOR [29,31–37]. Among of them, VIKOR is considered as particularly effective, as it can obtain a compromise solution that is closer to the ideal [38]. Moreover, to differentiate evaluation criteria, including universal and specific criteria, the VIKOR method can be modified to a group-based decision matrix that differs between different alternatives and different criteria [38]. Meanwhile, a hybrid model integrating the AHP into the VIKOR method was proposed, where the AHP was used to determine the weights of the evaluation criteria, and to improve the reliability and validity of the VIKOR model [39].

Building on this, this study develops a new, quantitative lean-based AHP–improved VIKOR framework to support the producer in selecting suppliers of PC production lines. The AHP is introduced to identify the weights of lean principles, including both the universal and specific criteria, to consider the multiple aspects of such decisions. After the weights are obtained, alterative suppliers are ranked by the improved VIKOR method for balancing conflicting and non-commensurable criteria. This offers a flexible system that not only ensures the lean aspects but also accounts for the distinctive characteristics of each PC production line. This framework reveals that it is important to consider both universal and specific evaluation criteria that highlight the advantages of each alternative supplier so to avoid focusing solely on one aspect.

2. Literature Review

Supplier selection is a critical element of the prefabrication production process and has been deeply investigated in the literature. This study closely reviews the following three streams to identify the gaps involved in supplier selection for PC production lines in prefabricated projects: (1) the lean principle, (2) supplier selection in PC production, and (3) MCDM methods for supplier selection. An in-depth review of these three streams in the literature is presented in the following sections.

2.1. Lean Principle

The "lean principle" began in the manufacturing environment and is known by a variety of synonyms, like lean manufacturing, lean production, the Toyota Production System, etc. [40]. It is mainly aimed at eliminating waste in process activities to reduce process cycles, improve the quality, and increase the efficiency, which is converted to a suitable form for use in construction [27,41]. Particularly, the rapid development of prefabricated construction has significantly increased the demand for PC production lines [42]. Concurrently, there is a growing trend towards adopting a lean culture and its principles in the construction sector [27,43–45]. Elmalky, et al. [46] and Rosli, Muhammad Tamyez and Zahari [44] discussed lean principles on construction waste to improve construction performance. Refocusing on prefabrication, it necessitates the specific molds, which are carried via various processing units to complete the entire production cycle [18]. This production process shares several characteristics, like modular production and sequential workflows, yet it also exhibits unique requirements and operational protocols [47]. With this in mind, Du, Xue, Sugumaran, Hu and Dong [25] shed light on the lean production scheduling of PCs using an improved biogeography-based optimization algorithm. In their study, lean principles, like value-based management and just-in-time, are incorporated into the production process of PCs. Similarly, Shabeen and Krishnan [48] employed value stream mapping in PC manufacturing to demonstrate lean principles so to increase production in a PC manufacturing unit. Moreover, the interactions of robotic systems and the lean principle were addressed in Gusmao Brissi, Wong Chong, Debs and Zhang [45], as well as the integration with the BIM [24,49,50]. However, most studies regarding lean principles focus on tackling phenomenal delays and budget over-runs [51]. The current lean research focus on production line evaluation and supplier selection remains limited.

2.2. Supplier Selection in PC Production

Notably, meticulous attention to production line evaluation and supplier selection is crucial for achieving PC production, not only because of cost-effectiveness, but also due to the higher quality and efficiency it ensures [15,52,53]. However, the multitude of diverse criteria must be considered. Among these, lean principles should have been prioritized [25,54,55].

The importance of evaluation in construction projects, which has been already highlighted by Yang, et al. [56], is that it builds the parameters to quantify the efficiency and effectiveness of the actions, and to reach the desired results. When introduced to PC production line evaluation, existing studies shed light on its supplier selection through various perspectives, like the BIM-based AHP method [15], sustainable development [53,57], the fuzzy TOPSIS model [16], and the neural network model [58]. In contrast to these studies, this paper introduces lean principles, which have been proven to be effective in industrial production processes and operations, to enhance the efficiency [59]. This becomes particularly imperative when selecting a PC production line, especially when facing challenges like low production efficiency, disorganized production processes, and poor production balance [42] at the initial stage of prefabrication development.

Though most studies regarding lean principles focus on tackling phenomenal delays and budget over-runs [51], several studies have concentrated on supplier selection based on lean principles. For instance, Rashidi, Noorizadeh, Kannan and Cullinane [29] determined seven lean manufacturing evaluation criteria for supplier selection, including the quality, cost, flexibility, customer service, just-in-time (JIT) performance, operational leanness, and labor leanness. Tsai [60] identified seven evaluation criteria based on lean concepts in the supplier selection model, i.e., reliability, capability, quality, distance, property, service, and economy. All the studies provided supportive and valuable information for evaluation and supplier selection through the lens of lean principles. However, specific selection criteria must be established based on practical considerations, including the context, the nature of the product, and the type of market [23,61]. In the context of PC production, there is a noted need for lean supplier evaluation metrics. These metrics should be selectively tailored from a wide range of criteria to specifically align with the uniqueness of PC production and industry perspective, reflecting the current state of prefabrication development.

2.3. MCDM Methods for Supplier Selection

PC production line evaluation and supplier selection is a problem of multi-criteria decision making (MCDM), wherein criteria such as cost, quality, and service will significantly influence the decision-makers' preferences regarding potential suppliers [30]. In this context, a series of methods associated with MCDM has been used for supplier selection [37]. These can be based on a single method and/or based on integration techniques.

The single methods include the Analytic Hierarchy Process (AHP), Analytic Network Process, Linear Programming, TOPSIS, Data Envelopment Analysis, and VIKOR [29,31–37,62]. For example, Pamucar, et al. [63] proposed a fuzzy decision-making approach for supplier selection. Among these single methods, VIKOR emerged as a particularly effective approach for addressing MCDM problems [64]. The reason is that VIKOR can identify a compromise solution for supplier selection that is the closest to the ideal [38]. Particularly, this method seeks to maximum group utility while minimizing individual regret [65]. To this end, Amiri, Hashemi-Tabatabaei, Ghahremanloo, Keshavarz-Ghorabaee, Zavadskas and Banaitis [38] developed an improved VIKOR decision-making method, which necessitates a group-based decision matrix that varies across different alternatives and norms, thereby broadening the applicability of this VIKOR method. Moreover, the improved VIKOR method facilitates the development of supplier selection in lean management [66]. This method integrates both universal criteria and the specific criteria of features, thereby establishing the groundwork for a decision-making model focused on the lean-based evaluation and selection of suppliers. However, it should be noted that the single method for supplier selection lacks flexibility, and has conflicting and incommensurable aspects, for instance, quality criteria conflicts with those centered on cost reduction [1].

As a result, integration techniques have become popular, like the AHP + Fuzzy-VIKOR. Compared to single methods, the integrated techniques provide a more comprehensive analysis by leveraging the strengths of different MCDM methods [37]. Considering the benefits of the improved VIKOR, the integrated AHP–improved VIKOR is adapted in this study as an effective and easy-going method [67,68]. The benefit is that the AHP can be used to determine the weight of criteria, while the improved VIKOR method can be applied to rank the alternatives, considering both universal and specific criteria, to offer a detailed and multi-faceted view of the decision problem [69,70]. This can be further attributed to the fact that the assigned weights significantly influence the accuracy of the results calculated using the improved VIKOR method [71]. While there is no evidence specifically showcasing the use of the AHP–improved VIKOR in our proposed context, the advantages of integrating the AHP with VIKOR can be found in Soner, et al. [72], Awasthi, et al. [73], and Luthra, et al. [74], who have applied this integrated method to optimize the selection process in ship construction and supply chains. This study develops the AHP–improved VIKOR method based on AHP-VIKOR, with a unique focus on evaluating PC production lines and supplier selection. This adaptation not only addresses a significant gap in the literature but enhances the methodological framework to better suit the unique challenges of this sector.

In summary, most single MCDM methods fail to provide balanced results, and some integrated techniques do not adequately consider the weights of criteria, which often involve a mix of subjective and objective conditions. Considering the shortcomings of the existing methods, the plurality of criteria, and the inherent uncertainty in PCs, this paper develops a lean-based AHP–improved VIKOR framework to address production line evaluation and supplier selection problems.

3. Lean-Based AHP–Improved VIKOR Evaluation Framework

The developed AHP–improved VIKOR framework can be considered as a three-phase MCDM model, which includes the identification of the lean criteria system, the weighting of lean criteria, and the selection of the supplier, as shown in Figure 1.

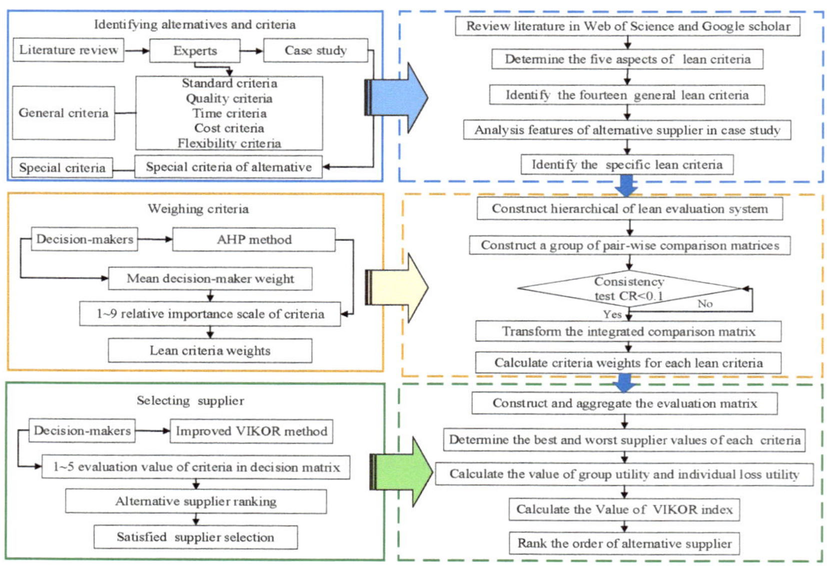

Figure 1. Lean-based AHP–improved VIKOR framework.

3.1. *Identifying the Universal Lean Evaluation Criteria*

Evaluating a PC production line from a lean perspective necessitates the application of lean principles, alongside universal evaluation standards, to determine the line's efficiency, resilience, and emergency responsiveness. However, applying a uniform set of evaluation criteria across diverse production lines poses challenges, as each line not only shares common features but also possesses its own unique attributes. Consequently, the evaluation of production line suppliers must incorporate both universal and specific criteria. Universal lean evaluation criteria constitute a foundational system for assessing PC production lines, encompassing the following five key aspects: standardization, quality, time, cost, and flexibility [75,76]. The universal second-level criteria of the five key aspects were identified by a literature review and interviews of experts.

Firstly, a systematic literature review was undertaken to identify a list of preliminary evaluation criteria. This study searched journal articles in *Web of Science* and *Google Scholar* with the following combination of keywords: "supplier selection and evaluation", "lean" and "production line", etc. The journal articles were further examined for whether they referred to the five aspects or not. Then, in-depth interviews were conducted with twenty experts (five lean experts, ten managers, and five PC experts) to validate the rationality and comprehensiveness of the preliminary criteria. These experts mainly had at least five years of experience in PC production lines. Based on their practical experiences, each criterion was reviewed, and suggestions were provided on whether certain preliminary criteria should be adjusted, added, or deleted. Based on these, the fourteen universal criteria of the five key aspects were identified, as shown in Table 1. Furthermore, the specific lean evaluation criteria, focusing on the unique advantages and capabilities of each PC production line, should be determined to build the holistic lean evaluation criteria.

Table 1. The universal lean evaluation criteria of PC production lines.

1st-Level Criteria	2nd-Level Criteria	Main Sources
A1: Standardization	Z1: Production process standardization degree	[77]
	Z2: Customized standard process design level	[78]
	Z3: The comprehensiveness of differentiated standard operating systems	[77]
	Z4: Standard operation level	[76]
A2: Quality	Z5: Total productive maintenance level	[17]
	Z6: Product quality and reliability level	[79]
	Z7: Equipment reliability or failure rate	[80]
A3: Time	Z8: Production Takt and turnround time	[81]
	Z9: Just-in-time and pull level	[82]
	Z10: Equipment utilization level	[77]
A4: Cost	Z11: Buying price	[83]
	Z12: Use and maintenance cost level	[84]
	Z13: Unit production cost	[83]
A5: Flexibility	Z14: Mass customization production level of the quick response to customers' demands	[85]

3.2. Weight Criteria Based on AHP

The AHP method is used to calculate the criterion weights, i.e., W_i. The calculation process of the AHP consists of the following four steps [31]: (1) constructing the pairwise comparison matrix, (2) calculating the priority vectors, (3) checking for consistency, and (4) determining the criterion weights.

3.2.1. Constructing Pairwise Comparison Matrices

Constructing a pairwise comparison matrix based on the relative importance scores among the criteria collected from experts is an important process for calculating the weights. The relative importance scores were evaluated by integrating the experts' evaluations of each pair of criteria *i* and *j* with the mean method [33], as shown in Table 2.

Table 2. Relative importance scale of the criteria.

Score	Relative Importance
1	Criteria *i* and *j* are of equal importance.
3	Criteria *i* is weakly more important than *j*.
5	Criteria *i* is strongly more important than *j*.
7	Criteria *i* is very strongly more important than *j*.
9	Criteria *i* is absolutely more important than *j*.

Note: 2, 4, 6, and 8 are intermediate values.

A pairwise comparison matrix A can be created based on the relative importance scores, shown as Equation (1). Each element a_{ij} of the matrix represents the result of comparing element *i* to element *j* in terms of their relative importance towards the upper hierarchy of the criteria.

$$A = (a_{ij})_{n\times n} \quad (1)$$

3.2.2. Calculating Priority Vectors

The priority vector, which means the relative weights of the elements being compared, is calculated from the pairwise comparison matrix A. The eigenvector method was used [86], where the priority vector W corresponds to the normalized principal eigenvector of the comparison matrix A. For the consistent reciprocal matrix, the largest eigenvalue λmax is equal to the number of comparisons, or $\lambda max = n$.

$$AW = \lambda_{max} W \quad (2)$$

The step to approximate W involves normalizing each column of A and then averaging across rows to obtain each criterion of W.

3.2.3. Checking for Consistency

To check the consistency of the pairwise comparisons, the Consistency Index (CI) and Consistency Ratio (CR) are calculated in Equations (3) and (4), as follows:

$$CI = \frac{\lambda_{max} - n}{n - 1} \quad (3)$$

$$CR = \frac{CI}{RI} \quad (4)$$

where n is the number of elements being compared and RI is the Random Index, which is an average CI for randomly generated pairwise comparison matrices of the same size n. The values of the RI vary with n and are tabulated based on simulations. If $CR < 0.10$, the judgments are considered acceptably consistent. If $CR \geq 0.10$, the judgments need to be reviewed.

3.2.4. Determining Criterion Weights

When determining the weights of the evaluation criteria corresponding to the upper hierarchy of criteria W_i, the lower hierarchy of the criteria corresponding to the evaluation goal can be finally calculated by Equation (5), where i is the number of hierarchies of the evaluation criteria, as follows:

$$W_{final} = w_1 \times w_2 \times \cdots \times w_i \quad (5)$$

3.3. *Supplier Evaluation and Selection Based on Improved VIKOR Method*

After determining the weights and scores for each criterion, the suppliers of PC production lines can be evaluated and selected. Using the steps of the improved VIKOR method, which requires modifications when deciding among suppliers, we can evaluate the suppliers based on both universal and specific criteria [66,87,88]. This approach leverages the strengths of the VIKOR method to ensure a comprehensive assessment. The steps of the improved VIKOR method are as shown below:

Step 1: Construct and aggregate the evaluation matrix.

We calculate the evaluation value f_{ij} of the group decision matrix D_i based on each alternative and each evaluation criteria belonging to the alternative, because each alternative has its own evaluation criteria. Every expert from the group of decision-makers makes a separate evaluation matrix, which consists of the evaluation of alternatives compared to the corresponding criteria, with a scale of 1–5 [66]. Then, we average the separated evaluation matrices by each expert into one decision matrix D, which is formed as in Table 3.

Table 3. Group decision matrix.

	Z_1	Z_2	Z_3	…	Z_n
S_1	f_{11}	f_{12}	f_{13}	…	f_{1n}
S_2	f_{21}	f_{22}	f_{23}	…	f_{2n}
S_3	f_{31}	f_{32}	f_{33}	…	f_{3n}
⋮	⋮	⋮	⋮	⋱	⋮
S_m	f_{m1}	f_{m2}	f_{m3}	…	f_{mn}

Step 2: Select the best value and the worst value of each sub-criterion.

We then set f_j^*, which represents the optimal evaluation value, and f_j^-, which represents the worst value. Universally, f_j^* and f_j^- can be separately determined by the largest and smallest ranking values f_{ij} in the decision matrix, or the theoretically largest and smallest evaluation values. Considering the sub-criteria Z_j, if Z_j belongs to the type of efficiency criteria

$$f_j^* = \overset{max}{i}(f_{ij},\ j = 1,2\ldots,\ n) \tag{6}$$

$$f_j^- = \overset{min}{i}(f_{ij},\ j = 1,2\ldots,\ n) \tag{7}$$

If Z_j belongs to the type of cost criteria

$$f_j^* = \overset{min}{i}(f_{ij},\ j = 1,2\ldots,\ n) \tag{8}$$

$$f_j^- = \overset{max}{i}(f_{ij},\ j = 1,2\ldots,\ n) \tag{9}$$

Step 3: Calculate the values of Q_i and G_i.

Q_i and G_i represent the group utility and individual loss utility. Q_i means the difference between the overall solution and the optimal solution in the ideal state. G_i is the difference between the actual value of a single criterion for the suppliers and its optimal value in each supplier. The group utility Q_i and individual utility G_i of the alternatives based on the corresponding evaluation criteria of every alternative are determined as follows:

$$Q_i = \sum_{j=1}^{n} \frac{W_j\left(f_j^* - f_{ij}\right)}{f_j^* - f_j^-} \tag{10}$$

$$G_i = \overset{max}{i}\left[\frac{W_j\left(f_j^* - f_{ij}\right)}{f_j^* - f_j^-}\right] \tag{11}$$

From the above, $Q^* = \overset{min}{i}Q_i$, $Q^- = \overset{max}{i}Q_i$, $G^* = \overset{min}{i}G_i$, $G^- = \overset{max}{i}G_i$. Q^*, and G^* are, respectively, the smallest values of the utility cost. The total utility of a supplier of PC production lines is the combination of both group optimal utility and individual loss utility.

Step 4: Computation of VIKOR index U_i.

$U_i(i = 1,2,\ldots,m)$ represents the total utility of an alternative supplier of PC production lines, r means the weight of the group utility, and $1 - r$ is the weight of the individual loss utility $(0 \leq r \leq 1)$; therefore, r shows the decision-making mechanism coefficient of alternative suppliers, as follows:

$$U_i = r\left[\frac{Q^* - Q_i}{Q^* - Q^-}\right] + (1 - r)\left[\frac{G^* - G_i}{G^* - G^-}\right] \tag{12}$$

It can be seen from the U_i formula that the comprehensive utility loss of the scheme considers the comprehensiveness of the whole and the uniqueness of one, the weights of which are variable. The decision-makers can reach a consensus through negotiation. If $r < 0.5$, then minimizing the individual loss utility plays a dominant role in the evaluation mechanism; if $r = 0.5$, then minimizing the individual regret utility plays the same role as maximizing the group utility, and decision-makers reach consensus through consultation; if $r > 0.5$, then maximizing the group utility plays a dominant role in the evaluation mechanism.

Universally, minimizing the individual loss utility is considered as equally important as maximizing the group utility, where U_i represents the expression of the total utility of decision making when using the improved VIKOR method to make decisions, upon which decision-makers reach agreement through discussion. The decision-making mechanism coefficient is 0.5 according to Valipour Parkouhi and Safaei Ghadikolaei [69], Anvari, Zulkifli and Arghish [70], Prasad, et al. [89], Chang [90].

Step 5: Rank the order of preference.

An alternative based on the calculation results is prioritized and the following sequence is formed according to the value of U_i from small to large: $S^{(1)}, S^{(2)}, \ldots, S^{(m)}$. The following two conditions must be satisfied before the alternative in the first position in the U_i ranking is suggested as the adjustment's solutions:

I. Acceptable advantage:

$$U\left(S^{(2)}\right) - U\left(S^{(1)}\right) \geq \frac{1}{m-1} \tag{13}$$

In which m is the number of alternative suppliers in the problem and S_2 in the U ranking list means the alternative with the second position.

II. Acceptable stability:

The alternative $S^{(1)}$ must also be the best in the Q or G ranking list. Go to the extra phase to obtain the compromise solution if either condition is not satisfied. When condition I is not satisfied, then the compromise solution is $S^{(1)}, S^{(2)}, \ldots, S^{(m)}$, and the maximum values of m need to be searched with the following relationship:

$$U\left(S^{(m)}\right) - U\left(S^{(1)}\right) < \frac{1}{m-1} \tag{14}$$

When condition II is not satisfied, then both $S^{(1)}$ and $S^{(2)}$ are adjustment solutions. So, condition I refers to the optimal terms of the acceptable alternative, and condition II is accepted as a stable decision-making alternative.

4. Case Study

4.1. Case Description

Yizhong Construction firm is a leading universal contractor in Tianjin, China. In response to China's PC development policy, this firm proactively embraced PC production and construction. This initiative included establishing PC factories for production to bolster industrialized construction capabilities. It also formed a collaborative partnership with PC experts for the evaluation and selection of PC suppliers. They emphasized the production line's critical role in determining the quality, efficiency, and cost of PCs. Meanwhile, they provided the valuable lean insights of selecting the supplier of PC production lines through the analysis of the site selection, factory, and workshop layout. The experts were selected and invited via the following snow sampling principle: (1) at least 5 years of related PC knowledge; (2) having undertaken important tasks in PC production; (3) holding senior positions in production teams [91]. These principles can ensure that the selected experts are qualified to answer questions pertaining to the supplier of PC production lines so to ensure the validity and accuracy. This expert group, comprised of professional individuals from

Yizhong, Tianjin University, Yuanda, and other PC factories, included five lean experts, ten managers, and five PC experts, totaling twenty members. The detailed information of these twenty experts is shown in Table 4.

Table 4. Expert profiles.

No. Expert	Stakeholder Group	No.	Main Position	Education Level	Experience
1	Tianjin University	3	Professor	Master's	$5\leq$
2	Tianjin University	2	Senior scholar	Bachelor's	$5\leq$
3	Yizhong	1	Universal manager	Master's	$5\leq$
4	Yizhong	4	Vice-Universal manager	Master's	$5\leq$
5	Yizhong	5	Project manager	Bachelor's	$5\leq$
6	Yuanda Factory	2	Factory director	Master's	$5\leq$
7	Other Factories	3	Universal manager	Master's	$5\leq$

In this case context, the following three suppliers participated in the survey: Tianyi, Zhongjian, and Xindadi. The PC production lines of each presented unique strengths and weaknesses. The experts conducted on-site studies and interviews with the three suppliers, summarizing their respective strengths and weaknesses, and extracting their key characteristics, as detailed in Table A1.

4.2. Data Analysis

4.2.1. Establishing Specific Lean Evaluation Criteria

In contrast, the specific lean evaluation criteria focus on the unique advantages and the capabilities particular to each PC production line, necessitating individual industrial considerations. For example, Zhongjian's production line is distinguished by its advanced digital technology and capabilities, while Tianyi excels in the layout efficiency and turnaround rate. The specific second-level criteria were determined by the twenty experts based the alternative suppliers of Zhongjian, Xindadi, and Tianyi. Then, in-depth interviews with the twenty experts were conducted to identify the specific lean evaluation criteria of three alternative suppliers so to construct the holistic lean evaluation criteria. First, the experts were invited by telephone or face-to-face to discuss whether the universal criteria identified from the literature exist or not in a real-world PC production situation. Additional criteria were also proposed and explained based on their experiences. Secondly, they identified the specific evaluation criteria based the character of alternative suppliers. Each interview lasted 1–2 h so to confirm the reliability of the identified criteria. When disagreements among interviewees existed, these experts were contacted for further discussion. After three rounds of discussion, the experts reached agreement for all the questions. The final evaluation criteria, combing universal and specific criteria, are detailed in Table A2.

4.2.2. Calculating Weights of Criteria

The weights of first-level criteria (standard, quality, time, cost, and flexibility) corresponding to the lean evaluation goals, WA1, WA2, WA3, WA4, and WA5, were calculated. The relative importance scores among the five criteria were determined by the twenty experts, according to Table 2, to construct a pairwise comparison matrix, as shown in Table 5.

The weights were calculated by Equations (1)–(4), based on the average scores of twenty experts, and are as follows:

$\boldsymbol{W} = [\boldsymbol{WA1}, \boldsymbol{WA2}, \boldsymbol{WA3}, \boldsymbol{WA4}, \boldsymbol{WA5}] = [0.102, 0.245, 0.102, 0.504, 0.046]$, $\boldsymbol{CI} = 0.0317$, $\boldsymbol{CR} = 0.0283 < 0.1$; thus, the judgments are considered acceptably consistent.

Then, the weights of the second-level criteria, corresponding to five first-level criteria, were calculated. Since the lean evaluation criteria of each supplier of a PC production line were different, which is shown in Table A2, the weights of the second-level criteria were calculated for each supplier in Zhongjian, Tianyi, and Xindadi. The calculating processes were the same as the calculations of the weights of the five first-level criteria. The weights of the second-level criteria, corresponding to five first-level criteria of Zhongjian, Tianyi, and Xindadi, are shown in Table A3. Then, the weights of the second-level criteria, corresponding to the lean evaluation goals, were further calculated with Equation (5).

Table 5. The pairwise comparison matrix of five first-level criteria.

1st-Level Criteria	A1	A2	A3	A4	A5
A1	1	1/3	1	1/5	3
A2	3	1	3	1/3	5
A3	1	1/3	1	1/5	3
A4	5	3	5	1	7
A5	1/3	1/5	1/3	1/7	1

4.2.3. Selecting Suppliers

After determining the weights of the evaluation criteria for each supplier, we used the improved VIKOR method to evaluate and select the supplier of the PC production line according to the steps shown in Section 3.3.

Step 1: The aforementioned experts from the set of twenty experts produced the evaluation value f_{ij} of the evaluation matrix through the comparison of three alternatives of PC production lines relative to each criteria by using a scale of 1–5 [66]. In the scale, one signifies 'tiny or extremely difficult', two denotes 'lesser or difficult', three represents 'medium', four indicates 'big or easy', and five corresponds to 'giant or very easy'. Then, we averaged every evaluation matrix by each expert into one decision matrix D, which was formed as shown below in Table A4.

Step 2: The next step was to select the best value f_j^* and the worst value f_j^- of each sub-criterion according to Equations (6) and (7). Considering that the criteria Z_j is the efficiency criteria, we set five as the highest-ranking value f_j^* and one as lowest-ranking value f_j^-.

Step 3: We then calculated the values of Q_i and G_i for the alternatives based on the corresponding evaluation criteria of every alternative using Equations (10) and (11), as shown in Table A5.

Step 4: This step included the calculation of the VIKOR index U_i, which represents the total utility of an alternative supplier of three PC production lines according to Equation (12). Here, the decision-making mechanism coefficient r was 0.5, according to Valipour Parkouhi and Safaei Ghadikolaei [69], Anvari, Zulkifli and Arghish [70], Prasad, Prasad, Rao and Patro [89], Chang [90]. Furthermore, the twenty experts were invited to evaluate the importance of minimizing the individual regret utility and maximizing the group utility. The experts thought that they played the same role, meaning that r was also 0.5. The results of U_i for Zhongjian, Tianyi, and Xindadi are shown in Table A5.

Step 5: We then ranked the order of three alternative suppliers of PC production lines based on the value of U_i from small to large. Originally, the ranking results of U_i were $U_1 < U_2 < U_3$, leading to the ranking results of S $^{(1)}$, S $^{(2)}$, and S $^{(3)}$ as Zhongjian, Tianyi, and Xindadi. According to Equations (13) and (14), two conditions should be further satisfied before the alternative supplier in the first position in the Ui ranking. Condition I: $U\left(S^{(2)}\right) - U\left(S^{(1)}\right) \geq \frac{1}{m-1}$, in which $\boldsymbol{U}\left(S^{(2)}\right) - U\left(S^{(1)}\right) = 0.160 - 0.156 = 0.04$, $\frac{1}{m-1} = \frac{1}{3-1} = 0.5$. Therefore, condition I is not satisfied. For condition II, the first-order value of U_1 is the first-order value of Q and of G. The ranking of the Q values from small to large is as follows: $Q_1 < Q_2 < Q_3$. The ranking of the G values from small to large is as follows: $G_1 < G_2 < G_3$. Therefore, the compromise solution is S $^{(1)}$, S $^{(2)}$, and S $^{(3)}$. Additionally, if condition I is not satisfied, the maximum values of m need to be searched with the relationship $\boldsymbol{U}\left(S^{(\mathbf{m})}\right) - \boldsymbol{U}\left(S^{(\mathbf{1})}\right) < \frac{1}{m-1}$, in which $U\left(S^{(3)}\right) - U\left(S^{(1)}\right) = 0.299 - 0.156 = 0.143 < 0.5$. Hence, the final ranking of the three alternatives, S $^{(1)}$, S $^{(2)}$, and S $^{(3)}$, is Zhongjian, Tianyi, and Xindadi, as shown in Table A5.

4.3. Case Result

The result means that the three suppliers are ranked as Zhongjian > Tianyi > Xindadi, where Zhongjian is the best supplier in this case study. The PC production line of Zhongjian offers significant advantages, including cost efficiency, ease of operation, balanced production, and the comprehensive management of information in practical scenarios. Furthermore, its proximity to the Yizhong Construction Co., Ltd. facilitates convenient services such as equipment installation, debugging, production guidance, and maintenance. As Yizhong Construction Co., Ltd. is currently undergoing a transition period, partnering with Zhongjian could lead to savings in investment and production costs while enhancing the production operability.

5. Discussion and Conclusions

This study examined the evaluation of production lines and supplier selection for prefabricated construction projects through the lens of lean principles. Unlike the existing selection models, like TOPSIS [53,57], the developed framework integrated the AHP [92] to assign weights to the lean criteria and employed the improved VIKOR method [93] to rank alternative suppliers, which was designed to facilitate the selection of suppliers for PC production lines, especially in developing countries.

To demonstrate the practical applicability the lean-based AHP–improved VIKOR evaluation framework, a case study was conducted in collaboration with experts from Tianjin University and three leading PC factories. The case study identified Zhongjian as the optimal supplier for the PC production line based on its superior performance in various lean-oriented criteria, such as cost efficiency, operational simplicity, and the effective management of information systems. These lean criteria are critical for fostering lean thinking in PC production and for optimizing prefabrication outcomes.

To be more specific, this study expands the scope beyond the limitations observed in the works by Abdollahi, Arvan and Razmi [23] by fully integrating these lean criteria into a comprehensive MCDM framework that includes both universal and specific criteria [15,66]. Notably, while the single methods, such as the AHP, DEA, VIKOR, and TOPSIS, have been explored in the literature, the integrated AHP–improved VIKOR method distinguishes itself by combining the weighted dimensions to synthesize diverse criteria into a coherent evaluation and ranking system [72]. This approach acknowledges the multifaceted nature of supplier selection, which aligns with the holistic objective of lean operations in prefabrication, thereby offering an advancement over traditional construction.

The contribution of this study is twofold. From a theoretical view, it expands the existing body of knowledge for the evaluation of production lines for PCs via a lean-based AHP–improved VIKOR framework, facilitating a deeper understanding of the combination of both universal and specific criteria in supplier selection. This developed framework enriches the project management practices of prefabrication development from an efficient procurement perspective. From a practical view, it provides the following actionable insights:

- For suppliers of PC production lines, it is vital to enhance and develop lean operation capabilities and management strategies based on the evaluation criteria so to improve the efficiency and success of PC production lines.
- For producers of PCs, they should consider the level of automaticity and information technology, and lean production requirements, while selecting the production line.
- For policymakers, it is necessary to formulate industry standards and conduct relevant activities, like adding lean principles into policies for PC development, so to improve the awareness of lean-based PC production lines.

6. Limitations and Future Directions

While the lean-based AHP–improved VIKOR framework provides a robust tool for supplier selection in prefabricated component (PC) production lines, this study has certain limitations. Firstly, the framework relies on comprehensive, high-quality data for accurate decision making, which can be difficult to obtain in resource-constrained environments. Future research could focus on adapting the framework specifically for suppliers operating under such constraints. Additionally, scalability remains a challenge, as adapting the framework to suppliers of varying sizes and operational complexities can be demanding. Future studies could aim to simplify the framework so to enhance the accessibility and to incorporate real-time data analysis tools for improved usability. Lastly, to adapt to the dynamic supply chain environment, a feedback mechanism, like the long-term performance evaluation of the framework, could be established. This would enable regular updates and refinements, ensuring that the criteria align with the current demands and requirements of the PC supply chain.

Author Contributions: Conceptualization, P.D. and H.G.; software, methodology, and validation, P.D.; resources, Z.N.; writing—original draft preparation, P.D. and C.S.; writing—review and editing, L.G. and F.K.P.H. All authors have read and agreed to the published version of the manuscript.

Funding: This research was funded by the Research on the Lean Construction Model and Implementation Path in the Context of Industrialization and Digital Transformation in the Construction Industry, grant number FHX2023-015.

Institutional Review Board Statement: Not applicable.

Informed Consent Statement: Informed consent was obtained from all subjects involved in the study.

Data Availability Statement: The raw data supporting the conclusions of this article will be made available by the authors on request.

Conflicts of Interest: The authors declare no conflicts of interest.

Appendix A

Table A1. The description of the suppliers.

Suppliers	The Main Description of the Production Line	Special Characteristic
Zhongjian	Automated production line with state-of-the-art automated equipment; requires little manual labor; simple equipment operation without the need for technical training; a high level of craftsmanship; strong expandability; extensive visual monitoring means of process control.	High level of automation and digitalization
Tianyi	Semi-automatic production line with partial automatic machinery, like a curing kiln and material feeder; well-designed logistics; free technical training provided; a variety of matching mold types, thereby improving the responsiveness to customer demands.	High level of mold utilization and optimization
Xindadi	Manual production line with less automatic equipment, with most of the operations carried out manually by workers; minimal fixed asset investment, using natural curing methods, requiring relatively more workers; the common solution without the need for specific criteria.	-

Table A2. The lean evaluation criteria of PC production lines with specific criteria.

1st-Level Criteria	2nd-Level Criteria	Suppliers		
		ZJ	Y	DD
A1: Standard	Z1: Production process standardization degree	✓	✓	✓
	Z2: Customized standard process design level	✓	✓	✓
	Z3: The comprehensiveness of differentiated standard operating systems	✓	✓	✓
	Z4: Standard operation level	✓	✓	✓
A2: Quality	Z5: Total productive maintained level	✓	✓	✓
	Z6: Product quality and reliability level	✓	✓	✓
	Z7: Equipment reliability or failure rate	✓	✓	✓
	Z8: * Automated monitoring level of quality and reliability	✓	-	-
A3: Time	Z9: Production Takt and turnround time	✓	✓	✓
	Z10: Just-in-time and pull level	✓	✓	✓
	Z11: Equipment utilization level	✓	✓	✓
	Z12: * One-piece flow level and line balance level	✓	-	-
	Z13: * Mold optimization level for improving the operation efficiency	-	✓	-
A4: Cost	Z14: Buying price	✓	✓	✓
	Z15: Use and maintenance cost level	✓	✓	✓
	Z16: Unit production cost	✓	✓	✓
	Z17: * Mold utilizing operation	-	✓	-
A5: Flexibility	Z18: Mass customization production level of the quick response to customers' demands	✓	✓	✓

Note: "*" indicates the specific evaluation criteria of a certain production line; "✓" indicates a certain production line has corresponding characteristics; "-" indicates a production line does not have corresponding characteristics.

Table A3. The lean evaluation criteria of PC production lines.

The Evaluation Goal	1st-Level Criteria	Weights of 1st-Level Criteria	2nd-Level Criteria	Weights Corresponding to 1st-Level Criteria			Weights Corresponding to the Evaluation Goal		
				ZJ	TY	XDD	ZJ	TY	XDD
The lean evaluation and selection of suppliers	A1	0.102	Z1	0.125	0.125	0.125	0.013	0.013	0.013
			Z2	0.375	0.375	0.375	0.038	0.038	0.038
			Z3	0.375	0.375	0.375	0.038	0.038	0.038
			Z4	0.125	0.125	0.125	0.013	0.013	0.013
	A2	0.245	Z5	0.390	0.637	0.637	0.096	0.156	0.156
			Z6	0.068	0.105	0.105	0.017	0.026	0.026
			Z7	0.152	0.258	0.258	0.037	0.063	0.063
			Z8	0.390	-	-	0.096	-	-
	A3	0.102	Z9	0.113	0.168	0.637	0.012	0.017	0.065
			Z10	0.064	0.075	0.258	0.007	0.008	0.026
			Z11	0.411	0.570	0.105	0.042	0.058	0.011
			Z12	0.411	-	-	0.042	-	-
			Z13	-	0.187	-	-	0.019	-
	A4	0.504	Z14	0.637	0.522	0.637	0.321	0.263	0.321
			Z15	0.258	0.200	0.258	0.130	0.101	0.130
			Z16	0.105	0.078	0.105	0.053	0.039	0.053
			Z17	-	0.200	-	-	0.101	-
	A5	0.046	Z18	1	1	1	0.046	0.046	0.046

Note: "-" means irrelevant evaluation criteria of PC production lines.

Table A4. The aggregation decision matrix of experts' opinions for alternatives.

	Z_1	Z_2	Z_3	Z_4	Z_5	Z_6	Z_7	Z_8	Z_9	Z_{10}	Z_{11}	Z_{12}	Z_{13}	Z_{14}	Z_{15}	Z_{16}	Z_{17}	Z_{18}
Zhongjian	4.50	3.83	4.67	3.33	3.83	3.50	3.33	4.33	4.00	3.83	3.50	4.50	-	4.83	2.67	3.83	-	3.50
Tianyi	4.00	3.67	4.17	4.00	4.33	3.50	3.50	-	4.33	4.17	4.33	-	4.33	4.00	3.50	3.83	4.00	4.17
Xindadi	3.67	3.00	3.83	4.00	4.67	3.50	4.17	-	3.67	3.50	2.83	-	-	2.33	4.00	4.33	-	3.50

Table A5. Results of the selection process based on the improved VIKOR.

	Z_j	f_{ij}^-	f_{ij}^*	f_{ij}	R_{ij}	W_{ij}	$W_{ij}{*}R_{ij}$	G_i	Q_i	U_i	Rank
Zhongjian	Z_1	1	5	4.50	0.125	0.013	0.002	0.066	0.235	0.156	1
	Z_2	1	5	3.83	0.293	0.038	0.011				
	Z_3	1	5	4.67	0.083	0.038	0.003				
	Z_4	1	5	3.33	0.418	0.013	0.005				
	Z_5	1	5	3.83	0.293	0.096	0.028				
	Z_6	1	5	3.50	0.375	0.017	0.006				
	Z_7	1	5	3.33	0.418	0.037	0.016				
	Z_8	1	5	4.33	0.168	0.096	0.016				
	Z_9	1	5	4.00	0.250	0.012	0.003				
	Z_{10}	1	5	3.83	0.293	0.007	0.002				
	Z_{11}	1	5	3.50	0.375	0.042	0.016				
	Z_{12}	1	5	4.50	0.125	0.042	0.005				
	Z_{14}	1	5	4.83	0.043	0.321	0.014				
	Z_{15}	1	5	2.67	0.583	0.130	0.076				
	Z_{16}	1	5	3.83	0.293	0.053	0.015				
	Z_{18}	1	5	3.50	0.375	0.046	0.017				
Tianyi	Z_1	1	5	4.00	0.25	0.013	0.003	0.076	0.254	0.160	2
	Z_2	1	5	3.67	0.333	0.038	0.013				
	Z_3	1	5	4.17	0.208	0.038	0.008				
	Z_4	1	5	4.00	0.250	0.013	0.003				
	Z_5	1	5	4.33	0.168	0.156	0.026				
	Z_6	1	5	3.5	0.375	0.026	0.010				
	Z_7	1	5	3.5	0.375	0.063	0.024				
	Z_8	1	5	4.33	0.168	0.017	0.003				
	Z_9	1	5	4.17	0.208	0.008	0.002				
	Z_{10}	1	5	4.33	0.168	0.058	0.010				
	Z_{11}	1	5	4.33	0.168	0.019	0.003				
	Z_{12}	1	5	4.00	0.250	0.263	0.066				
	Z_{14}	1	5	3.50	0.375	0.101	0.038				
	Z_{15}	1	5	3.83	0.293	0.039	0.012				
	Z_{16}	1	5	4.00	0.250	0.101	0.025				
	Z_{18}	1	5	4.17	0.208	0.046	0.010				
Xindadi	Z_1	1	5	3.67	0.333	0.013	0.004	0.214	0.384	0.299	3
	Z_2	1	5	3.00	0.50	0.038	0.019				
	Z_3	1	5	3.83	0.293	0.038	0.011				
	Z_4	1	5	4.00	0.250	0.013	0.003				
	Z_5	1	5	4.67	0.083	0.156	0.013				
	Z_6	1	5	3.50	0.375	0.026	0.010				
	Z_7	1	5	4.17	0.208	0.063	0.013				
	Z_8	1	5	3.67	0.333	0.065	0.022				
	Z_9	1	5	3.50	0.375	0.026	0.010				
	Z_{10}	1	5	2.83	0.543	0.011	0.006				
	Z_{11}	1	5	2.33	0.668	0.321	0.214				
	Z_{12}	1	5	4.00	0.250	0.130	0.033				
	Z_{14}	1	5	4.33	0.168	0.053	0.009				
	Z_{15}	1	5	3.50	0.375	0.046	0.017				

References

1. Lima, L.; Trindade, E.; Alencar, L.; Alencar, M.; Silva, L. Sustainability in the construction industry: A systematic review of the literature. *J. Clean. Prod.* **2021**, *289*, 125730. [CrossRef]
2. Kim, K.; Tiedmann, H.R.; Faust, K.M. Construction industry changes induced by the COVID-19 pandemic. *Eng. Constr. Archit. Manag.* 2024; *ahead of print*.
3. Hassan, N.M.; Alashwal, A. Developing a model for the implementation of designing out waste in construction. *Arch. Eng. Des. Manag.* **2024**, *1*, 17. [CrossRef]
4. Kim, Y.-W.; Rhee, B.-D. Incentive-based coordination for scheduled delivery in prefab construction. *Constr. Manag. Econ.* **2024**, *42*, 624–639. [CrossRef]
5. Sah, T.P.; Lacey, A.W.; Hao, H.; Chen, W. Prefabricated Concrete Sandwich and Other Lightweight Wall Panels for Sustainable Building Construction: State-of-the-Art Review. *J. Build. Eng.* **2024**, *89*, 109391. [CrossRef]
6. Dang, P.; Geng, L.; Niu, Z.; Chan, M.; Yang, W.; Gao, S. A Value-Based Network Analysis for Stakeholder Engagement through Prefabricated Construction Life Cycle: Evidence from China. *J. Civ. Eng. Manag.* **2024**, *30*, 49–66. [CrossRef]
7. Mao, C.; Shen, Q.P.; Pan, W.; Ye, K.H. Major Barriers to Off-Site Construction: The Developer's Perspective in China. *J. Manag. Eng.* **2015**, *31*, 04014038. [CrossRef]
8. Wu, Z.; Luo, L.; Li, H.; Wang, Y.; Bi, G.; Antwi-Afari, M.F. An analysis on promoting prefabrication implementation in construction industry towards sustainability. *Int. J. Environ. Res. Public Health* **2021**, *18*, 11493. [CrossRef]
9. Chippagiri, R.; Bras, A.; Sharma, D.; Ralegaonkar, R.V. Technological and Sustainable Perception on the Advancements of Prefabrication in Construction Industry. *Energies* **2022**, *15*, 7548. [CrossRef]
10. Gao, Y.; Tian, X.-L. Prefabrication policies and the performance of construction industry in China. *J. Clean. Prod.* **2020**, *253*, 120042. [CrossRef]
11. Liu, S.; Li, Z.; Teng, Y.; Dai, L. A dynamic simulation study on the sustainability of prefabricated buildings. *Sustain. Cities Soc.* **2022**, *77*, 103551. [CrossRef]
12. Luo, T.; Xue, X.; Tan, Y.; Wang, Y.; Zhang, Y. Exploring a body of knowledge for promoting the sustainable transition to prefabricated construction. *Eng. Constr. Archit. Manag.* **2021**, *28*, 2637–2666. [CrossRef]
13. Research&Market. Global Prefabricated Construction Industry Report 2023: Prefabricated Construction Startups Are Raising Funding Rounds to Support Their Growth. Available online: https://www.prnewswire.com/news-releases/global-prefabricated-construction-industry-report-2023-prefabricated-construction-startups-are-raising-funding-rounds-to-support-their-growth-301839783.html (accessed on 4 December 2024).
14. Zairul, M. The recent trends on prefabricated buildings with circular economy (CE) approach. *Clean. Eng. Technol.* **2021**, *4*, 100239. [CrossRef]
15. Zhao, L.; Liu, Z.; Mbachu, J. Optimization of the supplier selection process in prefabrication using BIM. *Buildings* **2019**, *9*, 222. [CrossRef]
16. Liang, R.; Li, R.; Yan, X.; Xue, Z.; Wei, X. Evaluating and selecting the supplier in prefabricated megaprojects using extended fuzzy TOPSIS under hesitant environment: A case study from China. *Eng. Constr. Archit. Manag.* **2023**, *30*, 1902–1931. [CrossRef]
17. Darwish, M.; Mohsen, O.; Mohamed, Y.; Al-Hussein, M. Integrated Simulation and Lean Approach for Production Line Improvement in a Prefabricated Panelized Homebuilding Facility. In Proceedings of the 28th Annual Conference of the International Group for Lean Construction (IGLC28), Berkeley, CA, USA, 6–12 July 2020.
18. Han, Z.; Yu, L.; Li, B.; Qi, Y. Research on prefabricated component production line mold platform combination method. *J. Phys. Conf. Ser.* **2022**, *2187*, 012059. [CrossRef]
19. Liu, C.; Wu, J.; Jiang, X.; Gu, Y.; Xie, L.; Huang, Z. Automatic assembly of prefabricated components based on vision-guided robot. *Autom. Constr.* **2024**, *162*, 105385. [CrossRef]
20. China Construction Association Certification Center. *Research Report on the Development of Prefabricated Construction in China's Building Industry*; China Construction Association Certification Center: Beijing, China, 2022.
21. Martin, H.; Ramjarrie, K. Cloud contractor selection model for design-build open tender. *J. Constr. Eng. Manag.* **2021**, *147*, 04021020. [CrossRef]
22. Nam, S.; Yoon, J.; Kim, K.; Choi, B. Optimization of Prefabricated Components in Housing Modular Construction. *Sustainability* **2020**, *12*, 10269. [CrossRef]
23. Abdollahi, M.; Arvan, M.; Razmi, J. An integrated approach for supplier portfolio selection: Lean or agile? *Expert Syst. Appl.* **2015**, *42*, 679–690. [CrossRef]
24. Barkokebas, B.; Khalife, S.; Al-Hussein, M.; Hamzeh, F. A BIM-lean framework for digitalisation of premanufacturing phases in offsite construction. *Eng. Constr. Archit. Manag.* **2021**, *28*, 2155–2175. [CrossRef]
25. Du, J.; Xue, Y.; Sugumaran, V.; Hu, M.; Dong, P. Improved biogeography-based optimization algorithm for lean production scheduling of prefabricated components. *Eng. Constr. Archit. Manag.* **2023**, *30*, 1601–1635. [CrossRef]
26. Zhang, S.; Zhongfu, L.; Tianxin, L.; Mengqi, Y. A holistic literature review of building information modeling for prefabricated construction. *J. Civ. Eng. Manag.* **2021**, *27*, 485–499. [CrossRef]
27. Akanbi, O.A.; Oyedolapo, O.; Steven, G.J. Lean Principles in Construction. In *Sustainable Construction Technologies*; Butterworth-Heinemann: Oxford, UK, 2019; pp. 317–348. [CrossRef]

28. Ho, C.T.T. Application of Optimization to the Production Planning of Construction Prefabrication Supply Chains. Ph.D. Thesis, University of Washington, Seattle, WA, USA, 2019.
29. Rashidi, K.; Noorizadeh, A.; Kannan, D.; Cullinane, K. Applying the triple bottom line in sustainable supplier selection: A meta-review of the state-of-the-art. *J. Clean. Prod.* **2020**, *269*, 122001. [CrossRef]
30. Lima-Junior, F.R.; Carpinetti, L.C.R. A multicriteria approach based on fuzzy QFD for choosing criteria for supplier selection. *Comput. Ind. Eng.* **2016**, *101*, 269–285. [CrossRef]
31. Chen, C.-H. A novel multi-criteria decision-making model for building material supplier selection based on entropy-AHP weighted TOPSIS. *Entropy* **2020**, *22*, 259. [CrossRef] [PubMed]
32. Dutta, P.; Jaikumar, B.; Arora, M.S. Applications of data envelopment analysis in supplier selection between 2000 and 2020: A literature review. *Ann. Oper. Res.* **2022**, *315*, 1399–1454.
33. Ecer, F. Multi-criteria decision making for green supplier selection using interval type-2 fuzzy AHP: A case study of a home appliance manufacturer. *Oper. Res.* **2022**, *22*, 199–233.
34. Ghasempoor Anaraki, M.; Vladislav, D.S.; Karbasian, M.; Osintsev, N.; Nozick, V. Evaluation and selection of supplier in supply chain with fuzzy analytical network process approach. *J. Fuzzy Ext. Appl.* **2021**, *2*, 69–88.
35. Göncü, K.K.; Çetin, O. A decision model for supplier selection criteria in healthcare enterprises with dematel ANP method. *Sustainability* **2022**, *14*, 13912. [CrossRef]
36. Hadian, H.; Chahardoli, S.; Golmohammadi, A.-M.; Mostafaeipour, A. A practical framework for supplier selection decisions with an application to the automotive sector. *Int. J. Prod. Res.* **2020**, *58*, 2997–3014. [CrossRef]
37. Schramm, V.B.; Cabral, L.P.B.; Schramm, F. Approaches for supporting sustainable supplier selection-A literature review. *J. Clean. Prod.* **2020**, *273*, 123089. [CrossRef]
38. Amiri, M.; Hashemi-Tabatabaei, M.; Ghahremanloo, M.; Keshavarz-Ghorabaee, M.; Zavadskas, E.K.; Banaitis, A. A new fuzzy BWM approach for evaluating and selecting a sustainable supplier in supply chain management. *Int. J. Sustain. Dev. World Ecol.* **2021**, *28*, 125–142. [CrossRef]
39. Siew, L.W.; Hoe, L.W.; Fai, L.K.; Bakar, M.A.; Xian, S.J. Analysis on the e-Learning Method in Malaysia with AHP-VIKOR Model. *Int. J. Inf. Educ. Technol.* **2021**, *11*, 52–58. [CrossRef]
40. Kilpatrick, J. Lean principles. *Utah Manuf. Ext. Partnersh.* **2003**, *68*, 1–5.
41. Sachdeva, G. Lean, Six Sigma, and Industry 4.0 Technologies Connection and Inclusion. In *Lean Six Sigma 4.0 for Operational Excellence Under the Industry 4.0 Transformation*; CRC Press: Boca Raton, FL, USA, 2023; pp. 47–57.
42. Wang, S.; Tang, J.; Zou, Y.; Zhou, Q. Research on production process optimization of precast concrete component factory based on value stream mapping. *Eng. Constr. Archit. Manag.* **2020**, *27*, 850–871. [CrossRef]
43. Potužák, B.J. Utilization of lean construction principles and techniques. *Supply Chain Manag.* **2017**, *1*, 38–54.
44. Rosli, M.F.; Muhammad Tamyez, P.F.; Zahari, A.R. The effects of suitability and acceptability of lean principles in the flow of waste management on construction project performance. *Int. J. Constr. Manag.* **2023**, *23*, 114–125. [CrossRef]
45. Gusmao Brissi, S.; Wong Chong, O.; Debs, L.; Zhang, J. A review on the interactions of robotic systems and lean principles in offsite construction. *Eng. Constr. Archit. Manag.* **2022**, *29*, 383–406. [CrossRef]
46. Elmalky, A.; Mohamed, S.A.; ELDASH, K. Impact of Adopting Lean Principles on Construction Waste in Developing Countries. *Eng. Res. J.* **2024**, *53*, 82–93. [CrossRef]
47. Wang, Y.; Bian, S.; Dong, L.; Li, H. Multiresolution Modeling of a Modular Building Design Process Based on Design Structure Matrix. *Buildings* **2023**, *13*, 2330. [CrossRef]
48. Shabeen, S.R.; Krishnan, K.A. Application of lean manufacturing using value stream mapping (VSM) in precast component manufacturing: A case study. *Mater. Today Proc.* **2022**, *65*, 1105–1111. [CrossRef]
49. Marzouk, M.; Elmaraghy, A. Design for deconstruction using integrated lean principles and bim approach. *Sustainability* **2021**, *13*, 7856. [CrossRef]
50. Goyal, M.; Gao, Z. Integration of building information modeling (BIM) and prefabrication for lean construction. In Proceedings of the International Conference on Construction and Real Estate Management 2018, Charleston, SC, USA, 9–10 August 2018; pp. 78–84.
51. Fahmy, M. Lean Principles Implementation in Construction Management: A One Team Approach. In Proceedings of the International Conference on Civil Infrastructure and Construction (CIC 2020), Doha, Qatar, 2–5 February 2020.
52. Zhang, Y. A Framework to Improve Modular Construction Manufacturing Production Line Performance. Master's Thesis, University of Alberta, Edmonton, AB, USA, 2017.
53. Song, Y.; Wang, J.; Guo, F.; Lu, J.; Liu, S. Research on supplier selection of prefabricated building elements from the perspective of sustainable development. *Sustainability* **2021**, *13*, 6080. [CrossRef]
54. Mostafa, S.; Tam, V.W.; Dumrak, J.; Mohamed, S. Leagile strategies for optimizing the delivery of prefabricated house building projects. *Int. J. Constr. Manag.* **2020**, *20*, 867–881. [CrossRef]
55. Sahu, A.K.; Sharma, M.; Raut, R.D.; Sahu, A.K.; Sahu, N.K.; Antony, J.; Tortorella, G.L. Decision-making framework for supplier selection using an integrated MCDM approach in a lean-agile-resilient-green environment: Evidence from Indian automotive sector. *TQM J.* **2023**, *35*, 964–1006. [CrossRef]
56. Yang, H.; Yeung, J.F.; Chan, A.P.; Chiang, Y.H.; Chan, D.W. A critical review of performance measurement in construction. *J. Facil. Manag.* **2010**, *8*, 269–284. [CrossRef]

57. Liang, R.; Li, R.; Chong, H.-Y. Sustainable supplier selection for prefabricated megaprojects during COVID-19: A novel large group decision-making model. *Build. Res. Inf.* **2023**, *51*, 533–554. [CrossRef]
58. Wang, H.; Guo, X. Selection of Prefabricated Components Supplier for Prefabricated Building Based on BP Neural Network Model. In *Applied Mathematics, Modeling and Computer Simulation*; IOS Press: Amsterdam, The Netherlands, 2022.
59. El Mokadem, M. The classification of supplier selection criteria with respect to lean or agile manufacturing strategies. *J. Manuf. Technol. Manag.* **2017**, *28*, 232–249. [CrossRef]
60. Tsai, W.C. Fuzzy measures of supplier evaluation under lean concepts. *J. Oper. Res. Soc.* **2017**, *60*, 1005–1011. [CrossRef]
61. Rezaei, A.; Rahiminezhad Galankashi, M.; Mansoorzadeh, S.; Mokhatab Rafiei, F. Supplier Selection and Order Allocation with Lean Manufacturing Criteria: An Integrated MCDM and Bi-objective Modelling Approach. *Eng. Manag. J.* **2020**, *32*, 253–271. [CrossRef]
62. Almasi, M.; Khoshfetrat, S.; Galankashi, M.R. Sustainable supplier selection and order allocation under risk and inflation condition. *IEEE Trans. Eng. Manag.* **2019**, *68*, 823–837. [CrossRef]
63. Pamucar, D.; Torkayesh, A.E.; Biswas, S. Supplier selection in healthcare supply chain management during the COVID-19 pandemic: A novel fuzzy rough decision-making approach. *Ann. Oper. Res.* **2023**, *328*, 977–1019. [CrossRef] [PubMed]
64. Zhao, J.; You, X.-Y.; Liu, H.-C.; Wu, S.-M. An Extended VIKOR Method Using Intuitionistic Fuzzy Sets and Combination Weights for Supplier Selection. *Symmetry* **2017**, *9*, 169. [CrossRef]
65. Opricovic, S.; Tzeng, G.-H. Compromise solution by MCDM methods: A comparative analysis of VIKOR and TOPSIS. *Eur. J. Oper. Res.* **2004**, *156*, 445–455. [CrossRef]
66. Jing, S.; Niu, Z.; Chang, P.-C. The application of VIKOR for the tool selection in lean management. *J. Intell. Manuf.* **2019**, *30*, 2901–2912. [CrossRef]
67. Xu, F.; Liu, J.; Lin, S.; Yuan, J. A VIKOR-based approach for assessing the service performance of electric vehicle sharing programs: A case study in Beijing. *J. Clean. Prod.* **2017**, *148*, 254–267. [CrossRef]
68. Rezaie, K.; Ramiyani, S.S.; Nazari-Shirkouhi, S.; Badizadeh, A. Evaluating performance of Iranian cement firms using an integrated fuzzy AHP–VIKOR method. *Appl. Math. Model.* **2014**, *38*, 5033–5046. [CrossRef]
69. Valipour Parkouhi, S.; Safaei Ghadikolaei, A. A resilience approach for supplier selection: Using Fuzzy Analytic Network Process and grey VIKOR techniques. *J. Clean. Prod.* **2017**, *161*, 431–451. [CrossRef]
70. Anvari, A.; Zulkifli, N.; Arghish, O. Application of a modified VIKOR method for decision-making problems in lean tool selection. *Int. J. Adv. Manuf. Technol.* **2014**, *71*, 829–841. [CrossRef]
71. Liu, H.-C.; You, J.-X.; You, X.-Y.; Shan, M.-M. A novel approach for failure mode and effects analysis using combination weighting and fuzzy VIKOR method. *Appl. Soft Comput.* **2015**, *28*, 579–588. [CrossRef]
72. Soner, O.; Celik, E.; Akyuz, E. Application of AHP and VIKOR methods under interval type 2 fuzzy environment in maritime transportation. *Ocean Eng.* **2017**, *129*, 107–116. [CrossRef]
73. Awasthi, A.; Govindan, K.; Gold, S. Multi-tier sustainable global supplier selection using a fuzzy AHP-VIKOR based approach. *Int. J. Prod. Econ.* **2018**, *195*, 106–117. [CrossRef]
74. Luthra, S.; Govindan, K.; Kannan, D.; Mangla, S.K.; Garg, C.P. An integrated framework for sustainable supplier selection and evaluation in supply chains. *J. Clean. Prod.* **2017**, *140*, 1686–1698. [CrossRef]
75. Li, X.K.; Wang, X.M.; Lei, L. The application of an ANP-Fuzzy comprehensive evaluation model to assess lean construction management performance. *Eng. Constr. Archit. Manag.* **2020**, *27*, 356–384. [CrossRef]
76. Barth, K.B.; Formoso, C.T. Requirements in performance measurement systems of construction projects from the lean production perspective. *Front. Eng. Manag.* **2021**, *8*, 442–455. [CrossRef]
77. Ahmad, S.; Soetanto, R.; Goodier, C. Lean approach in precast concrete component production. *Built Environ. Proj. Asset Manag.* **2019**, *9*, 457–470. [CrossRef]
78. Du, J.; Zhang, J.; Castro-Lacouture, D.; Hu, Y. Lean manufacturing applications in prefabricated construction projects. *Autom. Constr.* **2023**, *150*, 104790. [CrossRef]
79. Feldmann, F.G. Towards lean automation in construction—Exploring barriers to implementing automation in prefabrication. *Sustainability* **2022**, *14*, 12944. [CrossRef]
80. Liu, J.K.; Shi, G.S. Quality Control of a Complex Lean Construction Project Based on KanBIM Technology. *Eurasia J. Math. Sci. Technol. Educ.* **2017**, *13*, 5905–5919. [CrossRef]
81. Kong, L.; Li, H.; Luo, H.; Ding, L.; Zhang, X. Sustainable performance of just-in-time (JIT) management in time-dependent batch delivery scheduling of precast construction. *J. Clean. Prod.* **2018**, *193*, 684–701. [CrossRef]
82. Dave, B.; Kubler, S.; Framling, K.; Koskela, L. Opportunities for enhanced lean construction management using Internet of Things standards. *Autom. Constr.* **2016**, *61*, 86–97. [CrossRef]
83. Topaktas, R.; Altin, S.; Aydin, C.; Akkoc, A.; Urkmez, A.; Aydin, Z.B. Lean Tools Proposal to mitigate Delays and Cost Overruns in Construction Projects. *Cent. Eur. J. Urol.* **2020**, *73*, 336–341. [CrossRef]
84. Xu, Y.G.; Qian, C. Lean Cost Analysis Based on BIM Modeling for Construction Project. *Appl. Mech. Mater.* **2014**, *457–458*, 1444–1447. [CrossRef]
85. Pheng, L.S.; Gao, S.; Lin, J.L. Converging early contractor involvement (ECI) and lean construction practices for productivity enhancement: Some preliminary findings from Singapore. *Int. J. Product. Perform. Manag.* **2015**, *64*, 831–852. [CrossRef]

86. Saaty, T.L. Decision-making with the AHP: Why is the principal eigenvector necessary. *Eur. J. Oper. Res.* **2003**, *145*, 85–91. [CrossRef]
87. Rahardjo, B.; Wang, F.-K.; Lo, S.-C.; Chou, J.-H. A Hybrid Multi-Criteria Decision-Making Model Combining DANP with VIKOR for Sustainable Supplier Selection in Electronics Industry. *Sustainability* **2023**, *15*, 4588. [CrossRef]
88. Wu, Y.; Zhang, B.; Xu, C.; Li, L. Site selection decision framework using fuzzy ANP-VIKOR for large commercial rooftop PV system based on sustainability perspective. *Sustain. Cities Soc.* **2018**, *40*, 454–470. [CrossRef]
89. Prasad, K.G.D.; Prasad, M.V.; Rao, S.V.V.B.; Patro, C.S. Supplier Selection through AHP-VIKOR Integrated Methodology. *Int. J. Ind. Eng.* **2016**, *3*, 1–6. [CrossRef]
90. Chang, C.L. A modified VIKOR method for multiple criteria analysis. *Environ. Monit. Assess.* **2010**, *168*, 339–344. [CrossRef]
91. Yu, T.; Shi, Q.; Zuo, J.; Chen, R. Critical factors for implementing sustainable construction practice in HOPSCA projects: A case study in China. *Sustain. Cities Soc.* **2018**, *37*, 93–103. [CrossRef]
92. Petroutsatou, K.; Ladopoulos, I.; Nalmpantis, D. Hierarchizing the criteria of construction equipment procurement decision using the AHP method. *IEEE Trans. Eng. Manag.* **2021**, *70*, 3271–3282. [CrossRef]
93. Falch, L.; de Silva, C.W. Improvement of VIKOR method with application to Multi-Objective design problems. *Int. J. Inf. Technol. Decis. Mak.* **2023**, *22*, 777–802. [CrossRef]

buildings

Article

Cost–Benefit Framework for Selecting a Highway Project Using the SWARA Approach

Omar Shabbir Ahmed, Khalid S. Al-Gahtani * and Ayman Altuwaim

Department of Civil Engineering, King Saud University, P.O. Box 145111, Riyadh 4545, Saudi Arabia; 444106081@student.ksu.edu.sa (O.S.A.); aaltuwaim@ksu.edu.sa (A.A.)
* Correspondence: kgahtani@ksu.edu.sa

Abstract: The effective selection of highway projects is essential for driving economic growth and facilitating trade in Saudi Arabia's cities. However, current studies lack a comprehensive approach that assesses highways based on a full spectrum of economic, environmental, and social cost–benefit factors tailored to Saudi Arabia's construction industry. This study addresses this gap by developing a framework that incorporates the aspects. The methodology comprises five steps: (1) a literature review to identify benefit and cost criteria; (2) expert surveys to select significant criteria; (3) the application of the stepwise weight assessment ratio analysis (SWARA) method to determine criteria weights; (4) structured expert interviews to establish criteria quality weights; and (5) validation through application to three case studies, comparing the results with those obtained using the ANP method. The findings show that economically efficient road choices yield increased productivity and support industrial growth, while the most significant environmental benefit is reducing carbon emissions. Social benefits, as emphasized by experts, include accident reduction. Cost factors are also considered, with savings on vehicle operation costs identified as the most significant, as opined by the expert surveyed. Among the analyzed highways, Khurais Road, Riyadh, was the most efficient from the SWARA approach, with a value of 0.8, and the ANP case study conducted, with a normalized score of 0.045 and 0.230 for both benefits and cost criteria.

Keywords: cost; benefits; highway; Saudi Arabia construction industry; SWARA

Academic Editors: Pramen P. Shrestha, Wenxin Shen, Jin Xue and Wenzhe Tang

Received: 19 December 2024
Revised: 22 January 2025
Accepted: 28 January 2025
Published: 30 January 2025

Citation: Ahmed, O.S.; Al-Gahtani, K.S.; Altuwaim, A. Cost–Benefit Framework for Selecting a Highway Project Using the SWARA Approach. *Buildings* **2025**, *15*, 439. https://doi.org/10.3390/buildings15030439

1. Introduction

The construction industry in Saudi Arabia is poised for significant growth as the country diversifies its economy beyond oil [1]. This expansion presents numerous opportunities for both residential and large-scale transportation infrastructure projects. With an anticipated annual growth rate of 2.5% in 2021 [1], the sector is expected to continue expanding, driven by transportation-related megaprojects. Saudi Arabia's extensive road network, covering over 200,000 km, including 66,000 km of highways, plays a key role in connecting major cities, airports, ports, and railroads [2]. The development of over 5000 km of highways and bridges further enhances the country's transportation capacity, supporting both urban mobility and the movement of goods [3]. According to Aldagheiri [4], Saudi Arabia plans to construct over 3500 km of new roads, including 284 highways connecting major urban centers, as part of its growth strategy. Rahman et al. [5] indicate 2200 km of new roads are already under development, aimed at supporting the nation's economic expansion. The growth of e-commerce and industrial activity is expected to increase demand for land freight transportation [6]. To enhance road safety, the Transport Ministry is investing over USD 586.67 million in 23 new road safety measures [7]. Based on these developments, this study explores the cost and benefit factors influencing road selection in Saudi Arabia's construction industry. While previous studies have examined various methodologies for assessing the economic, environmental, and social impacts of road infrastructure projects, there remains a distinct gap in applying these frameworks to the Saudi context. Given the nation's focus on economic diversification and the growing emphasis on sustainable development, understanding how cost–benefit factors specifically influence road selection in Saudi Arabia is critical. Addressing this gap will ensure that road projects are evaluated comprehensively, considering the unique challenges and opportunities of the local construction landscape.

To address this gap, this study will conduct a comprehensive cost–benefit framework for road project selection. The framework consists of five steps. The first step is to identify criteria for benefits and costs by conducting a comprehensive literature review. Then, the significant criteria were identified by surveying the experts. After that, the criteria weights were determined using stepwise weight assessment ratio analysis (SWARA). Next, the criteria quality weight per criterion was established by conducting a structured interview with the experts. The framework was validated by applying the framework to the three case studies. The results were compared with the results of applying the ANP method in the same three case studies. The findings are expected to have significant implications for policymakers and project managers, guiding them toward decisions that maximize economic productivity, promote sustainable practices, and improve social outcomes in road infrastructure projects.

The remainder of this paper is structured as follows: Section 2 provides a comprehensive literature review; Section 3 describes the methodology and theoretical background of the SWARA approach, including data collection, statistical analysis, and computation of the criteria quality weight; Section 4 presents the results and the Analytic Network Process (ANP) case study validation as well as the conclusion.

2. Materials and Methods

This section provides an overview of road projects and how the cost–benefit analysis technique can be applied to efficient road selection. Numerous previous studies on cost-benefit methods were identified, and, at the end, research gaps were presented.

2.1. Road Projects Selection

The global expansion of road networks and the significant investments required for infrastructure highlight the need for effective project selection methods. Roads are growing in length and spatial expanse worldwide at an unprecedented pace [8,9]. Since 2000, the global length of legally permitted roads has grown by around 12 million kilometers, and by 2050, an additional 25 million kilometers of paved roads are anticipated [9]. The G20 industrial nations have stated that USD 70 trillion in spending will be required by 2030 [10]. For new roads and other infrastructure, they plan on more than doubling the global infrastructure investments to date to support this substantial expansion [11].

Various international studies have explored cost–benefit analysis, multi-criteria approaches, integrated strategies, and analytical network processes to evaluate and prioritize road projects, considering factors like economic, environmental, and social impacts. According to analysts, cost–benefit analysis (CBA) works well for resolving issues with multiple options, such as transportation projects at the same site. However, its use for the selection of road projects in the transport network is limited in several ways [12]. An infrastructure selection model was created by Macura et al. [13]. The model's primary features identify the group of projects for investment and consider the effects of external initiatives. The selection of public projects based on factors other than costs and monetary benefits is the subject of very few papers. Hinloopen et al. [14] applied ordinal and cardinal judgment criteria in the planning of public transport systems; the authors of [15] used the Data Envelopment Analysis (DEA) approach to measure target performance for traffic safety; the authors of [16] discussed a selection method for environmentally sustainable transport systems; and the authors of [17] used a multi-criteria sorting procedure to support public decisions. When it came to transportation projects, the authors of [18] contrasted the cost–benefit analysis with a multi-criteria approach.

An integrated strategy that uses the fuzzy Delphi technique, the analytical network process, and one-goal programming to solve the selection difficulties for road projects was proposed by [19]. The authors used the ANP approach for goal prioritizing and relationships between them to address the drawbacks of goal programming. The example given takes Taiwan's transportation project selection into account. Two years later, the BOCR criteria are used to improve this model [20]. The ANP approach is a useful tool for systems having linkages between objectives and stakeholders' environments [21]. The developed model's objective is to select Thailand's highway corridor. Economics, engineering, construction, transportation and traffic, environmental, social, and land use aspects are all taken into consideration.

2.2. Cost–Benefit Method

Due to the substantial economic and social contribution of roads to national development, governments must choose how to make the best possible use of limited funding for road safety legislation and initiatives. The key factors taken into consideration while making decisions about policies and budgets are suitability and legality. But in recent years, efficiency has become a common criterion for a good program. Efficiency assessment tools (EATs) are used to assess the efficiency of a system or process. The two most popular efficiency assessment tools are cost–benefit analyses (CBAs) and cost-effectiveness analyses (CEAs) [22]. While CEA is utilized for concerns involving partial efficiency and looks at the lives saved, CBA primarily examines the social output of a measure or a policy [23]. Cost–benefit analysis for highway selection quantifies project benefits, enhances transparency, informs decision-making, and prioritizes funding based on economic efficiency and potential safety improvements [24]. According to Coley [25], the cost–benefit analysis ensures that investments generate the maximum returns over their life cycle by quantifying agency costs and user benefits, which improves openness and accountability in the selection of transportation projects. Making decisions based on a variety of evaluation techniques results in better knowledge. Only a small number of studies have used multiple assessment techniques on the same project [26]. In the context of project financing for transportation infrastructure, the primary application areas for CBA pertain to the financial and economic feasibility of an infrastructure investment [27].

Typically, when several aspects are extremely speculative, CBA offers a rational framework for assessing potential courses of action. According to Matanhire [28], even though a monetary value cannot be readily assigned, it considers all the aspects that affect a project's costs or benefits. If a project's financial or economic rate of return is less than the required rate of return, the money would be better spent somewhere else. Shang et al. [29] discussed that the benefits that can be derived from roads include the following (economic, social, political, and technical benefits): regional development, employment, transportation industry, city importance, upward influence, living standard, transport quality, connection degree, and path density. Hence, when making highway investment decisions, cost–benefit analysis weighs capital costs against user and agency cost savings to determine the most economical roadway upgrades for systemwide deployment [30].

The CBA method was chosen for this research due to its ability to provide a comprehensive evaluation of both the economic and social impacts of highway projects. CBA is particularly effective in guiding resource allocation decisions in contexts with limited funding, allowing policymakers to prioritize projects that yield the highest returns on investment. This method not only quantifies agency costs and user benefits but also supports transparent decision-making by revealing the long-term economic efficiency and potential safety improvements of each project. Given that highway investments have widespread implications for regional development, social welfare, and economic growth, CBA enables a structured comparison of competing projects to determine the most advantageous options for system-wide implementation. In this study, CBA's analytical framework is instrumental in balancing capital costs against projected benefits, ensuring that selected projects align with Saudi Arabia's goals for efficient infrastructure development. The SWARA approach, along with other Multi-Criteria Decision-Making MCDM methods, has been applied in multiple contexts to facilitate more informed infrastructure choices. The application of CBA in various disciplines is shown in Table 1.

Table 1. Highway cost–benefit factors.

Field	Application	Authors
Transportation	Highway and railway selection	[31]
Construction	Road benefit evaluation	[32]
Construction	Road selection	[33]
Building	Decision-making process	[34]
Construction	Roofing selection	[35]
Environment	Water treatment options	[36]
Building	Green building	[37]
Agriculture	Resource allocation	[38]
Environment	Climate change mitigation	[39]
Public health	Social return on investment	[40]
Finance	Resource allocation	[41]
Transportation	Transportation method selection	[42]
Health	Health care evaluation	[43]

2.3. Gap Knowledge

Prior research underscores the need for robust frameworks that incorporate a comprehensive range of economic, social, and environmental factors when applying cost–benefit analyses to highway selection. Sarı and Şen [44] highlight that combining in-depth cost–benefit analysis with MCDA methods would strengthen decision-making frameworks in this field. Additionally, Pan et al. [45] point out limitations in current MCDM techniques, such as difficulties in integrating subjective decision-maker preferences with objective data analysis. In addition, there remains a critical gap in studies using the combined ANP and SWARA methods for highway selection. This underscores the need for future research to explore diverse MCDM methodologies and apply a cost–benefit approach to effectively identify optimal highway alternatives within infrastructure projects [46].

3. Methodology

The methodology consists of five steps which can be seen in Figure 1. The first step is to collect criteria for benefit and cost by conducting a comprehensive literature review. Then, the significant criteria were identified by surveying the experts. After that, the criteria weights were determined using SWARA. Next, the criteria quality weight per criterion was established by conducting a structured interview with the experts. The framework was validated by applying the framework to the three case studies. The results were compared with the results of applying the ANP method in the same three case studies.

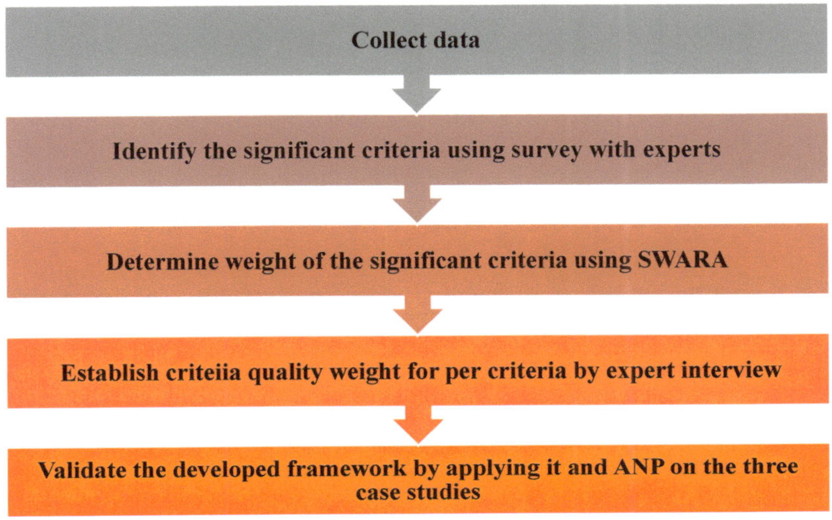

Figure 1. Research design.

3.1. Data Collection

The analyzed articles were taken from the 1995–2022 timeframe in the Web of Science database. Cost and benefit factors for highway selection were the search terms utilized. There were 1055 articles in all, consisting of original research journals, review papers, and conference proceedings. The first article on this subject was published in 2009, although the database contains studies dating back to 1995. Filters that adhered to preferred reporting items for systematic reviews and meta-analysis techniques were constructed to ascertain how the papers fit within the context of this research [47].

From the systematic literature review conducted, 46 cost–benefit factors were identified to be significant when considering road selection. The classification includes economic, environmental, social, technical, and political. The first classification, economic, includes economic benefits and cost factors such as improved economic activities, enhanced productivity, increased revenue generation, and capital investment [48]. The environmental classifications mainly address factors related to the environment such as reduced carbon emission, reduced congestion, less noise and smoke, ecology and pollution cost, environmental mitigation cost, and visual intrusion [49]. The social benefit and cost factors include increasing job opportunities, reducing accidents, marginal cost of public funds, and relocation cost [50]. The technical factors are technologically related, including smoother ride, optimized traffic flow, software and information system cost, and smart transportation system cost [51]. Items such as national unity, political stability, and infrastructural diplomacy are the elements that constitute political classification [52]. These factors were compiled from the literature and were validated using a qualitative approach. The identified cost–benefit factors were extracted from numerous studies, as can be seen in Table 2. The factors are categorized into five groups based on their functions: economic, environmental, social, technological, and political considerations.

Table 2. Highway cost–benefit factors.

Code	Economic Benefits	Economic Cost	References
ECB01/C01	Improve economic activities	Increase revenue generation	[48,53]
ECB02/C02	Enhanced productivity	Capital investment	[54,55]
ECB03/C03	Increased investment, tourism, and trade	Supply chain efficiency	[54]
ECB04/C04	Improve tax revenue generation Industry and commerce	Cost saving for business	[56]
C05		Savings on vehicle operation cost	[57]
C06		Reduce user cost	[58]
C07		Relative investment cost	[59]
	Environmental Benefits	**Environmental Cost**	
EVB01/EC01	Reduced carbon emission	Ecology and pollution cost	[49]
EVB02/EC02	Reduce congestion	Environmental mitigation cost	[60]
EVB03/EC03	Less noise and smoke	Visual intrusion	[49]
EVB04/EC04	Fuel saving	Noise/air pollution	[49]
EVB05/EC05	Erosion and sediment control	Reduced environmental cost	[61]
EVB06	Stormwater management		[62]
EVB07	Vegetation management		[62]
	Social Benefits	**Social Cost**	
SB01/SC01	Increase job opportunities	Marginal cost of public fund	[63]
SB02/SC02	Reduce accident	Relocation cost	[50]
SB03/SC03	Provide access to basic health care facility	Reduce accident and health cost	[50]
SB04	Improve the quality of life		[64]
SB05	Cultural and historic preservation		[65]
SB06	Crisis response		[66]
SB07	Disaster management		[66]
	Technical Benefits	**Technical Cost**	
TB01/TC01	Smoother ride	Software and information system cost	[51]
TB02/TC02	Optimized traffic flow	Smart transportation system cost	[67]
TB03/TC03	Efficient maintenance and repair	Design and engineering cost	[68,69]
TB04/TC04	Integration of intelligent transportation system	Construction cost Traffic control and safety measures	[26,70]
	Political Benefits		
PB01	National unity		[52]
PB02	Infrastructure diplomacy		[71]
PB03	Political stability		[26]
PB04	Fiscal responsibility		[26]
PB05	Enhanced government credibility		[26]

3.2. Identification of the Significant Criteria Using a Survey with Experts

This study utilized the expert survey to gather primary data on cost–benefit criteria considered for highway selection that are more suitable from the existing top-ranked factors initially discovered through an extensive literature review. Fourteen experts consisting of contractors, consultants, and experts working with government organizations who have indicated a medium knowledge of cost–benefits and participated in several road projects in Saudi Arabia are part of the survey. This is to further assess from the perspective of experts to come up with the most significant factors for efficient highway selection.

3.2.1. Expert Demography

This study involved fourteen experts from diverse backgrounds, including consultants, contractors, and government employees, all with extensive experience in road projects. Most participants have a medium-sized level of knowledge regarding cost–benefit analysis, with thirteen experts boasting over ten years of experience in the field and one expert having between five and ten years. The organizations represented vary in their implementation of cost–benefit analysis, with a majority applying it at a medium level and some at low or high levels.

The survey analysis indicates that 42.8% of respondents are contractors, while government employees and consultants each account for 28.6%. Among the experts, 42.86% report a high level of understanding of cost–benefit analysis, while 57.14% have a medium level of knowledge, demonstrating that all participants have at least moderate familiarity with the concept in the Saudi Arabian construction sector. Importantly, no respondents reported low knowledge or indicated being unaware of the concept.

In terms of implementation, 50% of experts indicated medium and low levels of the application of cost–benefit analysis within their organizations, while 36% reported a low level of implementation. Only 14% stated that their organizations utilize cost–benefit analysis at a high level.

3.2.2. Analysis of Expert Opinions

The cost–benefit factors found in this study may be applied to highway selection as shown in Tables 3 and 4. To determine the relative significance, the survey data were imported into Statistical Package for Social Science (SPSS) version 27 and evaluated using the RII formula as shown.

$$RII = \frac{\sum W}{AXN} \tag{1}$$

where W is the weight given to each factor by the respondent, A is the highest weight, and N is the total number of respondents. The RII number ranges from 0.0 to 1. As recommended by [72], the transformation matrix, which is a mathematical tool for transforming vectors or points in space in linear algebra, is utilized to compare the RII to the appropriate significance level. The cut-offs based on RII relevance indicates high (H) ranging from $0.8 < RII < 1.0$, high–medium ranging from $0.6 < RII << 0.8$, medium significance ranging from $0.4 < RII < 0.6$, medium–low ranges from $0.2 < RII < 0.4$, and low significant ranges from $0.0 < RII < 0.2$, [73]. The significance levels of the benefit factors are shown in Table 3, with increased productivity, reduced accidents, reduced carbon emission, and improved economic activities being the most significant with RII values of 0.92, 0.90, and 0.87, respectively. High–medium (H-M) criteria include improved quality of life and fuel saving as indicated by the respondents with RII values of 0.8 and 0.78, respectively. Political stability and fiscal responsibility as the least medium benefit factors for highway selection having RII values of 0.6 and 0.57.

Table 3. Highway benefit factors.

Benefit Factors	Codes	Frequency					RII	Ranking	Significance Level
		1	2	3	4	5			
Increased productivity	ECB02				5	9	0.92	1	High
Reduce accident	SB02			1	5	8	0.90	2	High
Reduced carbon emission	EVB01		1		6	7	0.87	3	High
Improve economic activities	ECB01		1		6	7	0.87	4	High
Less noise and smoke	EVB03	1			6	7	0.85	5	High
Increased investment, tourism, and trade	ECB03	1		1	5	7	0.84	6	High
Integration of intelligence transportation system	TB01	1		2	5	6	0.81	7	High
Improve the quality of life.	SB04		2		6	6	0.80	8	High–Medium
Fuel saving	EVB04		2		7	5	0.78	9	High–Medium
Crisis response	SB06		2		7	5	0.78	10	High–Medium
Improve tax revenue generation	ECB04	2		1	7	4	0.76	11	High–Medium
Optimized traffic flow	TB02	2		1	7	4	0.76	12	High–Medium
Provide access to basic health care facility	SB03	2		1	7	4	0.76	13	High–Medium
Increase job opportunities	SB01	2	1		8	3	0.73	14	High–Medium
Reduce congestion	EVB02	2	1	1	6	4	0.71	15	High–Medium
Erosion and sediment control	EVB05	1	2	2	7	2	0.70	16	High–Medium
Stormwater management	EVB06	3		2	6	3	0.68	17	High–Medium
Enhanced government credibility	PB05	3		2	6	3	0.68	18	High–Medium
Integration of intelligent transportation system	TB04	1	2	2	6	2	0.67	19	High–Medium
Disaster management	SB07	2	1	4	5	2	0.65	20	High–Medium
National unity	PB01	2	1	4	5	2	0.65	21	High–Medium
Efficient maintenance and repair	TB03	3		4	5	2	0.63	22	High–Medium
Infrastructure diplomacy	PB02	3		4	5	2	0.63	23	High–Medium
Cultural and historic preservation	SB05	3	1	4	4	2	0.61	24	High–Medium
Vegetation management	EVB07	4		4	4	2	0.60	25	High–Medium
Political stability	PB03	4		4	4	2	0.60	26	High–Medium
Fiscal responsibility	PB04	4		5	4	1	0.57	27	Medium

Table 4. Highway cost factors.

Cost Factors	Codes	Frequency					RII	Ranking	Significance Level
		1	2	3	4	5			
Savings on vehicle operation cost	C05			1	6	7	0.89	1	High
Increase revenue generation	C01		1		5	8	0.88	2	High
Reduce accident and health cost	SC03		1		6	7	0.87	3	High
Environmental mitigation cost	EC02		1		6	7	0.87	4	High
Reduce user cost	C06	1			6	7	0.85	5	High
Marginal cost of public fund	SC01	1			6	7	0.85	6	High
Noise/air pollution cost	EC04		2		6	6	0.79	7	High
Cost saving for business	C04		2		7	5	0.78	8	High–Medium
Relocation cost	SC02	2		1	6	5	0.77	9	High–Medium
Capital investment	C02	2		1	6	5	0.77	10	High–Medium
Design and engineering cost	TC03	2		1	7	4	0.76	11	High–Medium
Ecology and pollution cost	EC01	2	1		8	3	0.73	12	High–Medium
Reduce accident and health cost	SC03	2	1		8	3	0.73	13	High–Medium
Smart transportation system cost	TC02	2	1	1	6	4	0.71	14	High–Medium
Construction cost traffic control and safety measures	TC04	1	2	2	7	2	0.70	15	High–Medium
Software and information system cost	TC01	3		2	6	3	0.68	16	High–Medium
Supply chain efficiency	C03	1	2	2	6	2	0.67	17	High–Medium
Relative investment cost	C07	1	2	2	6	2	0.67	18	High–Medium
Visual intrusion cost	EC03	2	1	4	5	2	0.66	19	High–Medium

The cost factors, as shown in Table 4, indicate that savings on vehicle operation costs and increased revenue generation are the leading criteria with RII values of 0.89 and 0.88, respectively. Cost saving for business and relocation costs are high–medium factors with RII values of 0.78 and 0.77, respectively.

The significant benefit and cost criteria factors are presented in Tables 5 and 6, respectively, with the factors arranged in decreasing order of the relative importance index (RII) value. The selection of the significant factors was based on a high RII value from 0.80 to 1, as indicated by [72].

Table 5. Significant benefit criteria factors.

Benefit Factors	Codes	RII	Significance Level
Increased productivity	ECB02	0.92	High
Reduce accident	SB02	0.90	High
Reduced carbon emission	EVB01	0.87	High
Improve economic activities	ECB01	0.87	High
Less noise and smoke	EVB03	0.85	High
Increased investment, tourism, and trade	ECB03	0.84	High
Integration of intelligence transportation system	TB01	0.81	High

Table 6. Significant cost criteria factors.

Cost Factors	Codes	RII	Significance Level
Savings on vehicle operation cost	C05	0.89	High
Increase revenue generation	C01	0.88	High
Reduce accident and health cost	SC03	0.87	High
Environmental mitigation cost	EC02	0.87	High
Reduce user cost	C06	0.85	High
Marginal cost of public fund	SC01	0.85	High
Noise/air pollution cost	EC04	0.79	High

3.3. Determination of the Weight of the Significant Criteria Using SWARA

Keršuliene et al. [74] created the SWARA. When it comes to its application and the resolution of various issues, the SWARA method has garnered a lot of attention. These issues include the selection of machine tools, personnel [75], corporate social responsibility and sustainability, product design, and packaging design selection [76]. SWARA provides a clear and step-by-step procedure for weight determination, enhancing transparency in the decision-making process. This is vital when the decision involves multiple stakeholders who need to understand the rationale behind the assigned weights. In addition, SWARA leverages the knowledge and experience of experts who are familiar with the specific context of the problem. It relies on expert's preferences to prioritize different criteria, which ensures that the results are aligned with the domain knowledge. Regarding the rigorous method, SWARA systematically compares each criterion with the preceding one, thereby ensuring that every criterion is examined and rated for importance. This step-by-step approach helps eliminate potential biases in criteria importance as each comparison is made one step at a time, rather than rating them altogether. The SWARA techniques steps are as follows:

Step 1: The criteria are arranged according to their assessed significance in declining order.

Step 2: The response states the relative importance of criterion j concerning the preceding $(j-1)$ criterion, beginning with the second criterion and doing so for each criterion. This ratio is known as the comparative importance of the average value or s_j.

Step 3: The coefficient k_j is determined as follows:

$$k_j = \begin{cases} 1 & j = 1 \\ s_{j+1} & j > 1 \end{cases} \tag{2}$$

Step 4: The recalculated weight q_j is determined as follows:

$$q_j = \begin{cases} 1 & j = 1 \\ \frac{q_{j-1}}{k_j} & j > 1 \end{cases} \tag{3}$$

Step 5: The relative weights of the evaluation criteria are determined as follows:

$$w_j = \frac{q_j}{\sum_{k=1}^{n} q_k} \quad (4)$$

where w_j denotes the relative weight of the j-th criterion, and n denotes the number of the criteria.

The subjective weight s_j for the benefits and cost was obtained by conducting interviews with the seven experts from various sectors, such as consulting, contracting, and government, all with extensive experience in road projects, utilizing Delphi techniques. For the Delphi technique, experts were first selected based on knowledge, willingness, availability, and representation of a variety of backgrounds and positions so that the key viewpoints are provided. All experts know road selection and are conversant with multi-criteria techniques utilizing SWARA. Three rounds of questionnaire survey were conducted to establish the agreed criteria using Delphi. Consensus was reached by the expert, and the value of s_j was determined. Tables 7 and 8 show the weight of the evaluation criteria for the benefit and cost factors for highway selection in Saudi Arabia. The last column in Tables 7 and 8 represents the criteria weight for benefits and costs, respectively.

Table 7. The resulting weight of the benefit factors.

Codes	Benefit Factors	Rank	s_j	k_j	q_j	w_j
ECB02	Increased productivity	1		1	1	0.199
SB02	Reduce accident	2	0.15	1.15	0.870	0.174
EVB01	Reduced carbon emission	3	0.05	1.05	0.828	0.165
ECB01	Improve economic activities	4	0.1	1.1	0.753	0.150
EVB03	Less noise and smoke	5	0.3	1.3	0.579	0.115
ECB03	Increased investment, tourism, and trade	6	0.15	1.15	0.504	0.101
TB01	Integration of intelligence transportation system	7	0.05	1.05	0.480	0.096

Table 8. The resulting weight of the cost factors.

Codes	Cost Factors	Rank	s_j	k_j	q_j	w_j
C05	Savings on vehicle operation cost	1		1	1	0.203
C01	Increase revenue generation	2	0.1	1.1	0.909	0.184
SC03	Reduce accident and Health cost	3	0.2	1.2	0.757	0.153
EC02	Environmental mitigation cost	4	0.05	1.05	0.722	0.146
C06	Reduce user cost	5	0.25	1.25	0.577	0.117
SC01	Marginal cost of public fund	6	0.15	1.15	0.502	0.102
EC04	Noise/air pollution cost	7	0.08	1.08	0.465	0.094

3.4. Establishing Criteria Quality Weight per Criteria by Expert Interview

The criteria quality weight (CQW) was determined by a consulting expert. The criteria were assessed to determine if they were subjective or objective by the experts. The findings indicate that all the benefit and cost criteria were objective and measurable.

The type of criterion was determined in terms of its positivity or negativity by the experts. The benefit criteria were found to be positive on the following: increased productivity, improve economic activities, reduction in accidents, integration of intelligent transportation system, and reduced carbon emission. On the other hand, increased investment, tourism and trade, less noise, and smoke were found to be negative, as shown in Table 9. The measurement factors for each objective criterion was found by the experts, which are: speed; number of lanes, smart signage, barriers, and lighting; electronic signs; sensors and smart technologies; regular maintenance; green zones; smooth road surfaces; and noise recording technique of the sound pressure level (SPL) for the positive and negative benefit criteria, respectively.

Table 9. Benefit factor measurement.

Benefit Factors		Measurement	Type of Measurement	Optimum (Max/Min)	Unit	Standard
Economic	Increased productivity	Average speed	Positive	110 Km/h	km/h	Experts
	Improved economic activities	Number of road lanes	Positive	4 lanes	Number of lanes	Experts
	Increased investment, tourism, and trade	Smooth road surfaces	Negative	IRI < 2.0 m/km	m/km	IRI *
Social	Reduce accident	Smart signage, barriers, and lighting	Positive	5 smart signs per kilometer	Signs and lightings/km	Experts
Technical	Integration of intelligence transportation system	Electronic signs, sensors, and smart technologies	Positive	10–20 signs and sensors/km 70% of roads should be equipped with smart technologies	Signs and sensors/km Percentage of road coverage with smart technologies	Experts
Environmental	Reduced carbon emission	Smooth road surfaces	Negative	IRI < 2.0 m/km	m/km	IRI *
		Regular maintenance	Positive	4 times per year	Times/years	Experts
		Numbers of lanes	Positive	4 lanes	Number of lanes	Experts
		Green zones	Positive	50% highways	% coverage of road length	Experts
	Less noise and smoke	Noise recording technique Sound pressure level—SPL	Negative	SPL: 50–60 dB	dB (decibels)	EPA **

* International Roughness Index (IRI). ** European Union and United States Environmental Protection Agency (EPA) guidelines.

For the cost criterion, the positive criteria are savings on vehicle operation costs, increased revenue generation, reduced accident and health costs, marginal cost of public funds, environmental mitigation cost, and noise/air pollution cost. The negative criteria are reduced accidents and health costs. The measurement for the cost criterion includes average speed; number of road lanes, smart signage, barriers, and lighting; total budget allocation; installed noise barriers; and smooth road surfaces, as shown in Table 10. The optimal value for each measurement was found with its units used by reviewing the standards manual and interviewing the expert. The CQW can be easily determined for any case study depending on the measurement for each criterion. Then, the normalized CQW ($\overline{CQW}$) can be computed using Equations (5) and (6) for positive and negative measurements, respectively.

$$\overline{CQW} = \frac{CQW}{Max} \tag{5}$$

$$\overline{CQW} = \frac{Min}{CQW} \tag{6}$$

Table 10. Cost factor measurement.

Cost Factors		Measurement	Type of Measurement	The Optimum	Unit	Standard
Economic	Savings on vehicle operation cost	Average speed	Positive	110 Km/h	km/h	Experts
	Increase revenue generation	Number of road lanes	Positive	4 lanes	Number of lanes	Experts
	Reduce user cost	Smooth road surfaces	Negative	IRI < 2.0 m/km	m/km	IRI *
Social	Reduce accident and health costs	Smart signage, barriers, and lighting	Positive	5 smart signs per kilometer	Signs and lighting/km	Experts
	Marginal cost of public fund	Total budget allocation	Positive	20–50 billion SAR per year, based on historical data and global benchmarks	SAR	based on historical data and global benchmarks
Environmental	Environmental mitigation cost, noise/air pollution cost	Increase green zones	Positive	50% highways	% coverage of road length	Experts
		Install noise barriers		Aim for 60–80%	% coverage of road length	Experts

* International Roughness Index (IRI).

3.5. Validatation of the Developed Framework by Applying It and ANP to the Three Case Studies

The main objective of this section is to validate the developed framework with three (3) selected roads to serve as a case study using the Analytic Network Process (ANP). The roads selected include Khurais Road (Road 1), Al Urubah Road (Road 2), and King Fahd Road (Road 3), which are essential routes in Riyadh, with Khurais Road acting as a crucial east–west corridor for freight and logistics, Al Urubah Road supporting urban transport in an east–west direction, and King Fahd Road running north–south through the city's central business districts, serving as the primary artery for business and financial services, and facilitating the movement of goods and people across the city. The selection was based on their relevance to this study's focus on economic, social, environmental, and technological factors. These roads play crucial roles in the city's transportation network, directly influencing regional economic activities, public safety, environmental impacts, and technological advancements. The roads were chosen for their measurable and descriptive factors, such as traffic volume, the potential for economic growth, the need for safety features like smart signage and barriers, and the opportunities for environmental mitigation through smoother road surfaces and green zones. Additionally, these roads provide opportunities for integrating intelligent transportation systems (ITSs), making them ideal for evaluating the impact of technological improvements. By analyzing these three roads in Riyadh, this study ensures a comprehensive assessment of infrastructure interventions across all key factors. Figures 2–4 display the respective road networks on maps, providing a clear visual representation of the locations and routes under analysis.

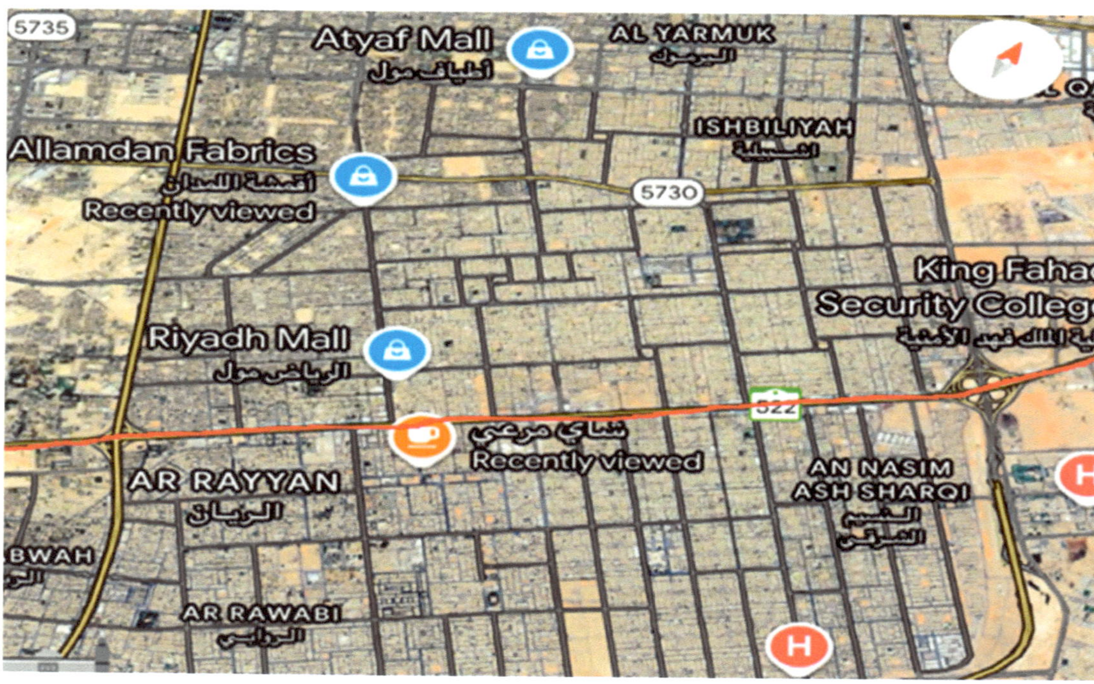

Figure 2. Khurais Road (Road 1).

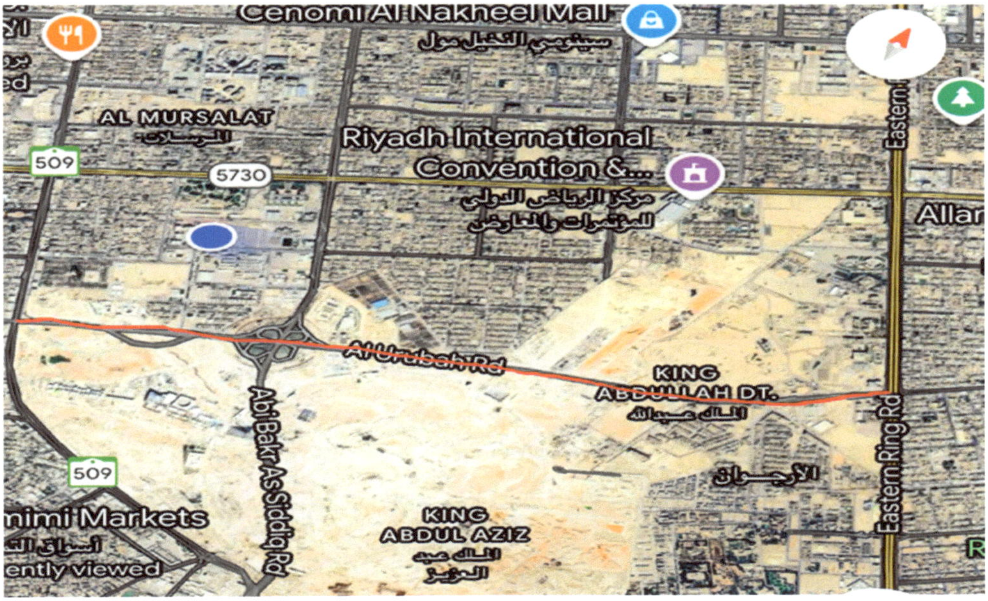

Figure 3. Al Urubah Road (Road 2).

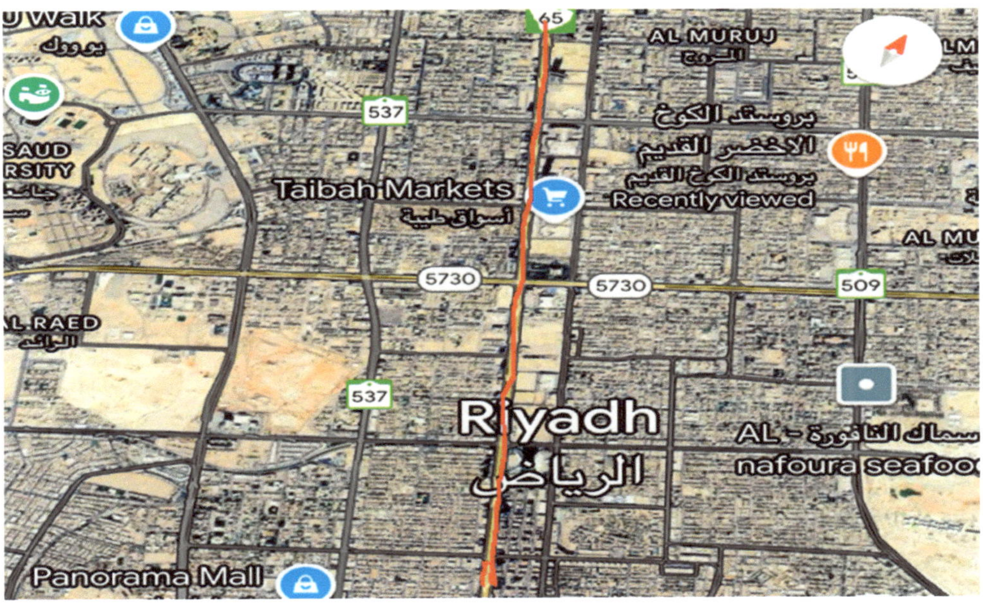

Figure 4. King Fahd Road (Road 3).

The results presented in Tables 11 and 12 were determined through a case study based on measurements for each criterion, which illustrates the initial assessment of the criteria quality weight for both benefits and costs.

Table 11. Matrix criteria quality weight of benefits CQW_B.

Name of Road	ECB02	SB02	EVB01	ECB01	EVB03	ECB03	TB04
Road 1	110	3	2	3	75	2.5	6
Road 2	90	6	5	3	65	2	10
Road 3	100	4	3	3	75	2.5	8

Table 12. Matrix criteria quality weight of cost CQW_C.

Name of Road	C05	C01	SC03	EC02	C06	SC01	EC04
Road 1	110	3	3	5%	2.5	20	10%
Road 2	90	3	6	5%	2	20	30%
Road 3	100	3	4	20%	2.5	20	10%

The results of the normalized matrix for the benefits are cost criterion presented in Tables 13 and 14, respectively, by using Equations (5) and (6) for positive and negative measurements, respectively.

Table 13. Matrix normalized criteria quality weight of benefits $\overline{CQW_B}$.

Name of Road	ECB02	SB02	EVB01	ECB01	EVB03	ECB03	TB04
Road 1	0.354	0.230	0.200	0.333	0.348	0.357	0.250
Road 2	0.290	0.461	0.500	0.333	0.302	0.285	0.416
Road 3	0.354	0.307	0.300	0.333	0.348	0.357	0.333

Table 14. Matrix normalized criteria quality weight of cost $\overline{CQW_C}$.

Name of Road	C05	C01	SC03	EC02	C06	SC01	EC04
Road 1	0.354	0.333	0.230	0.166	0.357	0.333	0.200
Road 2	0.290	0.333	0.461	0.166	0.285	0.333	0.600
Road 3	0.354	0.333	0.307	0.666	0.357	0.333	0.200

The quality of benefits Q_B and quality of cost Q_C can be determined using Equation (7) and Equation (8), respectively. The Q_B and Q_C of the three roads are shown in Table 15.

$$Q_B = \overline{CQW_B} \times w_B \quad (7)$$

$$Q_C = \overline{CQW_C} \times w_C \quad (8)$$

Table 15. Normalized benefit QBc and cost Qc.

Name of Road	QBc	Qc
Road 1	0.293	0.287
Road 2	0.374	0.338
Road 3	0.331	0.372

The quality criteria weights for both the benefit and cost factors for the three selected roads were evaluated and presented in the previous matrices. For the benefits factors, Khurais Road (Road 1), Al Urubah Road (Road 2), and King Fahd Road (Road 3) have normalized criteria weights of 0.293, 0.374, and 0.331, respectively. For the cost factors, the normalized quality criteria weight for the road (R1), (R2), and (R3) are 0.287, 0.338, and 0.372, respectively. The results (Table 16) were then subsequently utilized to compute the value engineering index as shown.

$$VE = \frac{Q_B}{Q_C} \quad (9)$$

Table 16. VE values for the roads.

Name of Road	VE
Road 1	0.084
Road 2	0.126
Road 3	0.125

Value engineering (VE) as obtained for the three roads is mainly concerned with the values of use, cost, and consideration. Value engineering aims to maximize function while decreasing cost and the ultimate formula for value is frequently described as function divided by cost. The VE results obtained indicate that Khurais Road (Road 1) having a VE value of 0.084 is the best alternative. This is then followed by King Fahd Road (Road 3) with a value of 0.125 and Al Urubah Road (Road 2) with a VE value of 0.126.

4. Results and ANP Case Study

To ensure the robustness of the decision process and validate the findings, this study used the ANP technique in the three case studies as it provides a different perspective for weighing the different criteria. While ANP has its limitations, its strength in analyzing complex, interconnected systems made it a suitable tool for validating the developed framework finding. The ANP model was developed as shown in Figures 5 and 6 to efficiently evaluate and select the best alternative from three (3) roads based on seven (7) different criteria as indicated in the cluster.

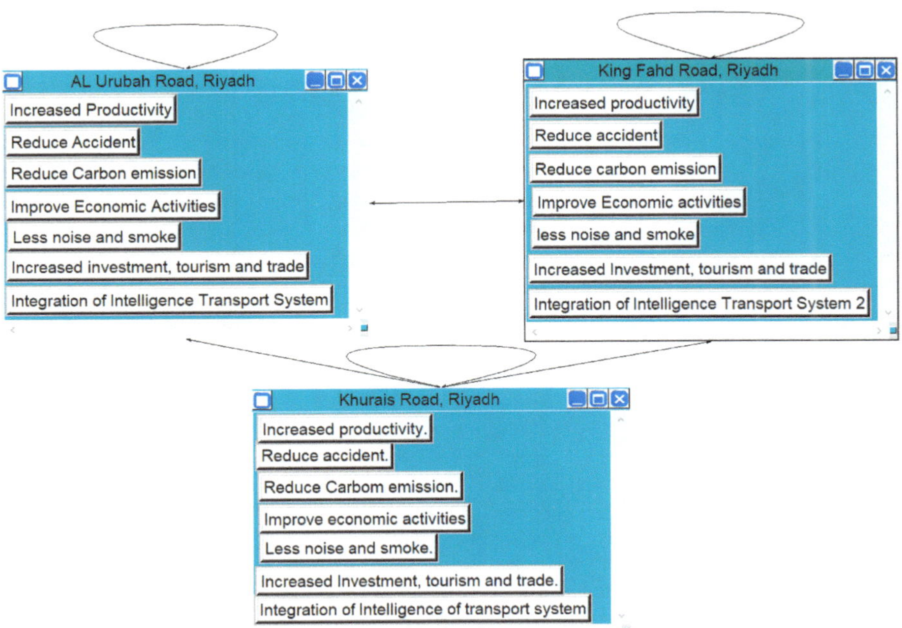

Figure 5. Road benefit factor ANP model.

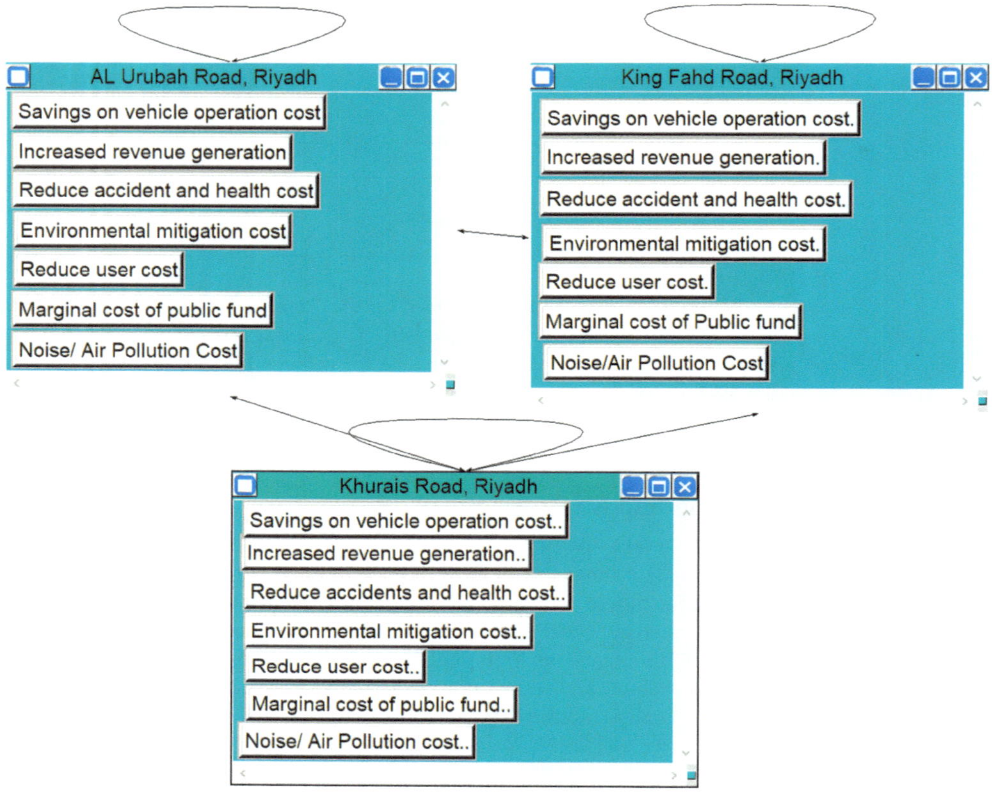

Figure 6. Road cost factor ANP model.

The road benefits cluster matrix, as shown in Table 17, indicates some similarities in the roads with a uniform value; however, each road has a unique value of its own, with Khurais Road (Road 1) having the highest cluster matrix value of 0.715, which was then subsequently followed by King Fahd Road (Road 3) and lastly Al Urubah Road (Road 2), with cluster values of 0.218 and 0.0668, respectively.

Table 17. Road benefit factor cluster matrix.

Cluster Node Labels	Al Urubah Road (Road 2)	Khurais Road (Road 1)	King Fahd Road (Road 3)
Al Urubah Road (Road 2)	0.067	0.333	0.333
King Fahd Road (Road 3)	0.218	0.333	0.333
Khurais Road (Road 1)	0.715	0.333	0.333

The priority matrix for benefit factors, as shown in Table 18, indicates that Khurais Road (Road 1), Riyadh, with a normalized value of 0.288, is ranked first in the expert opinion survey conducted. This is then followed by King Fahd Road (Road 3) with a value of 0.204, and Al Urubah Road (Road 2) ranked third based on the expert survey conducted considering the road benefit factors. The finding conforms with SWARA results.

The road cost cluster matrix, as shown in Table 19, indicates uniformity from the results of the matrix obtained. Al Urubah Road (Road 2), Riyadh, has a value of 0.637 all throughout, with King Fahd Road (Road 3) having a value of 0.258, while Khurais Road (Road 1) has a value of 0.105.

Table 18. Priority matrix for benefit factors.

Road Names	Normalized	Ideal	Rank
Khurais Road (Road 1)	0.288	0.099	1
King Fahd Road (Road 3)	0.204	0.071	2
Al Urubah Road (Road 2)	0.182	0.277	3

Table 19. Road cost factor cluster matrix.

Cluster Node Labels	Al Urubah Road (Road 2)	Khurais Road (Road 1)	King Fahd Road (Road 3)
Al Urubah Road (Road 2)	0.637	0.636	0.636
King Fahd Road (Road 3)	0.258	0.258	0.258
Khurais Road (Road 1)	0.105	0.105	0.105

The priority matrix for cost factors, as shown in Table 20, indicates that Khurais Road (Road 1), Riyadh, is ranked first with a normalized value of 0.230. King Fahd Road (Road 3) ranked second, and Al Urubah Road (Road 2) ranked third, with values of 0.221 and 0.05475, respectively. The ANP case study findings conform with the SWARA technique results.

Table 20. Priority matrix for cost factors.

Road Names	Normalized	Ideal	Rank
Khurais Road, (Road 1)	0.230	0.045	1
King Fahd Road, (Road 3)	0.221	0.043	2
Al Urubah Road, (Road 2)	0.055	0.011	3

Figure 7 shows the results of the VE, normalized, and cluster matrix for Khurais, Al Urubah, and King Fahad Roads. The results obtained from the value engineering (VE) analysis and the priority matrix for factors provide complementary insights into the evaluation of the three roads. According to the VE analysis, Khurais Road (Road 1) has the highest VE value of 0.084, indicating it as the most efficient alternative in terms of maximizing function while minimizing cost. This is followed by King Fahd Road (Road 3) with a VE value of 0.125 and Al Urubah Road (Road 2) with a VE value of 0.126. In contrast, the priority matrix ranks the roads, with Khurais Road (Road 1) again taking the top spot (ranked first) with a normalized value of 0.230. King Fahd Road (Road 3) is ranked second with a value of 0.221, while Al Urubah Road (Road 2) falls last with a much lower value of 0.055. Both analyses align in placing Khurais Road (Road 1) as the most favorable choice, with King Fahd Road (Road 3) coming second and Al Urubah Road (Road 2) ranking last. The findings obtained indicate that some roads function better than others, and this can be attributed to factors like better building materials, cutting-edge engineering technology, proactive maintenance, low accident rate, and economic contribution. Roads with advanced technologies, strategic prioritization, and adequate funding perform exceptionally well and last a long time. The findings from both the value engineering and the ANP-based priority matrix are consistent with each other, further reinforcing the superior performance of Khurais Road in terms of both cost-effectiveness and overall evaluation.

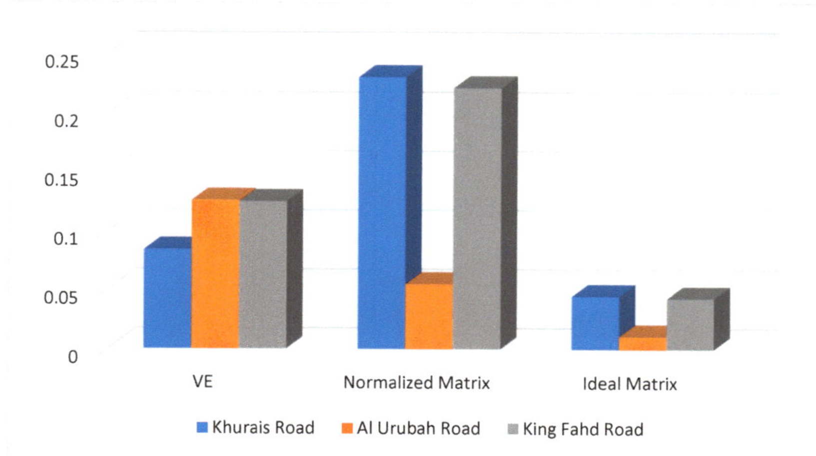

Figure 7. SWARA results for VE, normalized, and ideal matrix.

5. Discussion

From the stepwise weight assessment ratio analysis (SWARA) results obtained, as shown in Tables 4 and 6, the significant benefits factors for efficient road selection in Saudi Arabia's construction industry are ranked with a decreasing relative weight value. The SWARA results obtained indicate the most significant benefits factors towards effective road selection in Saudi Arabia's construction industry. The economic benefits include increased productivity, improved economic activities, and increased investment, tourism, and trade. The environmental benefits considered by the expert include reduced carbon emissions and less noise and smoke, also identified as the most significant factors by the expert. This is in line with the findings of Grace III [62].

The social benefits include reducing accidents, the most significant as indicated by the experts surveyed. This social benefit is regarded to aid in efficient road selection in Saudi Arabia's construction industry. This conforms with the studies conducted by Hoffmann et al. [77]. For the technical benefits, the integration of intelligence transportation is the most significant factor in effective road selection.

Economic cost, savings on vehicle operation cost, increased revenue generation, and reduced user cost are the most significant cost factors towards effective road selection in Saudi Arabia's construction industry. Environmental mitigation cost and noise/air pollution cost are the most significant environmental cost factors considered by the expert towards effective road selection.

One of the main theoretical contributions of this study is the identification of specific environmental benefits considered critical by the experts. These include reduced carbon emissions and less noise and smoke, which reflect the growing global focus on sustainability. As urbanization and traffic congestion increase, addressing environmental concerns like pollution and carbon emissions has become central to infrastructure planning, not only in Saudi Arabia but also worldwide. This study's emphasis on these environmental aspects contributes to the existing body of knowledge, further emphasizing the need for incorporating sustainability metrics into infrastructure decision-making processes [49].

On the social front, reducing accidents emerged as the most significant benefit in road selection, reinforcing the findings of several previous studies [50]. Road safety is a fundamental concern in many developing countries, including Saudi Arabia, where traffic-related fatalities are a major public health issue. This study contributes to the theoretical understanding of how safety considerations are integrated into infrastructure planning, aligning with global studies that highlight road safety as a top priority for transportation projects. The focus on social benefits in road selection demonstrates the increasing recognition that infrastructure projects should enhance public well-being beyond just economic gains.

From a technical standpoint, the integration of intelligent transportation systems was identified as a key benefit factor, highlighting a shift toward modern, technology-driven solutions in road construction and management. The adoption of intelligent transportation systems can lead to smarter, more efficient traffic management, and improved safety, supporting the findings of past studies [26,70] that explored the role of intelligent transportation systems in enhancing road safety and traffic flow in major cities worldwide. This study contributes to the theoretical understanding of how advanced technologies are becoming essential in the construction and management of roads, particularly in rapidly developing countries like Saudi Arabia.

The ANP model and findings from the three (3) roads evaluated indicate that Khurais Road, Riyadh, is the best alternative considering its benefits to road users compared to the two remaining roads. In comparison with other similar research, this study's focus on the unique context of Saudi Arabia's rapidly developing transportation infrastructure offers a novel perspective. While much of the existing literature on cost–benefit analysis in road selection is based on case studies from developed countries, this research emphasizes the challenges and opportunities faced by developing nations, where rapid growth, economic diversification, and sustainability concerns are key drivers for infrastructure development.

6. Conclusions

This paper demonstrated new efforts concerning the assessment of cost–benefit factors towards efficient road selection using the SWARA method. The main contribution of this research paper is that it assessed the cost–benefit factors that were derived from the literature. Hence, the validation of the factors was carried out through semi-structured interviews. An expert opinion survey was subsequently conducted, and responses received were further analyzed using the MCDM SWARA method and ANP technique. This study discovered that the beneficial factors for efficient road selection in Saudi Arabia's construction industry are increased productivity, promoted industries and commerce, and improved economic activities, and for social factors, less noise and smoke, fuel saving, and erosion and sediment control. Reducing accidents, providing access to basic health care facilities, and improving quality of life are regarded as the most significant social benefit factors.

The careful consideration of these factors during the road project's initial planning phase would enhance the road's benefit and performance to the end user. The findings will, in practice, improve the productivity of road construction. This study offered insights into current theories, including issues around cost–benefit factors related to road projects in academics and management.

Author Contributions: Conceptualization, O.S.A. and K.S.A.-G.; methodology, K.S.A.-G.; software, O.S.A.; validation, O.S.A., K.S.A.-G. and A.A.; formal analysis, O.S.A. and K.S.A.-G.; investigation, O.S.A., K.S.A.-G. and A.A.; resources, O.S.A. and K.S.A.-G.; data curation, K.S.A.-G.; writing original draft preparation, O.S.A.; writing review and editing, K.S.A.-G. and A.A.; visualization, K.S.A.-G. and A.A.; supervision, K.S.A.-G. and A.A.; project administration, K.S.A.-G. and A.A.; funding acquisition, O.S.A., K.S.A.-G. and A.A.; All authors have read and agreed to the published version of the manuscript.

Funding: Researchers Supporting Project number (RSP2025R280), King Saud University, Riyadh, Saudi Arabia.

Data Availability Statement: Dataset available on request from the authors.

Acknowledgments: The authors would like to thank King Saud University for their support in conducting this research for the master's graduation program.

Conflicts of Interest: The authors declare no conflicts of interest.

References

1. Abdel-Rahman, A. *Economic Diversification in the Kingdom of Saudi Arabia*; United Nations: New York City, NY, USA, 2002.
2. Almheiri, A.K. Is the United Arab Emirates Capable to Create a Green and Sustainable Future? Master's Thesis, The British University in Dubai, Dubai, United Arab Emirates, 2015.
3. Ramady, M.A. *The Saudi Arabian Economy: Policies, Achievements, and Challenges*; Springer: Berlin/Heidelberg, Germany, 2010.
4. Aldagheiri, M. The role of the transport road network in the economic development of Saudi Arabia. *WIT Trans. Built Environ.* **2009**, *107*, 275–286.
5. Rahman, S.M.; Khondaker, A.; Hasan, M.A.; Reza, I. Greenhouse gas emissions from road transportation in Saudi Arabia-a challenging frontier. *Renew. Sustain. Energy Rev.* **2017**, *69*, 812–821. [CrossRef]
6. Mubaraki, M. Predicting Deterioration for the Saudi Arabia Urban Road Network. Ph.D. Thesis, University of Nottingham, Nottingham, UK, 2010.
7. Al Turki, Y.A. How can Saudi Arabia use the Decade of Action for Road Safety to catalyse road traffic injury prevention policy and interventions? *Int. J. Inj. Control. Saf. Promot.* **2014**, *21*, 397–402. [CrossRef] [PubMed]
8. Laurance, W.F.; Balmford, A. A global map for road building. *Nature* **2013**, *495*, 308–309. [CrossRef] [PubMed]
9. Laurance, W.F.; Clements, G.R.; Sloan, S.; O'connell, C.S.; Mueller, N.D.; Goosem, M.; Venter, O.; Edwards, D.P.; Phalan, B.; Balmford, A. A global strategy for road building. *Nature* **2014**, *513*, 229–232. [CrossRef]
10. Chin, G.T.; Jiejin, Z. Infrastructure at the G20. In *The G20, Development and the UN Agenda 2030*; Routledge: London, UK, 2022; pp. 159–178.
11. Alexander, N. The Emerging Multi-Polar World Order: Its Unprecedented Consensus on a New Model for Financing Infrastructure Investment and Development. Heinrich Böll Foundation North America. 2014. Available online: https://us.boell.org/sites/default/files/alexander_multi-polar_world_order_1.pdf (accessed on 10 June 2024).
12. Tsamboulas, D.A. A tool for prioritizing multinational transport infrastructure investments. *Transp. Policy* **2007**, *14*, 11–26. [CrossRef]
13. Macura, D.; Nuhodžić, R.; Bojović, N.; Knežević, N. One model for rail infrastructure projects selection. In Proceedings of the Road and Rail Infrastructure II, Dubrovnik, Croatia, 7–9 May 2012.
14. Hinloopen, E.; Nijkamp, P.; Rietveld, P. Integration of ordinal and cardinal information in multi-criteria ranking with imperfect compensation. *Eur. J. Oper. Res.* **2004**, *158*, 317–338. [CrossRef]
15. Odeck, J. Identifying traffic safety best practice: An application of DEA and Malmquist indices. *Omega* **2006**, *34*, 28–40. [CrossRef]
16. Yedla, S.; Shrestha, R.M. Multi-criteria approach for the selection of alternative options for environmentally sustainable transport system in Delhi. *Transp. Res. Part A Policy Pract.* **2003**, *37*, 717–729. [CrossRef]
17. Norese, M.F.; Viale, S. A multi-profile sorting procedure in the public administration. *Eur. J. Oper. Res.* **2002**, *138*, 365–379. [CrossRef]
18. Tudela, A.; Akiki, N.; Cisternas, R. Comparing the output of cost benefit and multi-criteria analysis: An application to urban transport investments. *Transp. Res. Part A Policy Pract.* **2006**, *40*, 414–423. [CrossRef]
19. Wey, W.-M.; Wu, K.-Y. Interdependent urban renewal project selection under the consideration of resource constraints. *Environ. Plan. B Plan. Des.* **2008**, *35*, 122–147. [CrossRef]
20. Chang, Y.-H.; Wey, W.-M.; Tseng, H.-Y. Using ANP priorities with goal programming for revitalization strategies in historic transport: A case study of the Alishan Forest Railway. *Expert Syst. Appl.* **2009**, *36*, 8682–8690. [CrossRef]
21. Piantanakulchai, M. Analytic network process model for highway corridor planning. In Proceedings of the ISAHP 2005, Honolulu, HI, USA, 8–10 July 2005.
22. Riegg Cellini, S.; Edwin Kee, J. Cost-effectiveness and cost-benefit analysis. In *Handbook of Practical Program Evaluation*; Jossey-Bass & Pfeiffer Imprints, Wiley: San Francisco, CA, USA, 2015; pp. 636–672.
23. Bronsteen, J.; Buccafusco, C.; Masur, J.S. Well-being analysis vs. cost-benefit analysis. *DukE IJ* **2012**, *62*, 1603. [CrossRef]
24. Annes, B.A.; Carpenter, N.A.; Hashmani, A.; McGinnis, B.J.; Parrish, M.A.; Joshi, N.N.; Lambert, J.H. Prioritizing Highway Construction with a Benefits Analysis. In Proceedings of the 2006 IEEE Systems and Information Engineering Design Symposium, Charlottesville, VA, USA, 28 April 2006; pp. 118–123.
25. Coley, N. Spotlight on benefit-cost analysis. *Public Roads* **2012**, *75*, 24–29.
26. Henke, I.; Carteni, A.; Di Francesco, L. A sustainable evaluation processes for investments in the transport sector: A combined multi-criteria and cost–benefit analysis for a new highway in Italy. *Sustainability* **2020**, *12*, 9854. [CrossRef]

27. Harford, J.D. Congestion, pollution, and benefit-to-cost ratios of US public transit systems. *Transp. Res. Part D Transp. Environ.* **2006**, *11*, 45–58. [CrossRef]
28. Matanhire, B.B. Cost Benefit Analysis of a Regional Arterial PPP Toll Road Project A Case Study of the Proposed R-1 Toll Road. Master's Thesis, Eastern Mediterranean University (EMU)-Doğu Akdeniz Üniversitesi (DAÜ), Famagusta, Cyprus, 2019.
29. Shang, J.S.; Tjader, Y.; Ding, Y. A unified framework for multicriteria evaluation of transportation projects. *IEEE Trans. Eng. Manag.* **2004**, *51*, 300–313. [CrossRef]
30. Policy, U. *Guide to Cost-Benefit Analysis of Investment Projects*; The EU: Maastricht, The Netherlands, 2008.
31. Rattanakunuprakarn, S.; Jin, M.; Sussman, M.; Felix, P. Comparative evaluation of highways and railroads using life-cycle benefit-cost analysis. *Int. J. Sustain. Transp.* **2024**, *18*, 803–826. [CrossRef]
32. Martínez-Galán, E.; Leandro, F.J.B. A Qualitative Cost-Benefit Analysis of the Maritime Silk Road in Europe: Who Benefits from the Initiative and Who Does Not. *Asian Perspect.* **2024**, *48*, 13–39. [CrossRef]
33. Vagdatli, T.; Petroutsatou, K. Modelling approaches of life cycle cost–benefit analysis of road infrastructure: A critical review and future directions. *Buildings* **2023**, *13*, 94. [CrossRef]
34. Biancardo, S.A.; Gesualdi, M.; Savastano, D.; Intignano, M.; Henke, I.; Pagliara, F. An innovative framework for integrating cost-benefit analysis (cba) within building information modeling (bim). *Socio-Econ. Plan. Sci.* **2023**, *85*, 101495. [CrossRef]
35. Hekrle, M.; Liberalesso, T.; Macháč, J.; Silva, C.M. The economic value of green roofs: A case study using different cost–benefit analysis approaches. *J. Clean. Prod.* **2023**, *413*, 137531. [CrossRef]
36. García-Herrero, L.; Lavrnić, S.; Guerrieri, V.; Toscano, A.; Milani, M.; Cirelli, G.L.; Vittuari, M. Cost-benefit of green infrastructures for water management: A sustainability assessment of full-scale constructed wetlands in Northern and Southern Italy. *Ecol. Eng.* **2022**, *185*, 106797. [CrossRef]
37. Wu, Z.; Ma, G. Incremental cost-benefit quantitative assessment of green building: A case study in China. *Energy Build.* **2022**, *269*, 112251. [CrossRef]
38. Akinyi, D.P.; Karanja Ng'ang'a, S.; Ngigi, M.; Mathenge, M.; Girvetz, E. Cost-benefit analysis of prioritized climate-smart agricultural practices among smallholder farmers: Evidence from selected value chains across sub-Saharan Africa. *Heliyon* **2022**, *8*, e09228. [CrossRef]
39. Raihan, A.; Said, M.N.M. Cost–benefit analysis of climate change mitigation measures in the forestry sector of Peninsular Malaysia. *Earth Syst. Environ.* **2022**, *6*, 405–419. [CrossRef]
40. Edwards, R.T.; Lawrence, C.L. 'What You See is All There is': The importance of heuristics in cost-benefit analysis (CBA) and social return on investment (SROI) in the evaluation of public health interventions. *Appl. Health Econ. Health Policy* **2021**, *19*, 653–664. [CrossRef]
41. Mishan, E.J.; Quah, E. *Cost-Benefit Analysis*; Routledge: London, UK, 2020.
42. Henke, I.; Cartenì, A.; Molitierno, C.; Errico, A. Decision-making in the transport sector: A sustainable evaluation method for road infrastructure. *Sustainability* **2020**, *12*, 764. [CrossRef]
43. Brent, R.J. *Applied Cost-Benefit Analysis*; Edward Elgar Publishing: Cheltenham, UK, 2006.
44. Sarı, F.; Şen, M. Highway route planning via least cost path algorithm and multi criteria decision analysis integration, a comparison of AHP, TOPSIS and VIKOR. *Int. J. Environ. Geoinform.* **2022**, *9*, 27–38. [CrossRef]
45. Pan, X.; Xu, M.; Jia, J.; Xia, Y.; Chen, Q. Evaluation model of highway reconstruction and expansion scheme based on multiple attribute decision making. *Adv. Civ. Eng.* **2022**, *2022*, 3764557. [CrossRef]
46. Vlahogianni, E.I.; Karlaftis, M.G.; Golias, J.C. Short-term traffic forecasting: Where we are and where we're going. *Transp. Res. Part C Emerg. Technol.* **2014**, *43*, 3–19. [CrossRef]
47. Moher, D.; Liberati, A.; Tetzlaff, J.; Altman, D.G.; PRISMA Group*, t. Preferred reporting items for systematic reviews and meta-analyses: The PRISMA statement. *Ann. Intern. Med.* **2009**, *151*, 264–269. [CrossRef] [PubMed]
48. Forkenbrock, D.J.; Foster, N.S. Economic benefits of a corridor highway investment. *Transp. Res. Part A: Gen.* **1990**, *24*, 303–312. [CrossRef]
49. Luo, W.; Sandanayake, M.; Zhang, G.; Tan, Y. Construction cost and carbon emission assessment of a highway construction—A case towards sustainable transportation. *Sustainability* **2021**, *13*, 7854. [CrossRef]
50. Rahman, M.M.; Islam, M.K.; Al-Shayeb, A.; Arifuzzaman, M. Towards sustainable road safety in Saudi Arabia: Exploring traffic accident causes associated with driving behavior using a Bayesian belief network. *Sustainability* **2022**, *14*, 6315. [CrossRef]
51. Bridgelall, R. Inertial sensor sample rate selection for ride quality measures. *J. Infrastruct. Syst.* **2015**, *21*, 04014039. [CrossRef]
52. Morrison-Saunders, A.; Therivel, R. Sustainability integration and assessment. *J. Environ. Assess. Policy Manag.* **2006**, *8*, 281–298. [CrossRef]
53. Suksuwan, N.; Trangkanont, S. The conceptual framework of the government-sponsored rural road improvement project evaluation and selection. *Eng. J.* **2018**, *22*, 109–129. [CrossRef]
54. Acheampong, E.O.; Sayer, J.; Macgregor, C.J. Road improvement enhances smallholder productivity and reduces forest encroachment in Ghana. *Environ. Sci. Policy* **2018**, *85*, 64–71. [CrossRef]

55. Walls, J. *Life-Cycle Cost Analysis in Pavement Design: In Search of Better Investment Decisions*; US Department of Transportation, Federal Highway Administration: Washington, DC, USA, 1998.
56. Garber, N.J.; Hoel, L.A.; Sarkar, R. *Traffic and Highway Engineering*; Cengage Learning: Singapore, 2002.
57. Lingaitienė, O.; Merkevičius, J.; Davidavičienė, V. The model of vehicle and route selection for energy saving. *Sustainability* **2021**, *13*, 4528. [CrossRef]
58. Vadakpat, G.; Stoffels, S.; Dixon, K. Road user cost models for network-level pavement management. *Transp. Res. Rec.* **2000**, *1699*, 49–57. [CrossRef]
59. Bein, P. Cost Benefit Basis of Highway Investment Appraisal and Road Pricing. In Proceedings of the 29th Annual Canadian Transportation Research Forum, Vancouver, BC, Canada, 15–18 May 1994. 2020.
60. Daniel, J.I.; Bekka, K. The environmental impact of highway congestion pricing. *J. Urban Econ.* **2000**, *47*, 180–215. [CrossRef]
61. Perez, M.; Zech, W.; Donald, W.; Fang, X. Design methodology for the selection of temporary erosion and sediment control practices based on regional hydrological conditions. *J. Hydrol. Eng.* **2016**, *21*, 05016001. [CrossRef]
62. Grace III, J. Effectiveness of vegetation in erosion control from forest road sideslopes. *Trans. ASAE* **2002**, *45*, 681. [CrossRef]
63. Bener, A.; Abu-Zidan, F.M.; Bensiali, A.K.; Al-Mulla, A.A.; Jadaan, K.S. Strategy to improve road safety in developing countries. *Saudi Med. J.* **2003**, *24*, 603–608. [PubMed]
64. Khizindar, T.M Effects of tourism on residents' quality of life in Saudi Arabia: An empirical study. *J. Hosp. Mark. Manag.* **2012**, *21*, 617–637. [CrossRef]
65. Parumog, M.G ; Mizokami, S.; Cal, P. Using travel cost and contingent valuation methodologies in valuing externalities of urban road development: An application in valuing damages to cultural heritage. *J. East. Asia Soc. Transp. Stud.* **2003**, *5*.
66. Hall, S. *The Hard Road to Renewal: Thatcherism and the Crisis of the Left*; Verso Books: Brooklyn, NY, USA, 2021.
67. Weisswange, T.H.; Rebhan, S.; Bolder, B.; Steinhardt, N.A.; Joublin, F.; Schmuedderich, J.; Goerick, C. Intelligent traffic flow assist: Optimized highway driving using conditional behavior prediction. *IEEE Intell. Transp. Syst. Mag.* **2019**, *13*, 20–38. [CrossRef]
68. Kim, C.; Lee, E.-B.; Harvey, J.T.; Fong, A.; Lott, R. Automated sequence selection and cost calculation for maintenance and rehabilitation in highway Life-Cycle Cost Analysis (LCCA). *Int. J. Transp. Sci. Technol.* **2015**, *4*, 61–75. [CrossRef]
69. Kelle, P.; Schneider, H.; Raschke, C.; Shirazi, H. Highway improvement project selection by the joint consideration of cost-benefit and risk criteria. *J. Oper. Res. Soc.* **2013**, *64*, 313–325. [CrossRef]
70. Gordon, R. *Intelligent Transportation Systems*; Springer: Cham, Switzerland, 2016.
71. Lisinge, R.T. The Belt and Road Initiative and Africa's regional infrastructure development: Implications and lessons. *Transnatl. Corp. Rev.* **2020**, *12*, 425–438. [CrossRef]
72. Kineber, A.F.; Uddin, M.S.; Momena, A.F. Exploring the critical success factors of value management implementation for sustainable residential building project: A stationary analysis approach. *Sustainability* **2022**, *14*, 16215. [CrossRef]
73. Gündüz, M.; Nielsen, Y.; Özdemir, M. Quantification of delay factors using the relative importance index method for construction projects in Turkey. *J. Manag. Eng.* **2013**, *29*, 133–139. [CrossRef]
74. Keršuliene, V.; Zavadskas, E.K.; Turskis, Z. Selection of rational dispute resolution method by applying new step-wise weight assessment ratio analysis (SWARA). *J. Bus. Econ. Manag.* **2010**, *11*, 243–258. [CrossRef]
75. Zolfani, S.H.; Saparauskas, J. New application of SWARA method in prioritizing sustainability assessment indicators of energy system. *Eng. Econ.* **2013**, *24*, 408–414. [CrossRef]
76. Keršulienė, V.; Turskis, Z. An integrated multi-criteria group decision making process: Selection of the chief accountant. *Procedia-Soc. Behav. Sci.* **2014**, *110*, 897–904. [CrossRef]
77. Hoffmann, R.L.; Rohrer, W.M.; South-Paul, J.E.; Burdett, R.; Watzlaf, V.J. The effects of barriers on Health Related Quality of Life (HRQL) and compliance in adult asthmatics who are followed in an urban community health care facility. *J. Community Health* **2008**, *33*, 374–383. [CrossRef]

MDPI AG
Grosspeteranlage 5
4052 Basel
Switzerland
Tel.: +41 61 683 77 34

Buildings Editorial Office
E-mail: buildings@mdpi.com
www.mdpi.com/journal/buildings

www.ingramcontent.com/pod-product-compliance
Lightning Source LLC
Chambersburg PA
CBHW041421210926
51850CB00013B/3325